Materials and Design

THE ART AND SCIENCE OF MATERIAL SELECTION IN PRODUCT DESIGN

Materials and Design

The Art and Science of Material Selection in Product Design

Mike Ashby and Kara Johnson

ELSEVIER
BUTTERWORTH
HEINEMANN

AMSTERDAM • BOSTON • HEIDELBERG • LONDON • NEW YORK • OXFORD
PARIS • SAN DIEGO • SAN FRANCISCO • SINGAPORE • SYDNEY • TOKYO

Butterworth-Heinemann is an imprint of Elsevier
Linacre House, Jordan Hill, Oxford OX2 8DP, UK
30 Corporate Drive, Suite 400, Burlington, MA 01803, USA

First edition 2002
Reprinted 2003, 2004, 2005 (twice), 2006

British Library Cataloguing in Publication Data
A catalogue record for this book is available from the British Library

Library of Congress Cataloging-in-Publication Data
A catalog record for this book is available from the Library of Congress

ISBN–13: 978-0-7506-5554-5
ISBN–10: 0-7506-5554-2

For information on all Butterworth-Heinemann publications
visit our website at books.elsevier.com

Printed and bound in *Italy*

06 07 08 09 10 10 9 8 7 6

Preface

Books on material selection – and there are many – focus on finding a match between material properties and the technical requirements of a design. There are now well-developed methods for doing this, supported by sophisticated software tools. Together they form the basis for the teaching of materials selection in engineering programs around the world. But these programs frequently ignore, or at best devote little attention to what might be called the *art* of materials – the role they play in industrial design. This may be because the more technical aspects of engineering form a structured, analytical field that can be recorded and taught as a set of formal procedures. Industrial design cannot so easily be formulated as a method; it relies instead on "visual" thinking: on sketching and modeling; on color, texture and feel; on creating product personality. Product design combines both, requiring technical excellence to create a product that functions properly, safely and at acceptable cost, and excellence of industrial design to create its aesthetic appeal and perceived value.

This book is about the role of materials and processes in product design. It complements an earlier text that develops methods for choosing materials and processes to match the technical requirements of a product. Here, by contrast, the emphasis is on a wider range of the information about materials that designers need, the way they use it and the reasons they do so.

The book has two audiences: students and working designers. For students of engineering, the purpose is to introduce the role of materials in industrial design, using language and concepts with which they are already familiar; for students of industrial design, the purpose is to develop an understanding of materials helpful to the practice of product design. For working designers, the purpose is to present a concise reference source for materials and processes, profiling their characteristics. To this end, the book is divided into two parts. The first presents ideas about design and methods of material selection; the second is devoted to the profiles. For further reference, we have created a website that presents some of the ideas presented in this book (www.materialselection.com).

Many colleagues have been generous with their time and ideas. In particular, we are grateful for discussions, criticisms, contributions and constructive suggestions from Professor Yves Brechet of the University of Grenoble, Dr. David Cebon, Dr. John Clarkson, Dr. Hugh Shercliff, Dr. Luc Salvo, Dr. Didier Landru, Dr. Amal Esawi, Dr. Ulrike Wegst, Ms. Veronique Lemercier, Mr. Christophe LeBacq and Mr. Alan Heaver of Cambridge University, Dr. Pieter-Jan Stappers of the Technical University of Delft, Dr. Torben Lenau of the Technical University of Denmark, Patrick Hall and Sam Hecht of IDEO-London, Rickson Sun and Tim Brown of IDEO-Palo Alto, Julie Christennsen of Surface Design, San Francisco. We particularly wish to acknowledge the contribution of Willy Schwenzfeier and Patrick Fenton of Swayspace, New York, to the overall design of the book itself.

We wish to thank the people and organizations on the following page for permission to reproduce their images and photographs of their products.

Mike Ashby and Kara Johnson
December 2001

Acknowledgements

ALPA of Switzerland
Capaul & Weber
Neptunstrasse 96
PO Box 1858
CH-8032 Zurich
Switzerland

Antiques Collectors' Club
5 Church St
Woodbridge, Suffolk
UK

Apple Press
6 Blundell St
London N7 6BH
UK

Bang and Olufsen, UK
630 Wharfdale Rd
Winnersh Triangle
Berkshire RG41 5TP
UK

Cynthia Nicole Gordon
4108 Eastern Ave North
Seattle, WA 98103
USA

Dyson
20 Shawfield St
London SW3 4BD
UK

Ergonomic Systems Inc.
5200 Overpass Rd
Santa Barbara, CA 93111
USA

Arnoldo Mondadori Editore S.p.A.
Milan
Italy

Gisela Stromeyer
Architectural Design
165 Duane St #2B
New York, NY 10013
USA

Han Hansen
Admirilitätstrasse 71, 20459
Hamburg
Germany

MAS Design
Axis House, 77A
Imperial Rd
Windsor SL4 3RU
UK

Nokia Group
Keilalahdentie 4
FIN-02150, Espoo
Finland

Porsche Design GmbH
Flugplatzstrasse 29
A-5700 Zell am See
Germany

Sony Corp.
6-7-35 Kitashinagawa
Shinagawa-Kei
Tokyo
Japan

Vectra
1800 S. Great Southwest Parkway
Grand Prairie, TX 75051
USA

Vitra Management AG
Klünenfeldstrasse 22
CH-4127 Birsfelden
Switzerland

Yamaha Corporation
10-1, Nakazawa-cho
Shizuoka Pref., Hamamatsu
4308650
Japan

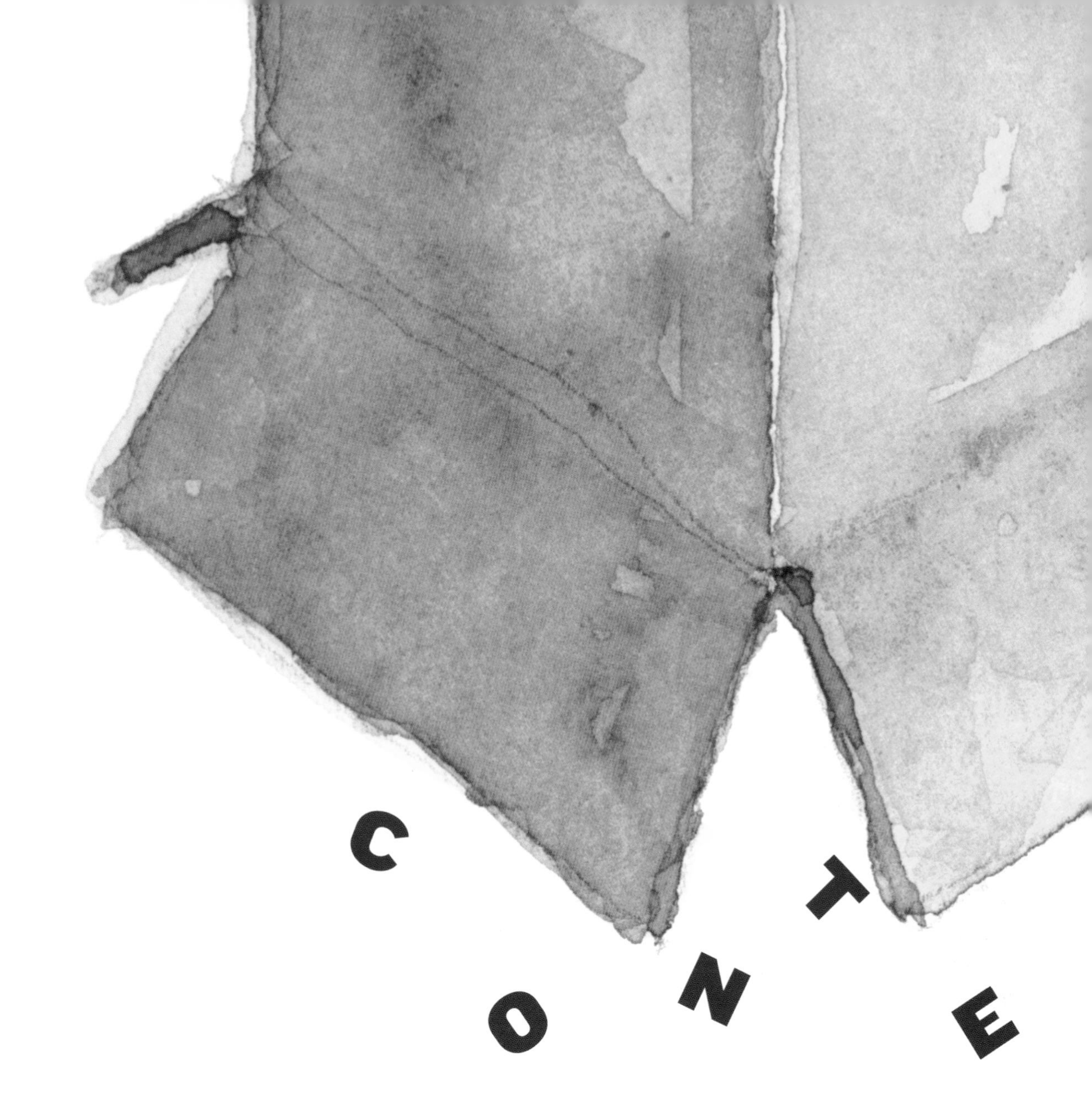

CONTE

Chapter 1 · Function and Personality 1
Chapter 2 · What Influences Product Design? 7
Chapter 3 · Design and Designing 27
Chapter 4 · The Stuff... Multi-dimensional Materials 49
Chapter 5 · Other Stuff... Shaping, Joining and Surfaces 89
Chapter 6 · Form Follows Material 99
Chapter 7 · A Structure for Material Selection 117
Chapter 8 · Case Studies in Materials and Design 141
Chapter 9 · New Materials – The Potential for Innovation 157
Chapter 10 · Conclusions 169

A Practical Reference for Inspiration

Material Profiles 174
Shaping Profiles 236
Joining Profiles 258
Surface Profiles 284

Appendices

Exercises for the Eye and Mind 311
Selected Material Maps 315

Index 329

Chapter 1

Function and Personality

We live in a world of materials; it is materials that give substance to everything we see and touch. Our species – *homo sapiens* – differs from others most significantly, perhaps, through the ability to *design* – to make things out of materials – and in the ability to see more in a material object than merely its external form. Objects can have meaning, carry associations or be symbols of more abstract ideas. Designed objects, symbolic as well as utilitarian, predate any recorded language – they provide the earliest evidence of a cultural society and of symbolic reasoning.

Some of these objects had a predominantly functional purpose: the water wheel, the steam engine, the gas turbine. Others were (and are) purely symbolic or decorative: the cave paintings of Lascau, the wooden masks of Peru, the marble sculptures of Attica. But most significantly, there are objects that combine the functional with the symbolic and decorative. The combination is perhaps most obvious in architecture – great architects have, for thousands of years, sought to create structures that served a practical purpose whilst also expressing the vision and stature of their client or culture: the Coliseum of Rome, the Empire State Building of New York, the Pompidou Centre of Paris, each an example of blending the technical and the aesthetic.

On a smaller scale, product designers seek to blend the technical with the aesthetic, combining practical utility with emotional delight. Think of Wedgewood china, Tiffany glass, Chippendale furniture – these were first made and bought to fulfil a functional purpose, but survive and are treasured today as much for their appeal as objects of beauty. Think, too, of musical instruments: the inlayed violin or harpsichord; of weapons of war: the decorative shields, the carved gun butt; or of the weapons of the mind: the gilded pen, the illuminated manuscript. All of these are tools made in forms that express aspects of their creator's imagination and desire to make objects of delight as well as of utility.

People – consumers – buy things because they like them – love them, even. To succeed, a product must, of course, function properly, but that is not enough: it must be easy and convenient to use, and it must have a personality that satisfies, gives delight (1.1). This last – personality – depends strongly on the industrial design of a product. When many technically equivalent products compete, market share is won (or lost) through its visual and tactile appeal, the associations it carries, the way it is perceived and the emotions it generates. Consumers now expect delight as well as function in everything they purchase. Creating it is a central part of design.

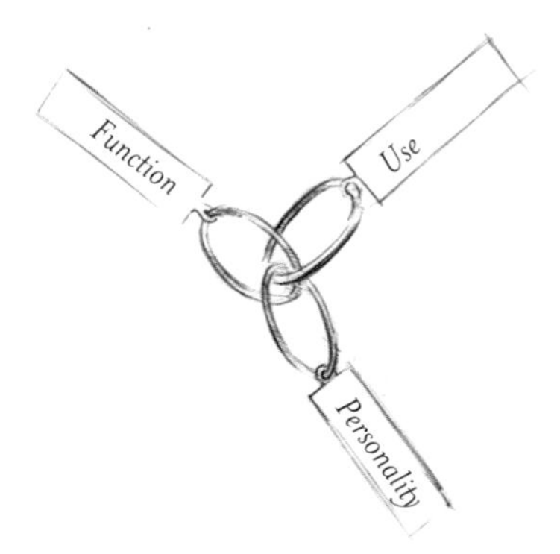

1.1 Function, Use and Personality
Balancing the requirements that a product must function, be easy to use and have a personality is key to innovative product design.

Advances in materials enable advances in industrial design, just as they do for technical design. And here we need a word that requires definition: "inspiration" – the ability to stimulate creative thinking. New developments in materials and processes are sources of inspiration for product designers, suggesting novel visual, tactile, sculptural and spatial solutions to product design. Examples drawn from the recent past are: the ability to color and mold polymers to make bright, translucent shapes; the co-molding of elastomers to give soft, tactile surfaces; toughened and textured glass to create transparent walls and flooring; surface coatings that reflect, refract or diffuse light; carbon fiber composites that allow exceptionally slender, delicate structures – and there are many more. In each of these examples innovative products have been inspired by the creative use of materials and processes.

Thus materials have two overlapping roles: that of providing technical functionality and that of creating product personality. It is here that an imbalance becomes apparent. Technical designers have ready access to information of the sort they need – handbooks, selection software, advisory services from material suppliers – and to analysis and optimization codes for safe, economical design. Industrial designers express frustration, both in print and in interviews, that they do not have equivalent support. In higher education the same discrepancy appears: the teaching of the science and technical application of materials is highly developed and systematized, supported by numerous texts, software, journals and conferences; there is no similar abundance of support for the teaching of materials in industrial design.

Bridging this gap in information and methods is not simple. The technical terms used by engineers are not the normal language of industrial designers – indeed they may find them meaningless. Industrial designers, on the other hand, express their ideas and describe materials in ways that, to the engineer, sometimes seem bewilderingly vague and qualitative. The first step in bridging the gap is to explore how each group "uses" materials and the nature of the information about materials that each requires. The second is to explore methods, and, ultimately, design tools, that weave the two strands of thinking into a single integrated fabric. That, in two sentences, is what this book is about.

Further Reading

There is a considerable literature on product design, some of it comprehensible, some not. Useful sources are listed, with ISB number and a brief commentary, at the end of each Chapter. Those listed below are a good starting point.

Clark, P. and Freeman, J. (2000) "Design, a Crash Course," The Ivy Press Ltd, Watson-Guptil Publications, BPI Communications Inc. New York, NY, USA. ISBN 0-8230-0983-1. *(An entertainingly-written scoot through the history of product design from 5000BC to the present day.)*

Dormer, P. (1993) "Design Since 1945," Thames and Hudson, London, UK. ISBN 0-500-20269-9. *(A well-illustrated and inexpensive paperback documenting the influence of industrial design in furniture, appliances and textiles — a history of contemporary design that complements the wider-ranging history of Haufe (1998), q.v.)*

Forty, A. (1986) "Objects of Desire – Design in Society Since 1750," Thames and Hudson, London, UK. ISBN 0-500-27412-6. *(A refreshing survey of the design history of printed fabrics, domestic products, office equipment and transport system. The book is mercifully free of eulogies about designers, and focuses on what industrial design does, rather than who did it. The black and white illustrations are disappointing, mostly drawn from the late 19th or early 20th centuries, with few examples of contemporary design.)*

Haufe, T. (1998) "Design, a Concise History," Laurence King Publishing, London, UK (originally in German). ISBN 1-85669-134-9. *(An inexpensive soft-cover publication. Probably the best introduction to industrial design for students (and anyone else). Concise, comprehensive, clear and with intelligible layout and good, if small, color illustrations.)*

Jordan, P.S. (2000) "Designing Pleasurable Products," Taylor and Francis, London, UK. ISBN 0-748-40844-4. *(Jordan, Manager of Aesthetic Research and Philips Design, argues that products today must function properly, must be usable, and must also give pleasure. Much of the book is a description of market-research methods for eliciting reactions to products from users.)*

Julier, G. (1993) "Encyclopedia of 20th Century Design and Designers," Thames & Hudson, London, UK. ISBN 0-500-20261-3. *(A brief summary of design history with good pictures and discussions of the evolution of product designs.)*

Manzini, E. (1989) "The Material of Invention," The Design Council, London, UK. ISBN 0-85072-247-0. *(Intriguing descriptions of the role of material in "invention" — here meaning creative design. The translation from Italian uses an interesting vocabulary, one unusual in a text on materials, and gives a commentary with many insights.)*

McDermott, C. (1999) "The Product Book," D & AD in association with Rotovison, UK. *(50 essays by respected designers who describe their definition of design, the role of their respective companies and their approach to product design.)*

Norman, D.A. (1998) "The Design of Everyday Things," MIT Press, London, UK. ISBN 0-385-26774-6. *(A book that provides insight into the design of products with particular emphasis on ergonomics and ease of use.)*

Chapter 2

What Influences Product Design?

Nothing is static. Today's designer seeks to optimize a design to best meet the needs of today's markets, but before the optimization is complete, the boundary conditions – the forces that influence the design decisions – shift, requiring re-direction and re-optimization. It is helpful to be aware of these forces; they create the context in which design takes place.

Figure 2.1 suggests five – the market, technology, investment climate, the environment and industrial design. It is a simplification, but a useful one. The central circle represents the Design Process, the workings and dynamics of which we explore in Chapter 3. It is subject to a number of external influences, indicated in the surrounding branches. A good designer is always alert to developments in technology, deriving from underlying scientific research. New technology is exploited in ways that are compatible with the investment climate of the company, itself conditioned by the economic conditions within countries in which the product will be made and used. Concern to minimize the ecological burden created by engineered products heightens the awareness of design for the environment, and, in the longer term, design for sustainability. And, in the markets of the 21st century, consumers want much more than a product that functions well and at an affordable price; they also want satisfaction and delight, making inputs from industrial design and aesthetics a high priority. There are of course many more influences, but discussions with designers suggest that these are currently among the most powerful. And, of course, they frequently conflict.

Before launching into the main development of the book, we briefly examine each of these influential forces, all of which thereafter lurk in the background, reappearing in later chapters when their importance makes itself felt.

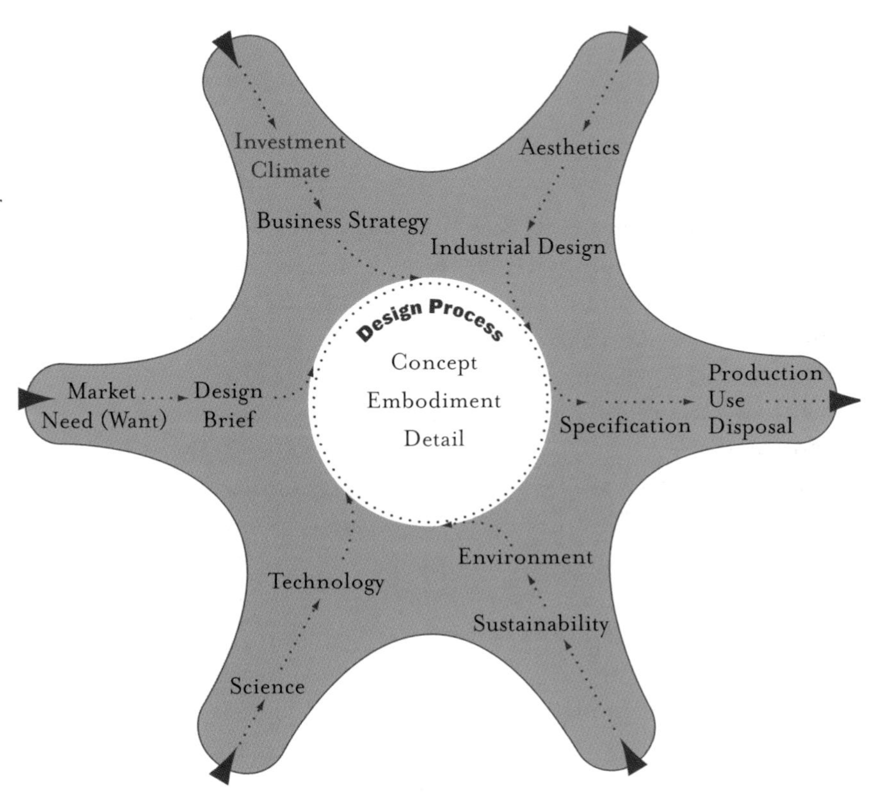

2.1 Inputs to the Design Process
The dominant inputs to product design derive from market need, science and technology, business strategy, concern for the environment and industrial design.

The Market

Present-day economic growth and prosperity, both national and personal, and the nature of free-market economies make the market a powerful driver in product design. In a developed country many products are technically mature and the market for them has saturated, in the sense that almost everyone who needs them, has them. Then it is desire ("want") not necessity ("need") that generates market forces. Much product design today is driven by desire, and one of the things consumers desire is greater functionality. This shifts emphasis from structural materials, which, after decades (centuries, even) of development are evolving only slowly, towards materials that behave in novel ways. These are largely provided by non-structural materials: it is the electrical, optical, magnetic and biological attributes that are important here. That is not to say that structural materials are dead – far from it: over 95% of all the engineering materials that are produced are used primarily to carry mechanical loads. Research and development of these emphasize more efficient processing, greater control of quality and more flexible manufacturing methods rather than a search for completely new materials. But while most cars – to take an example – are still made largely of steel, they also contain up to 30 small electric motors to position windows, seats, wipers and mirrors (magnetic and electrical materials), micro-processors for engine control and guidance systems (semi-conducting materials), toughened and non-reflective glass, liquid crystal and LED displays (optical materials), and flooring and paneling to insulate against heat and sound (thermal and acoustic materials). It is here that much current research and development is focused.

It is usual to suggest that designers respond to market needs, but sometimes it is the designer who creates the need. Revolutionary products take the market by surprise – few people felt the need for a walkman, or a digital watch, or a laser-pointer before these appeared in the marketplace. Here the designer anticipated a need, and by providing a solution, created it. The inspiration behind these concepts arose not from the market, but from advances in science and technology – the development of high-field magnets, of quartz oscillators, of solid-state lasers – to which we now turn.

Science and Technology

The least predictable of the driving forces for change is that of science itself (2.2). Despite periodic predictions[1] that science is "coming to an end" it continues to expose new technologies that enable innovation in materials and processes.

As already said, it is structural applications that, in terms of tonnage, overwhelmingly dominate the consumption of engineering materials. Cement and concrete, steels and light alloys, structural polymers and composites – all a focus for research in the last century – have reached a kind of technical maturity. What can new science and technology offer here?

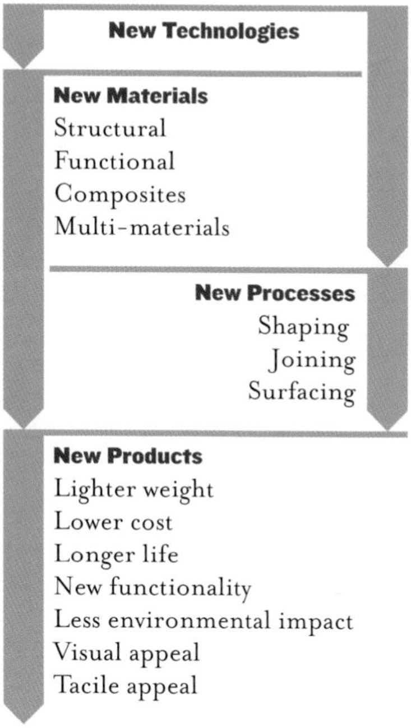

New Technologies

New Materials
Structural
Functional
Composites
Multi-materials

New Processes
Shaping
Joining
Surfacing

New Products
Lighter weight
Lower cost
Longer life
New functionality
Less environmental impact
Visual appeal
Tacile appeal

2.2 The Role of Science
Science reveals new technologies, from these technologies emerge new materials and processes. These, in turn, stimulate new concepts in product design.

1. *See, for example, Horgan (1996).*

Better materials for lightweight structures and for structural use at high temperatures offer such great potential benefits that research to develop them continues strongly. The drive towards miniaturization, too, creates new mechanical and thermal problems – the small size frequently means that the support structure must be exceptionally thin and slender, requiring materials of exceptional stiffness and strength, and that, while the power may be low, the power-density is enormous, requiring new materials for thermal management. The acknowledged future importance of MEMS devices (micro electro-mechanical systems) creates even greater challenges: materials for micron-scale beams, bearings, gears and chassis that must function properly at a scale in which the laws of mechanics operate in a different way (inertial forces become unimportant while surface forces become influential, for example).

Above all, there is the drive to discover and develop novel functional materials. Some examples of established functional materials and their application were given in the last section. Many more are under development and will, in time, inspire innovative products. Examples of such emerging materials are electro-active polymers, amorphous metals, new magnetic and super-conducting intermetallics and ceramics, metallic foams, and lattice materials made by micro-fabrication or 3-D weaving. Research on materials that mimic nature in subtler ways is stimulated by the deeper scientific understanding of cell biology, suggesting new approaches for developing bio-active and bio-passive surfaces that cells can recognize. Nanoscale assembly techniques allow the creation of two-dimensional devices that respond to the motion of a single electron or magnetic flux-quantum. If the 20th century is thought of as the era of bulk, three-dimensional materials, the 21st century will be that of surfaces, mono-layers, even single molecules, and the new functionality that these allow.

Sustainability and the Environment

All human activity has some impact on the environment in which we live. The environment has some capacity to cope with this, so that a certain level of impact can be absorbed without lasting damage. But it is clear that current human activities frequently exceed this threshold, diminishing the quality of the world in which we now live and threatening the wellbeing of future generations. The position is dramatized by the following statement: at a global growth rate of 3% per year we will mine, process and dispose of more stuff in the next 25 years than in the entire history of human civilization. Design for the environment is generally interpreted as the effort to adjust the present product design process to correct known, measurable, environmental degradation; the time-scale of this thinking is ten years or so, an average product's expected life. Concern for sustainability is the longer view: that of adaptation to a lifestyle that meets present needs without compromising the needs of future generations. The time-scale here is longer – perhaps 50 years into the future.

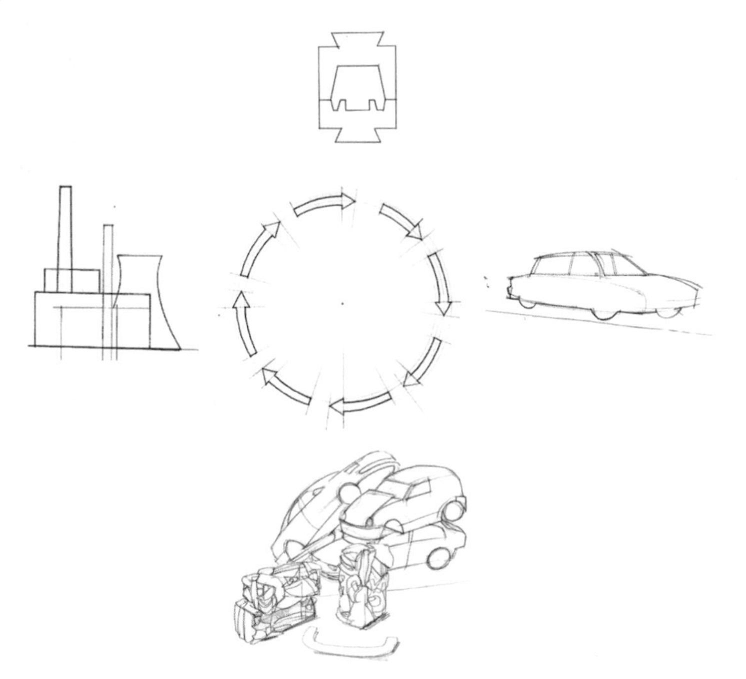

2.3 The Life of a Material
Ore and feedstock are mined and processed to give a material; this is manufactured into a product that is used and, at the end of its life, discarded or recycled. Energy and materials are consumed at each step, generating waste heat and solid, liquid or gaseous emissions.

The real considerations in design for the environment are brought into focus by examining the material lifecycle, sketched in 2.3. Ore and feedstock, most of them non-renewable, are processed to give materials; these are manufactured into products that are used and, at the end of their lives, discarded, a percentage perhaps entering a recycling loop, the rest committed to incineration or land-fill. Energy is consumed at each point in this cycle, with an associated penalty of CO_2 and other emissions – heat, and gaseous, liquid and solid waste. The problem, crudely put, is that the sum of these undesired by-products exceeds the capacity of the environment to absorb them. The visible injury is mostly local, and its origins can be traced and remedial action taken. Much environmental legislation aims at modest though continuous reductions in the damaging activity; regulation requiring a 20% reduction in – say – the average gasoline consumption of passenger cars is seen as posing a major challenge.

Sustainability requires solutions of a completely different kind. Even conservative estimates of the adjustment needed to restore long-term equilibrium with the environment envisage a reduction in the flows of 2.3 by a factor of four; some say ten.[2] Population growth and the growth of the expectations of this population more than cancel any 20% savings that the developed nations may achieve. It looks like a tough problem, one requiring difficult adaptation, and one to which no answer will be found in this book. But it remains one of the drivers and ultimate boundary conditions of product design, to be retained as background to any creative thinking.

2. The books by von Weizsäcker et al. (1997) and by Schmidt-Bleek (1997) make the case for massive reductions in the consumption of energy and materials, citing examples of ways in which these might be achieved.

How, then, to respond to the nearer-term problem of reducing the impact of usage? One obvious way is to do more with less. Material reductions are made possible by recycling, by the use of renewable materials made from things that grow, by miniaturization and by replacing goods by services. Energy reductions can be achieved by lightweight design of transport systems, by the optimized thermal management of buildings and by increased efficiency of energy conversion and utilization in industry. Probably the most effective measure of all is that of increasing product life: doubling life halves the impact of three out of the four stages of 2.3. And this refocuses the spotlight on industrial design – people don't discard possessions they love.

Economics and Investment Climate

Many designs have never reached the marketplace. Translating a design into a successful product (2.4) requires investment, and investment depends on confidence, on establishing economic viability. This relies in part on the design itself, but it also depends, importantly, on the nature of the market at which it is aimed, and the degree to which it can be protected from competition. The business case for a product seeks to establish its economic viability.

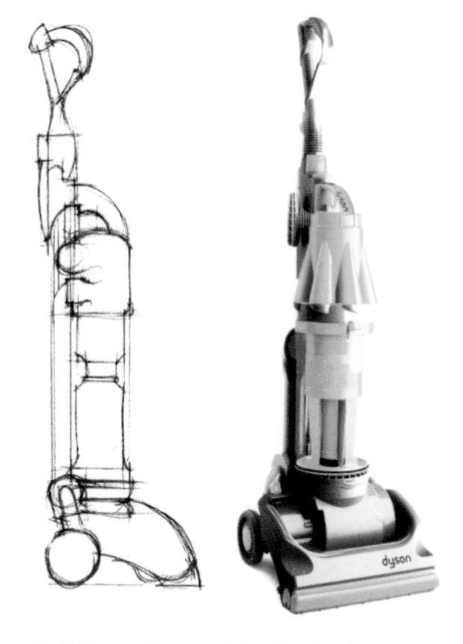

2.4 From Concept to Product
The movement of an idea from concept to product requires demonstration of technical and economic viability, a receptive market and capital investment. (Courtesy Dyson UK)

A product is economically viable if its value in the marketplace is greater than its cost by a sufficient margin to justify the investment required to make it; it is this that determines the potential revenue it can generate. Techniques of cost modeling allow estimates of the cost of production. Assessing value[3] is more difficult, requiring market research to establish the consumers' perception of the product, the importance they attach to performance, and the nature of the competition. Value depends on the market and industry at which the product is aimed: a titanium bicycle (to take an example) is attractive to mountain biking enthusiasts – to them, its value exceeds its cost; but it is not attractive to the average city shoppers, who perceive the cost to exceed the value.

3. See the article by Clark et al. (1997), listed under Further Reading, describing techniques of value analysis, utility and technical cost modeling.

Demonstrating economic viability is only one part of a credible business case. Establishing production requires investment, and this is forthcoming only if there is confidence in a significant level of value capture. The ability of the developers of the product to capture the value it offers relies on their ability to limit competition by retaining control of intellectual property and of key assets (people, trade secrets, licensing or partnership agreements). The investment required to commercialize a technically viable product will be forthcoming only if the technical, market and business case assessments are all attractive. And finally there is the question of attitude to risk. Some industries are slow to adopt new technologies, whether viable or not: the nuclear industry, the civil engineering industry and, increasingly, the aerospace industry are among them. Others are not: manufacturers of sports equipment and of many consumer products eagerly seize upon new materials and processes and enthusiastically promote products made from them, accepting the risk that, being new and imperfectly characterized, they may fail prematurely.

This scenario has significant impact on material development. Developing, certifying and commercializing a new structural material takes, typically, 15 years, and it is not always obvious that it will be technically or economically viable when that point is reached. In the last century much of the development costs of structural materials was underwritten by governments through defense, space and nuclear programs, willing to invest on a long time-scale – one that private industry is unwilling to accept. For functional materials, however, the time-scale for research and development can be shorter, and – partly because they can enable revolutionary products – their potential value can be higher, making them a more attractive investment. Thus the design process is influenced by the investment climate, judgements of expected profit and loss, sales volume and ease of production; each of these relates directly to the development and commercialization of materials.

Aesthetics and Industrial Design

Anaesthetics numb the senses, suppress feeling; anaesthesia is a lack of sensation. Aesthetics do the opposite: they arouse interest, stimulate and appeal to the senses, particularly the sense of beauty. Aesthetics (like "inspiration") is a difficult word, having too many shades of meaning to convey a sharp message, yet there seems to be no other that quite captures the sensory attributes of materials and products. Designers manipulate these senses – and the reactions to each sense – to create a product's personality.[4]

It has been argued in the past – and the argument is still sometimes used today – that a product designed to function

4. A designer's interaction with a material – or product – is probably better described as a synesthetic experience. Synesthesia is a psychological term that describes the physical experience of joined sensations. Many people say that they "see red" when angry, enjoy "sharp cheeses" with a glass of port or like listening to "cool jazz."

properly will automatically have aesthetic appeal; that "form follows function." This reasoning leads to the view that industrial design as an active pursuit is unnecessary, since good technical design will produce it as a by-product; industrial design, in this view, is mere packaging. And there is an opposing view: that products that are built around function alone have not been designed at all, but merely engineered. The reality, of course, is that both the technical and industrial design of a product influence its success and that of the company that makes it, and for very good reasons. Here are some.

Product Differentiation

Many products are now technically mature. Distinctions in technical performance are slight, and the prices of products with nearly the same performance are almost the same. As the market for a product saturates, sales can be stimulated by differentiation. This means creating product lines that are distinctive and have a personality that resonates with the tastes and aspirations of a particular user group: an elegance that appeals to women, a rugged character that can survive an athlete's use, a playful accessibility and tolerance of misuse that is appropriate for children.

Simple Interfaces

A product is safe, effective and pleasing if the way it works can be read from its design: the use of size, proportion, configuration and color to identify controls and what they do; the use of lights, sounds, displays and graphics to present its current state. Product functionality has increased and while consumers want this they also want a small size and a simple, easy to understand interface. The physical size that is reserved for an "interface" has decreased and the space with which the designer is free to express product use is limited. Good industrial design helps deal with this problem.

Corporate and Brand Identity

The image carried by a corporation and its products is one of its most valuable assets; indeed, the primary asset of some

2.5 A Product Family

The use of the same materials and surface finishes creates a brand-image that unites these four products. (Courtesy Porsche Design GmbH.)

companies is their brandname. Creating and presenting this image and carrying it from one generation of products to the next is a function of industrial design that touches every aspect of the company: its products, its advertising and even the architecture of its buildings. Empires, armies, religious orders, airlines, railways, manufacturing corporations – all of these use design to convey what they are, or how they would like to appear.

Product Life

Products have a "design life" – a time span after which replacement is anticipated, but need not occur. In reality, a product comes to the end of its life when the market or user no longer wants it. Cars have a design life of approximately 12 years, but classic cars survive far longer. A DC3 aircraft had a design life of 20 years – 60 years on they are still flying. The Eiffel Tower had a design life of only 2 years; more than a century later it stands as the symbol of Paris. "Classic" products are designs that are of such quality that they outlive their expected life – sometimes by many generations. Distinguished designs with elegant or imaginative use of materials survive because they are so endearing, symbolic or evocative that they are treasured and preserved. Unremarkable designs with dull or inappropriate materials are, deliberately or accidentally, transitory; we discard them without a thought.

A Necessary Balance

Products form part of the environment in which we live – in the home, in the workplace, on the street. Mass production now supplies products to a far larger market and in far larger numbers than in the earlier days of craft-based design, when few could afford them. In this sense mass production has enhanced the quality of life, but in others it has the capacity to diminish it. Our environment is enhanced by products that satisfy. By contrast if products create expectations that are not fulfilled, add nothing to (or even detract from) self-esteem or sense of place in society, or give no sense of satisfaction, then the quality of life has suffered.

Successful products depend on a balanced mix of technical and industrial design. This fact has been more readily accepted in architecture than in product design – architects[5] speak of the three "ideals" of efficiency, economy and elegance. There is an increasing awareness[6] that similar ideals apply to product design. For these reasons and others, industrial design is now as important an aspect of the total design process as any other. The examples that conclude this chapter illustrate these points further.

5. See, as an example, Billington (1983).

6. Pye (1997), particularly, makes this point.

Some Examples...

Product Differentiation – Wristwatches

Wristwatches have been commercially available since 1850, though for most of that time only the rich could afford them. As little as 40 years ago, a watch was a valuable and valued possession – a present for life given on a significant occasion (2.6, top). Once you had one watch, you didn't need another.

With increasing wealth after 1945, the market for watches broadened and saturated – meaning that practically everyone had one. The need to stimulate sales became a priority, and how better than to make watches an item of fashion, not just for the rich, but for everyone. Watches have become a commodity, mass-marketed, like clothes. As an example, a mail-order catalogue published in 2001 (in which watches are only a tiny part) lists and illustrates 312 watches, most priced below $50, some below $10. All use digital technology and thus have a precision far higher than most people need – marketing on performance ceases to be relevant. If a watch is to sell for more than $50 (and many, of course, do) it must appeal in some way to the consumer's sense of style, taste and individuality.

If further demonstration of the recent dominance of industrial design in marketing is needed, consider the Swatch. Prior to 1975, Switzerland had a monopoly of the world manufacture of watches, sustained by the national brand-image of quality. The electronic watch, invented in the United States and then developed in Japan, was a catastrophe for the Swiss watch industry. By 1984, the number of watch-making houses, their output,

2.6 Product Maturity: Wristwatches
Above, watches from the era when they were expensive heirlooms. Below, extreme styling from Swatch. (Courtesy Apple Publishers, London)

and the workforce they employed, had all fallen to less than one-third of their peaks a few years earlier. Desperate for recovery, Swiss watchmakers collaborated to design a watch that was as accurate, better made and thinner than those from the East Asian countries, and was competitively priced. More significantly, they devised the concept of the "collection," a series of products, technically identical, but differing in color, texture, pattern and styling (2.6, bottom). Artists were commissioned to design one or a series; and each new design was given a name ("Calypso Beach," "Lolita," "Graffiti"). By 1989 the image was established and successful, not only in broadening the market, but encouraging individuals to buy not one watch, but two, ten..., a hundred, even. The brand has now launched some 1,000 designs of a single technical innovation.

2.7 Product Differentiation: Mobile Phones

One of the first Nokia phones to some of the latest designs.

Small and Simple Interfaces – Mobile Phones

Mobile (cellular) phones typify recent and successful electronic innovations. Like the portable CD player, the Walkman, and even the personal computer, mobile phones now sell as much for their aesthetics as for their technical merit. Cellular communication systems require an established network of repeater stations; initial industry growth, around 1988, depended on a minimum network of these stations. The technology was new, taken up by "early adopters" – the relatively small group of consumers that love new technology and will buy it even before it works well or is easy to use. Mobile phones of that era were brick-shaped, black and heavy, with a "technical" image like that of military radiophones (2.7, top). Increasing the market required technical advances: lighter phones (meaning more efficient batteries and lower power-drain) and, with increasing concern about health risks, lower microwave intensity and thus higher sensitivity. By 1995, coverage by repeater stations in developed countries was nearly universal and take-up of the technology had grown faster than anticipated. By 1999, the European and American markets were approaching initial saturation.

Most mobiles now have more organic forms and make extensive use of color and pattern, and come with snap-on covers to match clothes, mood or occasion. Contemporary mobiles exploit the visual and tactile attributes of materials, to produce products that appeal to a wider range of social groups, and to create shifting design trends that makes a one-year-old model appear out of date. Translucent cases with dimly visible innards, pop-art or high-art décor like the "Mona Lisa" and limited editions – all are ways of stimulating markets. Perceived value has now supplemented technical value as a driver in design. Other recent innovations have followed the same patterns, with differing time-scales. The Walkman (launched 1970), the portable CD player (launched 1980), the computer mouse (launched 1984) have transmuted from functional, dull and flat objects into objects that are translucent or brightly colored, sculptured and emotional.

Corporate Identity – Bang & Olufsen

Anyone with the smallest interest in stereo and audio equipment can recognize a Bang & Olufsen (B&O) product even if they have never seen it before (2.8). In the late 1960s, this company set out to establish a recognizable corporate identity through the design of their products. Their radios, televisions and phones were – and are now – made recognizable by high-tech shapes, sharp edges, flat surfaces, careful choice of finish and neat, clean graphics. The combination of wood veneer, satin finished aluminum and stainless steel combine to suggest quality and attention to detail. Knobs, moving lids and complex controls are eliminated, replaced by infrared sensors and simplified remote controls. The designers have sought to combine the discrete with the remarkable.

2.8 Corporate Identity: Bang & Olufsen

These images show a consistent and powerful corporate and product identity. (Courtesy of Bang & Olufsen, Denmark)

Bang & Olufsen express their philosophy[7] in the following way. "Design is nothing unless it is used to unite form and function... We have never tried to be the first to introduce new technology... Our aim is different: we want our products to make sense and to make you feel special in their company... We believe that such products are built by combining unrivaled technology with emotional appeal; products with personality that make people feel special and create emotional bonds... Most of our products come in a range of colors to present you

7. *Bang and Olufsen (2002).*

with a choice, just as if you were buying a new sofa... Our colors don't follow fashion, but they will match your furniture... Surface textures are another design element which some would regard as secondary, but which we care about deeply... We have but one rule: Bang & Olufsen products must create an immediate understanding of their performance... We aim to surprise with unusual initiatives, to escape the sense of boredom of mass product development..."

Here is a company that sees perceived value, created by focused industrial design, as central to product recognition and appeal, and to corporate identity. Their continuing profitability and success suggest that they are right.

Material Evolution

The influences on product design discussed earlier – technology, the environment, the investment climate and industrial design – result in an evolution in the use of materials for a given product or product class. Here are two examples – the camera and the hairdryer.

2.9 Material Evolution: the Camera
From metal to polymer to paper to metal-coated polymers, the evolution of materials is shown in these selected cameras.

The Camera

Roughly 140 years ago, when the camera was invented, changes in the design came slowly. Early cameras were made from wood, brass, leather, and, of course, glass. They do have a certain style: that of the cabinet-maker, reflecting their craft-based manufacture. Bit by bit they became tools of art, business and science, and evolved to become more like scientific instruments, the emphasis more on functionality than on visual appeal. The materials chosen for these cameras were mostly metal – they were durable and put an emphasis on "engineered" quality. But as the market for the standard product started to saturate new variants appeared: the disposable camera (made of paper), the miniature spy camera (metalized polymers) and waterproof cameras (elastomers over-molded on a range of polymers) – differentiation through technical innovation is often coupled with changes in material. Technical development continues today:

the zoom lens and digital imaging are examples. But this has gone hand-in-hand with differentiation of a different kind: cameras for children, cameras for under-water tourism, instant cameras, cameras to be worn as fashion accessories. Figure 2.9 shows some examples.

Here it is worth noting how products make evolutionary jumps, redefining their role, and capturing new markets. The jump from film-based photography to digital imaging is an example. The evolution of phones from traditional landline connection to mobile technology is another. The jump from balance-wheel to quartz-oscillator time-keeping is a third. These are what are called "revolutionary" innovations, meaning that they frequently disrupt existing markets because the new technology is controlled by a different manufacturing sector than the old, setting off a new evolutionary process.

Hairdryers

Electric hairdryers first became widely available about 1945. Little has changed in the way that their function is achieved since then: an electric motor drives a fan which propels air through heating elements, whence it is directed by a nozzle onto the hair. Early hairdryers (2.10a) had a power of barely 100 watts. They were made from zinc die castings or pressed steel sheet, and they were bulky and heavy. Their engineering was dominated by the "metal mentality": parts that could be easily cast, pressed and machined were held together by numerous fasteners. Metals conduct both electricity and heat, so internal insulation was necessary to prevent the user getting electrocuted or burned. This, together with inefficient motors and fans, made for a bulky product. The only technical development since has been that of the centrifugal fan, allowing a slimmer design by putting the motor at the hub of the fan itself (compare 2.10a and b).

The development of polymers led to hairdryers that at first used Bakelite, a phenolic resin, then other polymer materials, for the casing and handle (2.10c,d and e). The early versions are plastic imitations of their metal counterparts; the Bakelite model (2.10c) has the same shape, the same number of parts but even more fasteners than the metal one. Polymers were at

2.10 Material Evolution: the Hairdryer
From cast iron to pressed steel to injection molded polymers.

first attractive because of the freedom of decorative molding they allowed. Hairdryers like that of 2.10d have lost some of the industrial form; they were aimed at a more fashion-conscious public. But these designs did not exploit the full advantages offered by polymers: brighter colors, more complex moldings which interlock, snap-fasteners for easy assembly and the like. There were some gains: the unit was lighter, and (because the thermal conductivity of polymers is low) it didn't get quite so hot. But if the fan stalled, the softening point of polymers was quickly exceeded; most old hairdryers that survive today from this era are badly distorted by heat. Nonetheless, smaller motors and better thermal design pushed the power up and the weight down, and allowed a reversion to the more efficient axial flow design (2.10e).

The changes in design described above derive largely from the introduction of new materials. The tiny motor of a modern hairdryer uses ceramic magnets and a ceramic frame to give a high power density. Sensors detect overheating and shut the unit down when necessary. The higher velocity of airflow allows a heater of high power density and reduces the need for insulation between the heating element and the casing. The casing is now designed in a way that exploits fully the attributes of polymers: it is molded in two parts, generally with only one fastener. An adjustable nozzle can be removed by twisting – a snap fit, exploiting the high strength/modulus ratio of polymer. The design is now youthfully attractive, light and extremely efficient. Any company left producing pressed metal hairdryers when a unit like this becomes available finds that its market has disappeared.

Hairdryers have, for some years now, converged technically. Increased market share relies on creating demand among new social groups (children, athletes, travellers...) and in new models that differ in their visual, tactile and associative appeal: a Snoopy hair dryer for children (2.10f); a miniature sports/travel dryer weighing just under 400g; a translucent dryer within which the heating element and other parts are dimly visible through a frosty blue exterior that glows tantalizingly. In each case, materials and design are used to enable technological advances and capture the consumer's attention, and increase market share.

Conclusions

Today industrial design is integrated into the design process of most new products, and in a world of global competition, it makes good business sense to do so. In some products – sunglasses, skis, coffee makers – the influence of the industrial designer is more obviously central than in others – light bulbs, machine tools, jet engines – but even here it can be found. Next time you step off a plane, glance for a moment at the discretely noticeable Pratt and Whitney or Rolls Royce logo on the engine cowling – its styling is no accident. In short: technical and industrial design are the essential, complementary, parts of the design of *anything.*

This book seeks to explore the role that materials can play in balancing technical and industrial design and the selection of materials in product design itself.

Further Reading

Bang and Olufsen (2002) Product catalogue and website. www.bang-olufsen.com *(Illustrations of B & O products and statements of company design philosophy.)*

Billington, D.P. (1985) "The tower and the bridge, new art of structural engineering," Princeton University Press, Princeton, New Jersey, ASM. ISBN 0-69102393-X. *(Professor Billington, a civil engineer, argues that great structures – the Brooklyn Bridge, the skyscrapers of Chicago, the concrete roof-shells of Pier Luigi Nervi not only overcome technical challenges, but also achieve the status of art. Much is biographical – the achievements of individual structural architects – but the commentary in between is forceful and enlightening.)*

Braun (2002) Product Catalogue and website. www.braun.de. *(Illustrations of Braun products and statements of company design philosophy.)*

Clark, J. P., Roth, R. and Field, F. R. (1997) "Techno-economic Issues in Materials Science," p. 255, ASM Handbook Vol 20, "Materials Selection and Design," ASM International, Materials Park, Ohio, USA. ISBN 0-87170-386-6. *(The authors explore methods of cost and value analysis, and environmental issues in material selection.)*

Edwards, F. (1998) "Swatch, a guide for connoisseurs and collectors," Apple Press, London, UK. ISBN 1-85076-826-9. *(A well written and lavishly illustrated documentation of the design history of the Swatch, with over 300 examples, each with a brief commentary.)*

Horgan, J. (1996) "The End of Science," Abacus Books, Little, Brown Co. London, UK. ISBN 0-349-10926-5. *(Hogan argues that, in almost every branch of science, the developments are those of detail, not of revolutionary discovery. Many disagree.)*

Jordan, P.W. (2000) "Designing Pleasurable Products," Taylor and Francis, London, UK. ISBN 0-748-40844-4. *(Jordan, Manager of Product Aesthetics at Philips Design, argues that products today should give pleasure. Much of the book is a description of market-research methods such as interviews, questionnaires, field studies and focus groups for eliciting reactions to products from users.)*

Pye, D. (1997) "The Nature and Aesthetics of Design," Cambium Press, Connecticut, USA. ISBN 0-9643999-1-1. *(A book devoted to understanding design, with a particular focus on mechanical design. The role of aesthetics is described but no particular references to materials are given.)*

Schmidt-Bleek, F. (1997) "How much environment does the human being need – factor 10 – the measure for an ecological economy," Deutscher Taschenbuchverlag, Munich, Germany. *(The author argues that true sustainability requires a reduction in energy and resource consumption by the developed nations by a factor of 10.)*

Von Weizsäcker, E. Lovins, A.B. and Lovins, L.H. (1997) "Factor Four," Earthscan Publications, London, UK. ISBN 1-85383-407-6. *(The first publication to take the disturbing, but defensible, point of view that sustainability requires massive changes in human behavior and consumption.)*

Chapter 3

Design and Designing

In Italian there is one inclusive word for design and engineering – la progettazione;[1] a designer or an engineer is "il progettista." Translated literally, "il progetto" is the plan. In English, too, the word "design" is defined as "the plan," with an even wider spectrum of meanings: design with regard to fashion – like hats; or with regard to aerodynamics and fluids – like that of turbine blades; or even with regard to states of mind.[2] Such breadth of meaning creates opportunities for confusion. To avoid this, we will speak of *technical design, industrial design and product design,* using these words with the following sense.

Technical (or engineering) design includes the aspects of design that bear on the proper technical functioning of the product: its mechanical and thermal performance, cost, and durability. We shall call these, collectively, the technical attributes of the product – attributes that describe how it works and how it is made. Industrial design includes the aspects of design that bear on the satisfaction afforded by the product: the visual and tactile attributes, associations and perceptions, historical antecedents – attributes that describe its personality or character. Product design, in the sense we use it here, means the synthesis of technical and industrial design to create successful products.

There is a risk in drawing these distinctions that technical and industrial design will be seen as separate, unconnected activities. A more balanced view is that they form a continuum, that they are simply parts of the overall design process. But while technical design utilizes well-established methods and sophisticated computer-based tools, industrial design cannot so easily be made systematic or quantitative. Unlike the technical attributes, which are absolute and precisely specified, many attributes of industrial design depend on culture and on time: the Japanese ideal of beauty differs from that of the European; and what is beautiful to one generation can appear ugly to the next. Scientific and technical language and thinking work well when ideas can be expressed explicitly and precisely, but these break down when the ideas are imprecise, or involve subjective appreciation or taste; then, other ways of thinking are needed.

1. Adapted from Manzini's (1989) remarkable book "Materials of Invention."

2. "I don't design clothes, I design dreams" – Ralph Lauren, NY Times, 19 April 1986.

Ways of Thinking

Technical design relies on deductive reasoning – thinking based on logic and analysis. Deductive reasoning, applied to the selection of materials, is described more fully in Chapter 7. It lends itself to formulation as a set of steps, often involving mathematical analysis. Industrial design, by contrast, relies on inductive reasoning – synthesis, drawing on previous experience. Inductive methods for selecting materials, also explored in Chapter 7, use perception and visualization. These we need to explore more fully since they are central to the discussion that follows.

3.1 Virtual Violin?
The form of the violin is an essential part of its personality. In this electronic violin, the ghost-like form both makes the connection to the original and suggests the transmutation that has taken place. (Courtesy Yamaha Corp.)

Observation and Perception

Imagine yourself to be standing in a motorcycle tradeshow behind two men who are looking at a Harley Davidson. The Harley has technical attributes, listed in its specification: weight, number of cylinders, power, maximum speed, the material of which the frame is made – these and many other attributes can be precisely defined and accurately measured. The Harley also has aesthetic attributes – it is black, metallic and loud. The two men see the same motorcycle but they perceive it in different ways. In the mind of one is an ideal image of a smooth, yellow, urban scooter, without visible mechanical parts, clean lines and trendy styling; he perceives the Harley as heavy, extravagant and dangerous. The ideal in the mind of the other is an image of an open road, a black leather bodysuit, a helmet with darkened visor, twin-exhausts; he perceives the Harley as powerful, authoritative, an expression of freedom.

Perception is the result of interpreting what is observed. Two observers of the same product will perceive it in different ways, ways that derive from their reaction to the physical object they see and the accumulated mental images and experiences they carry with them. Both observation and perception contribute to creativity in design, and here it is necessary that we sharpen the definition of four terms we will use to describe, in increasing order of abstraction, the attributes of products –

particularly those relating to industrial design and the personality of a product.

- *Aesthetic attributes* are those that relate directly to the senses: sight, touch, taste, smell, hearing; those of sight include the form, color and texture of a material or product.

- *Attributes of association* are those that make a connection to a time, place, event or person – thus a jeep has military associations, gold has associations of wealth, the color black, in some cultures but not all, of death.

- *Perceived attributes* describe a reaction to a material or product – that it is sophisticated, or modern, or humorous, for instance.

- *Emotional attributes* describe how a material or product makes you feel – happy, sad, threatened perhaps – "emotional ergonomics," in the words of Richard Seymour of SeymourPowell, London.

To these we add the word *style*. Styles have names: Art Nouveau, Art Deco, Modernist, Post-Modern, etc. Each is shorthand, so to speak, for a particular grouping of aesthetic, perceived and emotional attributes and associations – one about which there is general agreement. Styles, sometimes, are linked to certain materials, but it cannot be said that a material has a style, only that it acquires one when it becomes part of a product. Examples developed later in this and the next chapter will make these distinctions clearer.

3.2 Left Brain, Right Brain
Thinking from the left or right – the first seeking solutions by logic and analysis, the second seeking solutions by synthesizing elements from recalled or imagined images or analogies.

Verbal-Mathematical and Visual Thinking

Writers such as McKim,[3] discussing ways in which the human brain manipulates information in order to reason, distinguish two rather different processes (3.2). The first, the domain of the left-hemisphere of the brain, utilizes verbal reasoning and mathematical procedures. It moves from the known to the unknown by analysis – an essentially linear, sequential path. The second, the domain of the right-hemisphere, utilizes

3. McKim's book Visual Thinking (1980) explains better than most the ways in which images enter the creative thought process.

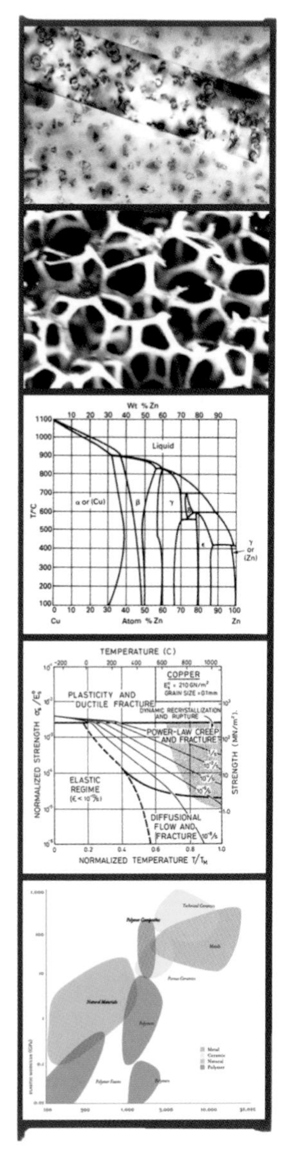

3.3 Images from Material Science
Some of the ways visual images occur in material science (top to bottom): Electron micrographs; Optical micrographs; Phase diagrams; Deformation mechanism maps; Material property charts.

images – both remembered and imagined. It creates the unknown from the known by synthesis – by dissecting, recombining, permuting, and morphing ideas and images. The first way of thinking, the verbal-mathematical, is based on learned rules of grammar and logic. The second way of thinking, the visual, makes greater use of the imagination; it is less structured but allows greater conceptual jumps through free association.

Think for a moment about the following example of the way you store visual information. You probably know and recognize several hundred people, perhaps many more. Could you, if asked to draw a recognizable picture of any one of them, do so? Most people can't; many can't even conjure up a picture of the face in their "mind's eye" (their imagination). This suggests that the visual image is stored only in a very crude way. Yet if you unexpectedly encountered a person that you know in – say – Los Angeles International Airport, you would instantly distinguish them from the thousands of other people there. Recognition of a face or a place requires a detailed comparison of a visual image with an image stored in the mind, seeking a match of a very subtle kind – and the average person can store enough information to recognize and distinguish not one, but hundreds of these. The way the mind stores images is not well understood but it is clear that its image database is very large and, when triggered, capable of very rapid access and great precision.

Creativity in design (both technical and industrial) involves the free association and combination of images to achieve a desired set of attributes. The images may be visual – observed objects, photographs, sketches and drawings – or mental – stored in the memory and imagination of the designer.

Visual Thinking in Material Science

The word "science" immediately suggests deductive reasoning – analysis. But creative scientists from Leonardo da Vinci and Newton to Einstein and Crick/Watson testify that their moments of great insights arose as much from synthesis – visual thinking – as analysis. Material science, particularly, makes use of images for communication and as a way of thinking (3.3). Venn diagrams and flowcharts illustrate relationships and procedures; bar charts and graphs show magnitudes and numerical trends. Schematics display molecular structures, show how mechanisms work or how equipment functions.

Information can be more densely packed in diagrams and images that show relationships. Phase diagrams relate the regimes of stability of competing alloys. Micrographs reveal structural similarities between differing materials, suggesting, perhaps, that a heat-treatment used for one might be effective for another. Deformation mechanism maps relate the regimes of dominance of competing deformation mechanisms. Material property charts[4] relate a population of materials in material-property space, a space with many dimensions. Each of these captures a vast amount of information and compresses it into a single image, revealing patterns in the data that words and equations do not. It is here that they become a tool not only for communication, but also for reasoning.

4. These charts, used extensively in the book, are introduced in Chapter 4.

The power of the visual image lies in the ease with which it can be manipulated by the mind and its ability to trigger creative thought. A picture of a car tail-light made of acrylic, taken to show its transparency and ability to be colored, reveals much more: that it can be molded to a complex shape, that it can withstand water and oil, and that it is robust enough to cope unprotected with use in the street. Diagrams showing relationships, particularly, have the power to trigger new ideas – examples in Chapter 4 will show how plotting material information can suggest novel material composites and combinations. Without the visual image, these ideas do not so easily suggest themselves. Visual communication and reasoning, then, have a long-established place in the world of material science. But how are they used in design? To answer that we must examine the design process itself.

The Design Process

First, a word about types of design. Original design starts from a genuinely new idea or working principle: the lamp bulb, the telephone, the ball-point pen, the compact disc, the mobile phone. More usually, design is adaptive, taking an existing concept and seeking an incremental advance in performance through refinement.

The starting point of a design is a market need or a new idea; the endpoint is the full specification of a product that fills the need or embodies the idea (3.4). It is essential to define the need precisely, that is, to formulate a need statement or design brief, listing product requirements, an expected environment of use and possible consumers. Writers on design emphasize that the statement should be solution-neutral (that is, it should not imply how the task will be done) to avoid narrow thinking limited by pre-conceptions.

Between the design brief and the final product specification lie many steps. One way of modeling this design process is shown in the left-hand column of 3.4. Design, in this view, has three broad stages: conceptual design, development and detailed design.[5] The concept presents the way the product will meet the need, the working principle. Here the designer considers the widest possible range of ideas, both technical and aesthetic. The choice of concept has implications for the overall configuration of the design, but leaves decisions about material and form largely unanswered.

5. Here we are using the words of a respected design text (Pahl and Beitz, 1997); many others follow this influential book on technical design.

The next stage, development, takes each promising concept and develops it, analyses its operation, and explores alternative choices of material and process that will allow safe operation in the anticipated ranges of loads, temperatures and environments. In parallel with this, alternative forms, colors and textures are explored, seeking, in ways described in Chapter 6, materials and processes capable of creating them.

Development ends with a feasible design that is then passed on for detailing. Here specifications for each component are drawn up; critical components are subjected to precise mechanical or thermal analysis; optimization methods are

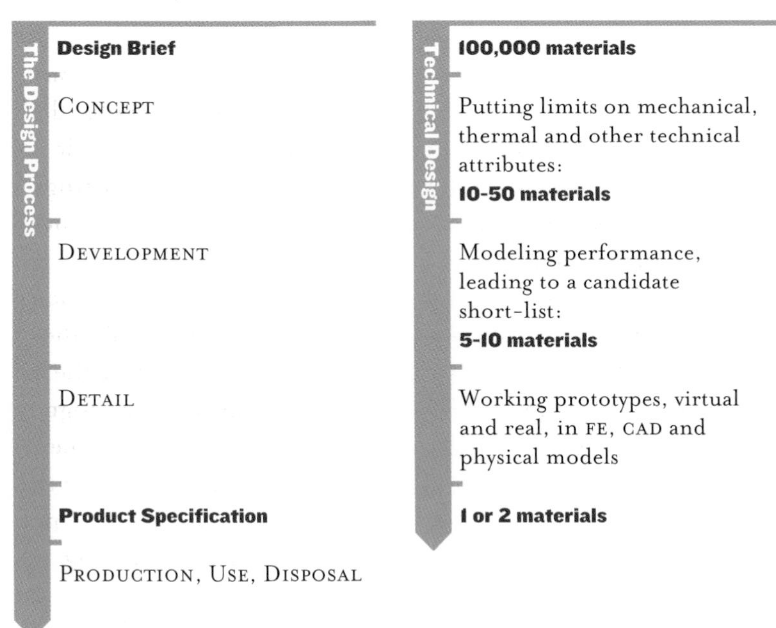

3.4 Materials in the Design Process
The design of a product proceeds from an identification and clarification of the task through concept, development and detailed design to a final product specification. Initially all materials (of which there are perhaps 100,000 in all) are candidates. Technical constraints (center) and constraints of industrial design (right) narrow the choice, leading to a small number that can be explored in detail.

applied to components and groups of components to maximize performance, and costs are analyzed. 3D surface models are used to develop form, and a final choice of geometry, material, manufacturing process and surface is made. The output of this stage is a detailed product specification.

In this way of thinking, materials information is required at each stage of the design (3.4,center and right). The nature of the information needed in the early stages differs greatly in its level of precision and breadth from that needed later on. In conceptual design, the designer requires generic information – broad character sketches – for the widest possible range of materials. All options are open: a polymer may be the best choice for one concept, a metal for another, even though the required function is identical. The need, at this stage, is not for precision; it is for breadth and ease of access: how can the vast range of materials be presented to give the designer the greatest freedom in considering alternatives?

The next step of development requires information for a subset of these materials, but at a higher level of precision and

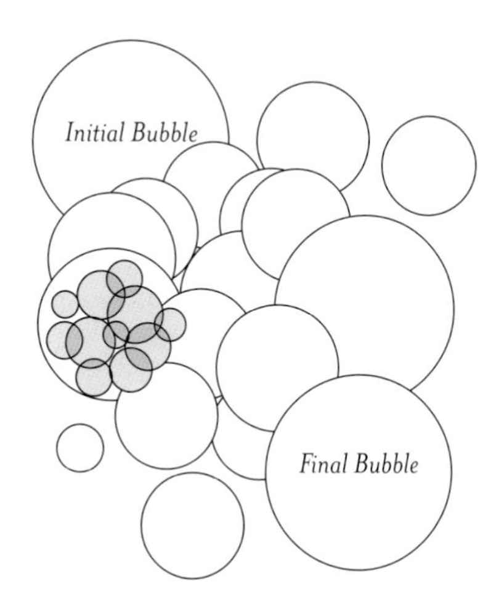

3.5 Bubbles in the Design Process
A bubble analogy for the design process. Adapted from Wallace (1991), in "Research in Design Thinking," a conference at the Technical University of Delft in 1991.

detail. Technical attributes are found in more specialized handbooks and software which deal with a single class of materials and property – corrosion resistance of metals for instance – and allow choice at a level of detail not possible from the broader compilations which include all materials. The material attributes relevant to industrial design are assembled in a different way – ideas gleaned from other designers and products, the study of material collections, the assembly of mood boards, the use of creativity aids, sketching and model-building – we return to all of these in a moment.

The final stage of detailed design requires a still higher level of precision and detail, but for only one or a very few materials – information best found by contacting the material supplier. A given grade of a material (polypropylene, for instance) has a range of properties that derive from differences in the way different suppliers make it. And sometimes even this is not good enough. If the component is a critical one (meaning that its failure could, in some sense or another, be disastrous) then it may be prudent to conduct in-house tests to measure the critical properties, using a sample of the material that will be used to make the product itself. The final step is to make and test full-scale prototypes to ensure that the design meets both the technical and aesthetic expectations of the customer.

The materials input into design does not end with the establishment of a product specification and final production. Products fail in service, and failures contain information. It is an imprudent manufacturer who does not collect and analyze data on failures. Often this points to the misuse of a material, one which redesign or re-selection can eliminate.

There is much to be said for this structured model of design. Its formality appeals to technical designers, trained in systematic methods for tackling problems of stress analysis or heat flow. But the degree of interdependence in design far exceeds that in stress analysis, so that design requires additional skills, more nearly like those of the experienced lawyer or politician – practiced in assembling and rearranging facts and in judging similarities, differences, probabilities and implications. For design – and particularly for the role of materials in design – this model is perhaps too structured. It does not allow for the variety of paths and influences that lie between the market need and the product specification.

Ken Wallace, translator of the concept/development/detail design model from its original German, suggests another analogy – that of bubbles (3.5). Here each bubble represents a step in the design process or the output of such a step. There is no linear path from the initial "design brief" to the final "product specification;" instead, many paths link the thousands of bubbles that lie between them. It is important to get into the initial bubble – to confront the problem. But there is no identified path from that bubble to the final bubble and it is always possible that other bubbles will have significant influence. The bubble model is more representative of a random and unstructured design process. The most efficient designers can immerse themselves in the bubbles and have the appropriate resources to answer any challenges that may arise. Getting caught in one bubble – the weakness of the specialist – is fatal. Progress requires the ability to step back, seeing the bubbles as pattern as well as sources of detailed information, exercising control of the pattern and only unpacking the detail when it is needed. This is closer to our view of material selection as it relates to the design process and is expanded in Chapter 7 when we detail methods of selection.

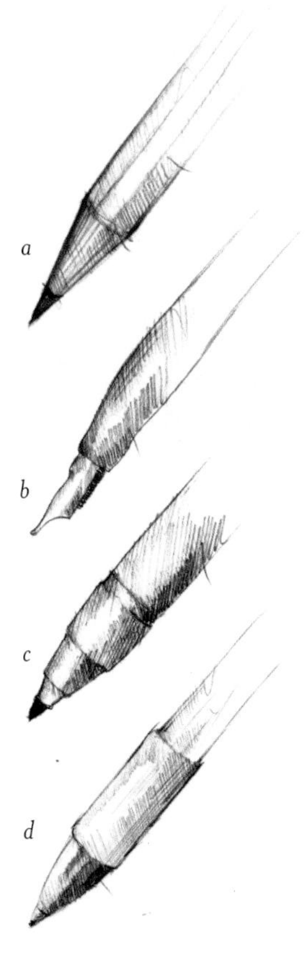

3.6 Concepts for Writing
Sketches for four concepts for writing instruments: (a) transfer of solid pigment by wear; (b) marking by flow of liquid ink from a nib; (c) marking by transfer from a porous felt or nylon, and (d) transfer of oil-based ink via a rolling ball.

The Design of Pens – an Example

To illustrate these ideas, consider the design of pens. Pens are objects both of utility and of desire. It is possible to buy, for $2, a pen that writes perfectly well; yet it is also possible to market with success one that costs more than $2000. Pens exemplify products that combine technical and aesthetic aspects in a particularly effective way. Here we look briefly at their design.

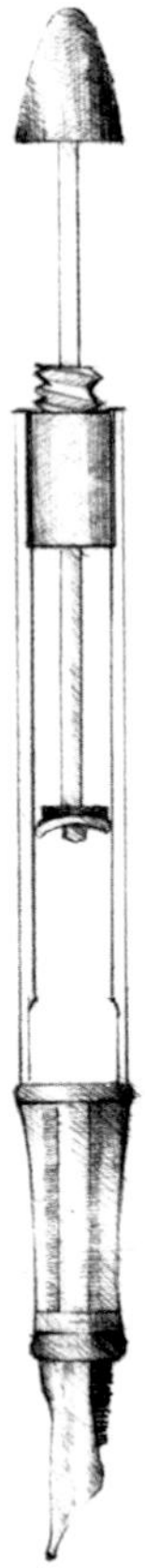

The Concepts

Ever since man devised a written language, a need existed for making accurate, durable marks on flat surfaces. Marks can be made by scratching or carving, but this is slow; the real need is for a compact instrument for marking quickly. Marking with pigments is the answer.

Figure 3.6 shows four concepts for making marks with pigments. In the concept at (a), marks are made by frictional transfer of solid pigments such as charcoal, graphite or crayon (colored pigment in a wax or chalk base). In the concept at (b) a water-based pigment or ink is caused to flow onto the surface from a nib – a cut goose-quill, or a steel or gold equivalent – using surface tension to hold the ink in the nib and duct it to its tip. The concept at (c) also relies on surface tension; it uses a porous felt or nylon tip to suck up pigment from a reservoir where it is deposited on any surface that it touches. The final concept, shown at (d), rolls an oil-based pigment onto the surface with a ball, relying on the viscosity and the good wetting of the oil to transfer it to the ball.

All of these could acquire their pigment by dipping in a well of ink or powdered pigment. But to be practical today, a pen must be capable of writing continuously for hours without recharging, and it must tolerate changes in temperature and (modest) changes in pressure without exuding ink where it is not wanted.

Development

Figure 3.7 shows a development of a concept (b): it is a reservoir-fed ink pen. The diagram is largely self-evident. The reservoir is filled by pumping a piston (which works like the piston in a bicycle pump) attached to the tail of the pen. After filling, the tail is "parked" by screwing it down onto the pen body. Ink reaches the nib by capillary action through channels cut in the plug that slots into its underside (not shown).

3.7 Development of a Fountain Pen
Development of a fountain pen, showing working principles and layout.

The Detail

To make the pen, much more must be specified. Figure 3.8 shows part of the information developed during the detailed design stage: the analysis of stresses where necessary, optimization of weight and balance, precise dimensioning, costing and, of course, the choice of materials.

The invention of ebonite (1841) – heavily vulcanized rubber – coincided with the invention of the fountain pen, and proved a near-ideal material for its body, still occasionally used today. It can be machined, molded and polished, and its rich black color makes it a sophisticated setting for gold and silver ornamentation. Its drawback is that colors other than black are difficult to achieve. New entrants in the now-competitive market (1880 onwards) switched to celluloid (a compound of nitro-cellulose and camphor) because it is easily dyed, flexible and tough, and allows special effects simulating ivory, mother of pearl and other finishes. Bakelite (1910) proved too brittle for making pens, so it was not until after 1950 that much changed. But since then, synthetic thermoplastics that can be dyed, patterned and molded quickly – notably acrylic – have dominated pen design. Special editions with metal bodies (stainless steel, titanium, silver) are all possible, but even these have polymeric inserts.

The dominance of gold for nibs arose because early inks were acidic and attacked most other metals. But gold, even when alloyed, is relatively soft, and wears when rubbed on paper, which contains abrasive ceramic fillers. This led to the inclusion of granules of osmium or iridium – both much harder than gold – into the tip, and this practice continues today. The development of neutral inks now allows the use of other materials, notably steel, alloyed or coated to prevent rusting.

Why pay $2000 for a pen when you can buy one that works very well for $2 – well, $5? (This book was written with a $5 pen.) The history of writing carries many strings of associations. Here are a few. Writing ↔ reading ↔ love of books ("a home without books is like a house without windows"). Writing ↔ learning ↔ wisdom, leadership ("knowledge is power"). Writing

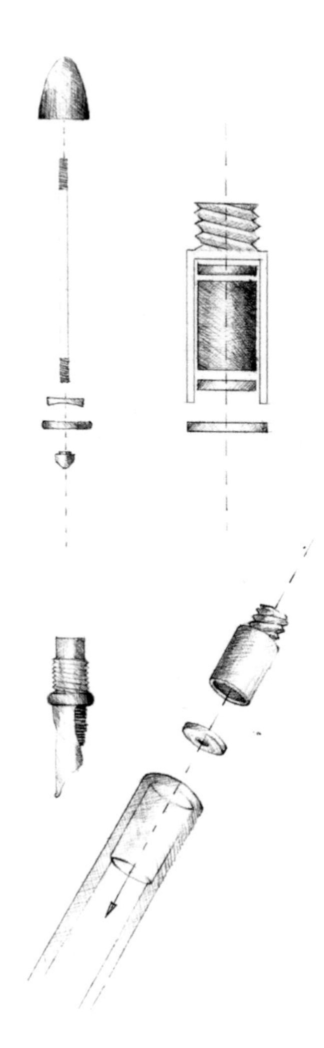

3.8 Detail of a Fountain Pen

Detailed design: materials, processes and dimensions are specified for each component.

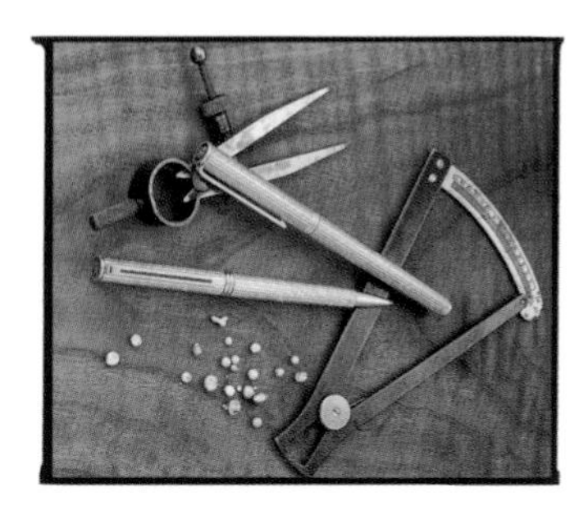

3.9 Visual Prompts

The advertising message is clear: these pens are to be thought of as precise scientific instruments, tools for intellectual achievement. (Courtesy Arnoldo Mondadori S.p.A.)

3.10 A Pen and its Utility

Ladislas Biro, a Hungarian, patented the ball-point pen in 1935. But he ran out of money and sold the idea to Marcel Bich who, in 1950, launched the BIC pen. Though cheap, thought has gone into the shape, color and texture. The clear shell allows you to see how much ink is left, its hexagon shape keeps the pen from rolling off your desk and the color of the caps matches that of the ink. The Bic Pen Company now sells over 3 million pens per day.

↔ creativity ↔ artistic acclaim ("a man of letters"). Writing ↔ political persuasion ↔ political eminence ("the pen is mightier than the sword"). Writing ↔ books and libraries ↔ personal wealth (public libraries are a comparatively recent innovation). Writing ↔ letters ↔ friendship, love ("letters – the elixir of love!").

Here we have an unusually rich brew. The linked word-strings tap into the aspirations of large and diverse groups of people. For those to whom the links are apparent, a pen that discretely suggests the individuality of the owner is worth a lot more than $5. Pen manufacturers have used this reasoning (or their own version of it) to great effect. The pen as an intellectual statement, or as an indicator of wealth, or good taste, or individuality, or youth, or fun: these are deliberate attempts to attract consumers through the personality of the product.

And now we see a sort of method at work, involving the following chain of reasoning. The product has a function. Here: writing. What associations does that function carry? Here: creativity, artistry, refinement, power, riches, sophistication. Which of these associations are seen as desirable attributes of an individual? Here: intellectual refinement, artistic achievement, sophisticated taste, leadership, wealth. What features – visual prompts, stylistic references – can be applied to the product to suggest one or another of these associations? Expensive materials, discretely noticeable silver, gold or enamel ornamentation, classical form, graphics and color suggesting recognizable artistic movements, form and surface hinting at youthful vigor and vision. The same reasoning motivates advertising: the pen seen in the hand of the Pope or of a well-known actor, the pen in opulent surroundings next to bejeweled objects, the pen lying on a sheet of classical music. Figures 3.9 and 3.10 bring out some of the associations of pens.

Product design in the 21st century must deliver a blend of function, use and personality. The first requires good technical design – products must work properly. The second is a question of human factors or ergonomics – the matching of the product to the physical and mental abilities of the user. The last is a compound of both of these and something more – the delight afforded by good industrial design. It is the role of designers to create this many-sided personality. But where do they get their ideas?

Sources of Inspiration

Newspaper reporters, it is said, never relax. Half their mind is always alert to the possibility that a news story might be about to break just round the corner. Relax and they'll miss it – worse,

Country	Design Museum
Australia	· The Powerhouse Museum, Sydney
Britain	· The Design Museum, London · The Victoria and Albert Museum, London · The Science Museum, London
Czechoslovakia	· Musée National des Techniques, Prague
Denmark	· The Danish Museum of Decorative Art, Copenhagen · Danish Design Centre, Copenhagen
France	· Musée Nationale des Techniques, Paris · Musée des Arts Décoratifs, Paris · Musée National d'Art Modern, Paris. · The Musée d'Orsay, Paris · Fondation National d'Art Contemporain, Paris
Germany	· Das Deutsche Museum, München · Vitra Design Museum, Weil am Rhein
Holland	· The Stedljk Museeum, Amsterdam · The Booymans van Beumijen Museeum, Rotterdam
Italy	· Sandretto's Plastics Museum, Pont Canavese
Sweden	· The Form Design Center, Malmo
Switzerland	· Design Collection, Museum für Gestaltung, Zurich
USA	· The Smithsonian Museum, Washington DC · Museum of Modern Art (MOMA), New York · The National Academy of Design, New York · Cooper-Hewitt, New York

3.11 Museums of Design and Applied Art
Design museums are a rich source of inspiration for designers.

another reporter may get it first. Conversations with designers reveal a similar syndrome: they are always alert to the possibilities offered by a shape, a texture, a material, a surface, an image that might be adapted in an innovative design. It is a characteristic not confined to product designers: fashion designers and architects also use what they have observed and their ability to manipulate it as a creative tool.

Industrial designers and engineers, like fashion designers and architects, describe their work as "creative," implying that their best ideas are self-generated, the result of a sort of inspiration. But even inspiration has its sources and methods.[6] Discussions with designers suggest several, and emphasize the central role of materials and processes in each. Here are some.

6. "Where do architects and designers get their ideas? The answer, of course, is mainly from other architects and designers." Stephen Bayley, British Design Critic, Commerce and Culture, Chapter 3 (1989).

Design Magazines and Annual Reviews, Museum Exhibitions and Tradeshows

Design magazines[7] and annual reviews[8] illustrate innovative products, attaching a brief description listing designer and – sometimes – materials. Design yearbooks[9] do much the same, but devote more space to the design history and antecedents, to motivation, and to descriptions of features, often augmented by interviews with the designers themselves. Browsing through these can provide inspiration, but the largely unstructured nature of the information that they contain makes their efficient use difficult. Most designers also seek ideas from a wider range of sources encompassing fashion, architecture and design publications such as Vogue, Domus and Wallpaper. Design exhibitions[10] expose the viewer to rich sources of ideas drawn from diverse product sectors, often suggesting the use of materials in a product to achieve a particular feel, texture, color or association. Museums of Design and Applied Art (3.11) record design history and innovation, combining permanent displays with temporary exhibitions and educational programs. Finally, there are tradeshows such as the annual Hanover Trade Fair, the Polymer Tradeshow – K – in Dusseldorf and Milan's Furniture Show; each acts as a showcase for contemporary materials and design.

7. The ID Magazine is a good example.

8. See, for example, Byars (1995, 1997a, b, 1998).

9. The International Design Year Book (1998, 1999); The ID Magazine Annual Design Review (1998, 1999, 2000).

10. See for example, the catalog of the MOMA exhibition "Mutant Materials" (Antonelli, 1995), or that mounted by Material Connexion called "Materials and Ideas for the Future" (Arredo, 2000).

Material Sample Collections

Many individual designers and design houses assemble collections of materials – not just metals, plastics and ceramics, but also weaves, finishes, coatings and shapes – materials that have, in various ways, been processed. The physical nature of the samples is the key point: new ideas – inspiration – can come more readily from handling (not just visualizing) a material. Familiar materials carry associations that derive from their traditional uses: polished wood – the sense of warmth, civilization, discrete luxury; brushed aluminum – the sense of clean mechanical precision, and so forth. But the use of familiar materials in an unfamiliar way is also a creative step. New materials, particularly, act as a trigger to inventive thinking, offering the potential for novel design. This is where maintaining a material collection becomes challenging. Most material development is driven by technical need, not by motives of industrial design, with the result that information does not readily reach the designer. Material information services[11] exist that carry large sample collections and offer web-based access to images and suppliers, but not much more. There is a need for an accessible sample collection with links from the image and supplier to a larger file of aesthetic and technical data. We return to this point later in the book.

11. An example is the subscription service of Material Connexion, www.materialconnexion.com.

Mood Boards or Collages

A "mood board" is, in part, a personalized, project-focused image collection, enhanced by a collection of material samples. Samples of materials, colors and textures are chosen because they have features – an ability to be formed, to accept a finish, to evoke an association or emotion – that might contribute to the design. Images of products that have features like those the designer seeks, and images of the environment or context in which the product will be used, act as prompts for creative thinking. It is not uncommon for a designer, confronted by the challenge of molding the character of a product, to first buy and dissect examples of any product that has a feature – a surface finish, an association, a style – that might be exploited in a fresh way. Visually, these are assembled

3.12 Mood Boards
A mood board for a pen that will be used on the beach.

into a collage or mood board (3.12), acting as a trigger for ideas both about choice of materials and about their juxtaposition. Here, too, it is the material and the way it is used that is the starting point; the creative step is that of transforming it into the context of the new product.

Creativity Aids

Techniques exist to inspire creativity, though it must be said that views differ about their value. Brainstorming relies on the group dynamics that appear when participants express their ideas, however wild, deferring all value judgement until the process is over. One person's wild idea can stimulate a practical solution in another: "golf clubs are made of titanium, golf clubs swing, could playground swings be made of titanium?" In synectics a familiar design problem is made strange by placing it in another context and looking for solutions, a few of these solutions are then transferred back to the original problem: "we need a strong, flexible and lightweight material solution – how does bamboo grow so that it is strong, flexible and lightweight – can we mimic these solutions?" Mind-mapping is a sort of personalized brainstorming in which ideas are placed on a page and linked as appropriate; these links are used to stimulate further thinking: "wood... cells... porous solids... foams... metal foams... titanium foam...?" And finally,

Improvisation, a dramatic technique, can be used to inspire new ideas; on the stage of design a random association of materials and products can sometimes provide innovative result.

Sketching and CAD

The form of a new product first takes shape by sketching – freehand drawing, freely annotated, that allows the designer to explore alternatives and record ideas as they occur. Sketching is a kind of image-based discussion with oneself or with others – a way of jotting down ideas, rearranging and refining them. Only at a later stage is the design dimensioned, drawn accurately and coded into a surface modeling package. Modeling software[12] allows the display of projections of the product and experimentation with visual aspects of color and texture. But designers emphasize that it is not these but the act of sketching that stimulates creativity.

12. Pro-engineer's SolidWorks 2001 is an example of a solid modeling program for mechanical engineers and designers with the ability to download files for rapid prototyping.

Model-making and Prototyping

A design evolves through models; these models are an important means of communication between industrial and technical designers, and between the designers and the client for whom the product is to be made. Preliminary models, often made of polymer foam, plaster, wood or clay, capture the form of the product; later models show the form, color, texture, mechanisms and weight. Models allow the product to be handled (if small) and viewed from multiple angles (particularly important when large). Rapid prototyping has transformed the later stages of model-making, allowing a CAD file from a modeling package to be downloaded and converted directly into an wax or polymer model.[13]

13. Profiles of rapid prototyping systems will be found in the Reference Profiles at the end of this book.

Nature as Inspiration

To "inspire" is to stimulate creative thinking. Inspiration can come from many sources. That for product design comes most obviously from other products, and from materials and processes, particularly new ones. After that, nature is perhaps the richest source. The mechanisms of plants or animals – the things they can do, and the way they do them – continue to

3.13 Nature
Throughout history nature has provided a rich source of inspiration for designers.

mystify, enlighten and inspire technical design: Velcro, non-slip shoes, suction cups, and even sonar have their origins in the observation of nature (3.13). Nature as a stimulus for industrial design is equally powerful: organic shapes, natural finishes, the use of forms that suggest – sometimes vaguely, sometimes specifically – plants and animals: all are ways of creating associations, and the perceived and emotional character of the product.

Conclusions: a Creative Framework

The picture that emerges is that of the designer's mind as a sort of melting pot. There is no systematic path to good design, rather, the designer seeks to capture and hold a sea of ideas and reactions to materials, shapes, textures and colors, rearranging and recombining these to find a solution that satisfies the design brief and a particular vision for filling it. Magazines and yearbooks, material collections, museum exhibitions, tradeshows and the designer's own experience provide the raw ingredients; the designer permutes, modifies and combines these, stimulated by brainstorming or free discussions and astute observation of what other designers are doing. This involves drawing freely on memories or images, combining aspects of each, imagining and critically examining new solutions – a sort of self-induced virtual reality. The brain is good at storing an enormous number of images, but is poor at recalling them in detail without prompts or specific triggers.

These prompts can be created by using stored images and visual information about materials and products, provided these are indexed in ways that allow rapid recall. Indexing, then, is key to ordering material and process information to provide a design tool. To do this effectively, indexing must be sufficiently abstract to capture relationships, yet be sufficiently concrete to be easily understood by a novice user. If this can be achieved, a framework for organizing and manipulating the attributes of materials and their role in technical and industrial design becomes possible. The ultimate aims of this creative framework might be:

· to capture and store material and process information of the sort found in magazines, yearbooks, and exhibition-based publications.

· to present information about materials and processes in a creative format that is relevant to product design.

· to allow browsing of potential materials or processes via hypertext links or free-text searching.

· to allow retrieval of "bits" of information about the technical and perceived attributes of materials, processes and the products they create.

These all lie within the scope of present day software engineering, but they require more thought about the best way to organize information for designers. We will consider them further in Chapter 7.

Further Reading

Antonelli, P. (1995) Mutant Materials in Contemporary Design, Museum of Modern Art, New York, USA. ISBN 0-87070-132-0 and 0-8109-6145-8. *(A MOMA publication to accompany their extensive 1995 review of materials in product.)*

Arredo, F. (2000) "Materials and Ideas for the Future," Material Connexion, 4, Columbus Circle, New York, NY, USA. *(The catalogue of an exhibition at the Salon Internatazionale del Mobile, Milan, 2000.)*

Ashby, M.F. (1999) "Materials Selection in Mechanical Design," 2nd edition, Butterworth Heinemann, Oxford, UK. ISBN 0-7506-4357-9. *(A text that complements this book, presenting methods for selecting materials and processes to meet technical design requirements, and presenting a large number of material property charts.)*

Beylerian, G.M. and Osborne, J.J. (2001) "Mondo Materialis," the Overlook Press, Woodstock and New York, NY, USA. ISBN 1-58567-087-1. *(Creative use of materials from 125 architects and designers, presented as collages, in large format and full color. The original collection was created for a 1990 exhibition.)*

Byars, M. (1995) "Innovations in Design and Materials: 50 Chairs;" (1997) "50 Lights"and "50 Tables;" (1998) "50 Products," RotoVision SA, Switzerland. ISBN 2-88046-264-9, 2-88046-265-7 and 2-88046-311-4. *(Lavish color illustration with minimal text cataloging the designer, material and processes surrounding groups of 50 contemporary products.)*

Dragoni, G. and Fichera, G. (1998) editors, "Fountain Pens, History and Design," Antique Collectors' Club, 5, Church Street, Woodbridge, Suffolk UK. ISBN 1-85149-289-5. *(A lavishly illustrated compilation of the history and characteristics of pens from the great pen-makers: Parker, Sheaffer, Montblanc, Waterman, Pelikan.)*

Ferguson, E.S. (1992) "Engineering and the Minds Eye," MIT Press, Cambridge Mass USA. ISBN 0-262-06147-3. *(A largely historical sweep through design history since 1700, highlighting past and present deficiencies, and emphasizing visual imagination above pure analysis in the act of design.)*

ID Magazine (1998, 1999), Pearlman, C. editor, 440 Park Avenue South, Floor 14, New York, NY. *(The International Design Magazine carries reviews of contemporary and experimental products and graphics.)*

McKim, R.H. (1980) "Experiences in Visual Thinking," 2nd edition, PWS Publishing Company, Boston, USA. ISBN 0-8185-0411-0. *(A very readable introduction to creative thinking and design.)*

Pahl, G. and Beitz, W. (1997) "Engineering Design," 2nd edition, translated by K. Wallace and L. Blessing, The Design Council, London, UK and Springer Verlag, Berlin, Germany. ISBN 0-85072-124-5 and 3-540-13601-0. *(The Bible – or perhaps more exactly the Old Testament – of the technical design field, developing formal methods in the rigorous German tradition.)*

The International Design Yearbook (1998), Sapper, R. editor, and (1999), Morrison, J. editor, Laurence King, London, UK. *(An annual review of innovative product designs.)*

Tufte, E.R. (1983) "The Visual Display of Quantitative Information," Graphics Press, PO Box 430, Cheshire, CT 06410 USA. ISBN 0-9613921-0-X. *(The bible of graphical methods for representing – and misrepresenting – information. The book makes the case that graphics, at their best, are instruments for reasoning about information. The book itself is a masterpiece of graphic design.)*

Chapter 4

The Stuff... Multi-dimensional Materials

Materials are the stuff of design. When we speak of "information" for materials what do we mean? Figure 4.1 illustrates the steps involved in moving a material from the laboratory into a successful product. Tests yield raw data. These are distilled, via appropriate statistical analyses, into data for "allowables": values for properties on which design can safely be based (typically, 3 standard deviations below the mean). A material may have attractive "allowables" but to make it into a product requires that it can be shaped, joined and finished. The characterization of the material is summarized in a table of such information (4.1, center). It enables safe, technical design of the product – the engineering dimension, so to speak.

That is only the start. The product will be used – the choice of material is influenced by the nature of the user: children, perhaps, or travelers, or the elderly. Does it comply with legislative requirements that the product must meet (FDA approval, perhaps, or requirements limiting flammability, noise, vibration or bio-compatibility)? Would it survive the use and misuse it will encounter in service? Is it toxic? Information for this – the dimension of use – is as important in guiding selection.

There is more. Product manufacturers today strive for ISO 9000 or 14000 qualification – meaning that they have established quality standards and procedures of environmental auditing and responsiveness. This, the environmental dimension, requires yet another layer of materials information, one relating to product manufacture, use and disposal. And even that is not the end. The industrial design of a product is as much a part of its creation as any other. The characteristics of a material that contribute to industrial design, though harder to pin down, are as important as those for technical properties. This – creating satisfaction – requires a fourth dimension of materials information: that relating to aesthetics and personality. Given these, rational design with the material becomes possible. Carrying the design into production depends – as we said in Chapter 2 – on investment, attracted by a successful business case. But we have enough to discuss without that. Here we explore the four dimensions of materials information: engineering, use, the environment and perceptions.

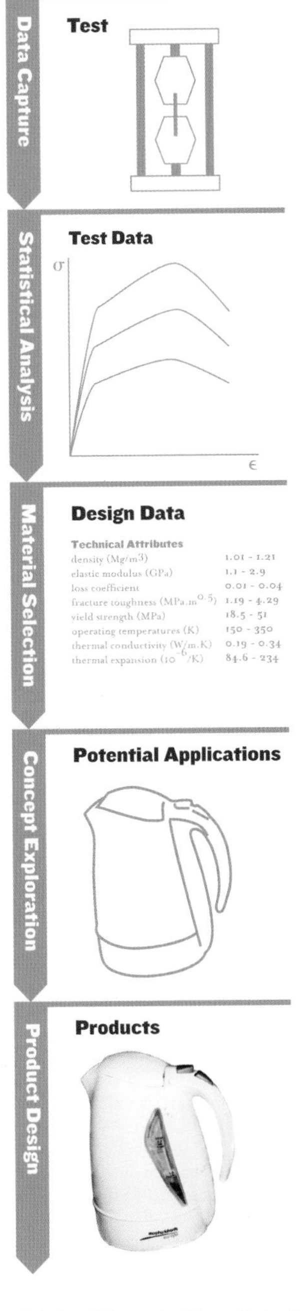

4.1 Confidence in New Materials
The steps in moving a material from the laboratory to a successful product.

The Engineering Dimension: Technical Attributes

The scientific study of materials – materials science – seeks to understand the fundamental origins of material properties, and, ultimately, to manipulate them. It has had remarkable success in doing both. The origins of many material properties derive directly from the atomic and electronic structure of the material: among these are density, stiffness, thermal and electrical conductivity, optical transparency and many others. These are now well understood, and can, within the limits imposed by the laws of physics, be manipulated. Composites, one of the great technical advances of the last 40 years, combine the properties of two very different materials: polymers and carbon fibers in sports equipment, elastomers and steel in car tires, metals and ceramic fibers in aerospace components. Here, too, the scientific understanding is deep and the ability to "design" materials is considerable.

Material science has developed a classification based on the physics of the subject (4.2). It is not the only one – architects, for instance, think about materials in other ways.[1] The science-based classification emerges from an understanding of the ways in which atoms bond to each other, and (in the case of composites) how mixtures of two different materials, each with its own attributes, behave. But science is one thing, design is another. Is the science-based classification helpful to the technical designer? To explore this further we must first look at the technical attributes of materials. The classification of 4.2 – Family, Class, Member – is based, at the first level, on the

1. *Cardwell et al. (1997), for example, suggest a radically different classification – one based on familiarity rather than physics: unfamiliar, familiar, contemptible (!), unknown, and unknowable.*

4.2 Classification of Materials
A classification of materials based on a scientific understanding of the nature of the atoms they contain and the bonds between these atoms. The final column shows a list of possible attributes for a specific material.

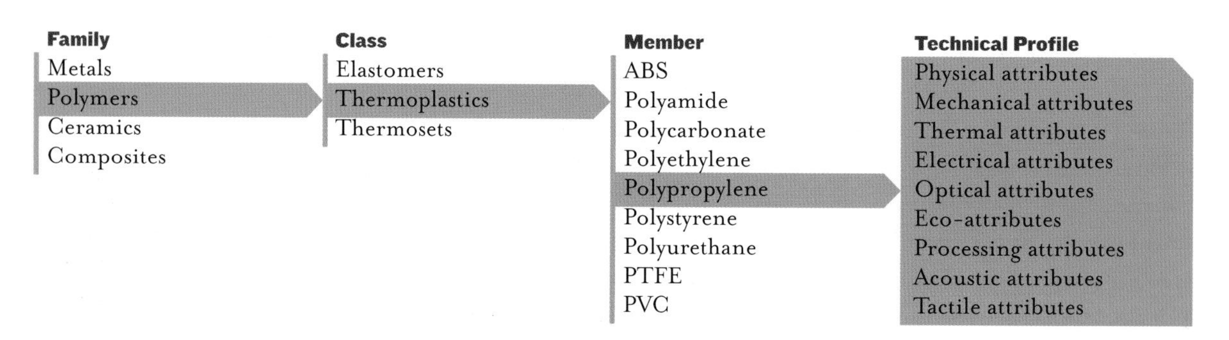

nature of the atoms of the material and that of the bonding between them (e.g. "polymer"), at the second level on its variants (e.g. "thermoplastic"), and at subsequent levels, the details of its composition. Each member has a set of attributes that quantify its physical, mechanical, thermal, electrical and optical behavior – what we will call its technical profile.

Open a handbook or search the web, seeking a material by name, and you will find technical data. This information is largely numeric: values for modulus, strength, toughness, hardness, thermal conductivity and expansion coefficient, electrical resistance and so forth (4.3), backed up with a little text-based information on corrosion resistance and wear. These are the data needed for technical design – for the calculation of safe loads, temperatures, heat fluxes, and operating life. The most useful classification of materials for technical design is one that groups materials that have similar profiles of these technical attributes. But what do we mean by "similar"?

4.3 Technical Profiles

Selected technical attributes for polypropylene.

Physical Attributes	
Density, kg/m³	900–910
Mechanical Attributes	
El. modulus, GPa	1.14–1.55
Yld. strength, MPa	31–35
Tensile strength, MPa	33–36
Comp. strength, MPa	37–45
Elongation, %	100–350
Toughness, kJ/m²	10–11
Fatigue limit, MPa	11–15
Hardness, Vickers	9.2–11
Thermal Attributes	
Max use temp., C	90–105
Th. conductivity, W/m·C	0.11–0.12
Th. expansion, /C · 10^{-6}	145–180
Molding temp., C	210–250
Electrical Attributes	
Dielectric constant	2.2–2.3
Dielectric loss, %	0.05–0.08
Resistivity, ohm·cm	$3 \cdot 10^{22}$–$3 \cdot 10^{23}$

Mapping Technical Attributes

Asking if two colors are "similar" can be answered by comparing their wavelengths. But if, by "similar" you mean a larger set of properties, you are asking for something more difficult: recognition of a pattern of behavior. The brain is better at pattern-recognition when the input is visual rather than text-based. So: how can we make technical attributes visible? One way is by plotting them in pairs, to give a sort of map of where they lie. Figure 4.4 is an example. Here the first two of the attributes from 4.3 – elastic modulus, E, measuring stiffness, and density, ρ, measuring weight – are mapped, revealing the layout of the E–ρ landscape. The dimensions of the little bubbles show the range of modulus and density of individual classes of material; the larger envelopes enclose members of a family. Metals cluster into one part of the map, polymers into another, ceramics, woods, foams, elastomers into others. Like maps of a more conventional kind, it condenses a large volume of information into a single, easily readable, image.[2]

What do we learn? That members of each of the families of 4.2 cluster in a striking way. They do, when judged by these two

2. The labels that are used in these maps are shortened for the sake of space: fPU is flexible polyurethane foam, rPU is rigid, ocPU is open cell, ccPU is closed cell, tsPolyester is thermo-setting polyester, tpPolyester is thermoplastic.

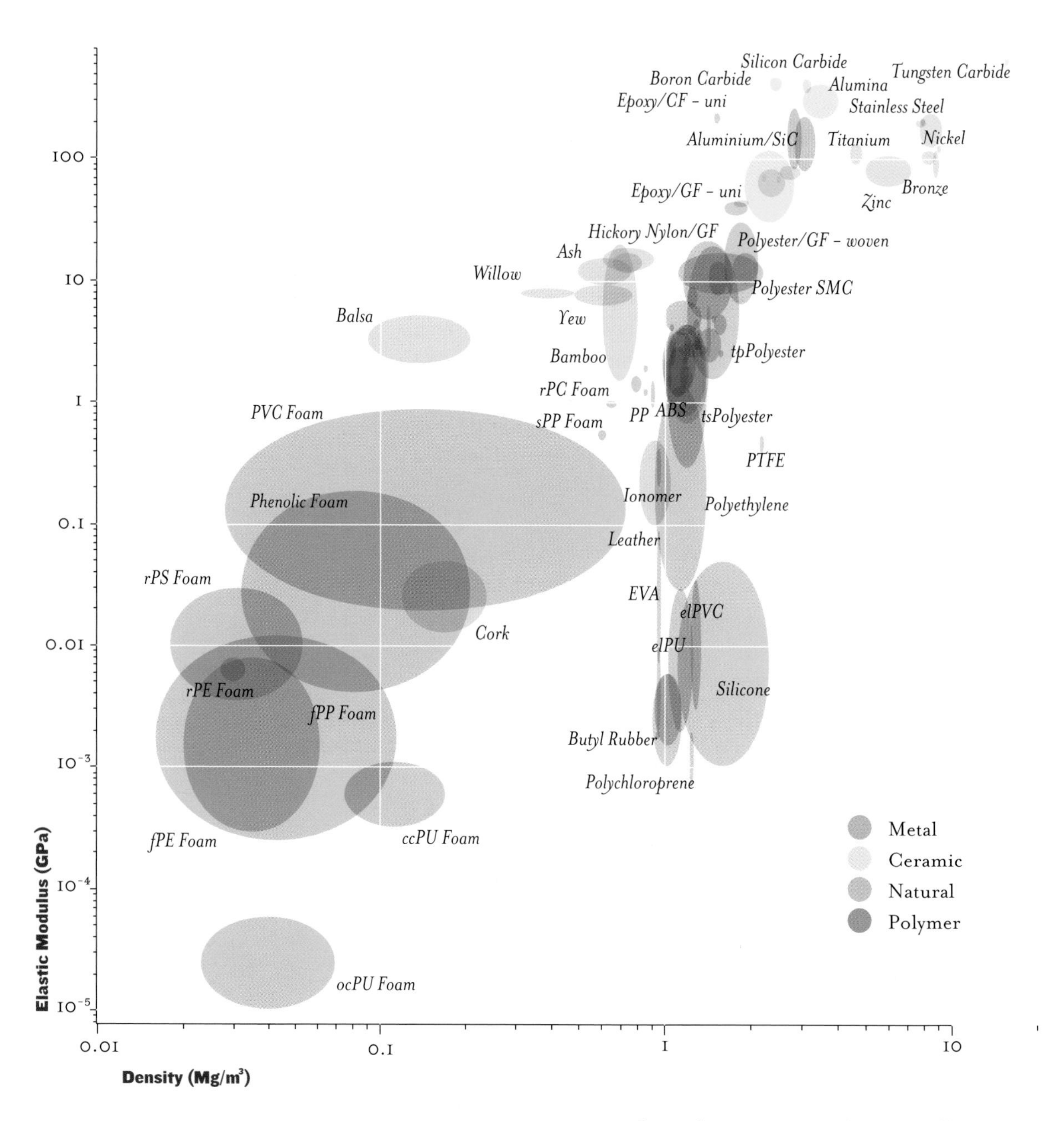

4.4 Elastic Modulus and Density

A chart of Elastic Modulus and Density for materials, showing the clustering of metals, polymers, ceramics, foams, woods, etc. (CES 4, 2002).

engineering properties, have distinctive similarities, allowing them to be grouped and to be distinguished from other families that group elsewhere. There is, however, overlap – the lightest, stiffest, metals overlap with technical ceramics and composites; elastomers overlap with polymers – here the separation is ambiguous.

But that was only two technical attributes. Materials have many more. Plotting them,[3] you find the same groupings, but the overlaps that are not the same. If the classification you want is one based on technical, "engineering" attributes then it looks as though the right one is the one the scientists propose.

Maps like that shown in 4.4 can suggest composites and blends of two materials. The most common composites are fiber or particulate-reinforced polymers. Glass fiber is stiffer and denser than nylon; adding glass fibers to the nylon gives a material with stiffness and density that lies somewhere in between. Polymer blends, similarly, have properties intermediate between those of the pair of polymers that were blended. The maps allow this to be visualized, suggesting combinations that might meet a specific need.

So visual presentation of data can reveal similarities that are hard to see in other ways. And that makes you wonder if you could go one step further and, instead of mapping many separate pairs of attributes, find a way of combining them all to make a single super-map – a global overview. Well, yes and no. Yes, there is a way of doing this. But what it reveals is not so clear. Still, it's worth looking at. It is called multi-dimensional scaling, or MDS.[4]

MDS is a way of revealing similarities and differences between members of a group, using information about many attributes, not just two. Applied to materials, it works like this. First, calculate the "distance" between each material and every other in the group, taken in pairs. "Distance" is a measure of dissimilarity: if all the attributes of two materials are identical, the distance between them is zero. If all but one are identical, and the odd one out differs by ten units, the distance might be given the value 10. If all are different, and by varying degrees, each difference is given a value and all the values are aggregated to give a single number: this is the overall "distance" between the two materials. The output of this step is a table, showing the distances between each material and every other.

You can learn something by looking at this table, picking out groups of materials that are separated by the shortest distances. But when the number of materials is large, this is hardly a visually

3. There are maps like 4.2 for many other engineering properties of materials. They can be copied without restriction from Ashby (1999), or downloaded free of charge from www.grantadesign.com.

4. The methods of multi-dimensional scaling are well described in the little book by Kruskal and Wish (1987).

inspiring task. So there is a second step – that of manipulating this table to make a "distance-map." It involves a kind of optimization technique that is best explained as follows. Suppose you had only three materials, and had calculated, for as many attributes as you wished, the three distances between them. You could then make a picture on which you placed a dot and called it "material 1," and then a second dot, at a length equal to the material 1 – material 2 distance that you called "material 2," and then a dot for material 3, making a triangle, so positioned that it was just the right distance from 1 and 2.

Fine. But now add a fourth material. In general, you can't find a place where it exactly fits – where its distances from 1, 2 and 3 are all correct; it's a consequence of the two-dimensionality of the picture. But suppose you could stretch some of the distances and squeeze others, you could go on doing so until there was a position in which the fourth material would fit – although the map is now distorted, an approximation, no longer an exact visualization of the distances. And you could go on adding more and more materials, each time doing a bit more stretching and squeezing. This, you might think, would lead to an ever more distorted map, but if you had an algorithm that minimized, at each step, the total stretch and squeeze, you could keep the distortion to a minimum. And that is exactly what multi-dimensional scaling does.[5]

5. Many arbitrary decisions are involved. "Distance" can be measured and aggregated in more than one way. An example: normalize all data to a set range – say -10 to 10 – calculate the individual distances for each attribute of a material, then aggregate by forming the root-mean-square of these (Euclidean aggregation). Minimizing the stretch and squeeze, too, allows many alternatives. An example: minimize the root-mean-square of the deviations of the stretched or squeezed distances from the values they should, ideally, have. Sounds sensible, but both examples contain implicit assumptions about the relative "weights" of small and big differences, and about the extent of acceptable distortion. If you want to use the method, read Kruskal and Wish (1987) first.

The maps that emerge have to be interpreted with caution – they are approximations. The "distance" measure on which they depend rolls many attribute into a single number; and in doing so it throws away a great deal of information ("does a distance of 20 mean that all attributes differed a little, or that most were identical and one was very different?"). But the method does produce interesting pictures, in which clustering is clearly evident.

Figure 4.5 is an example. It is an MDS map for engineering materials based on 15 of their mechanical and thermal attributes. Ceramics, metals and polymers each fall into separate groups; and within polymers, thermoplastics and elastomers form overlapping groups. The analysis has led to groupings that in most ways resemble the families and classes of material science.

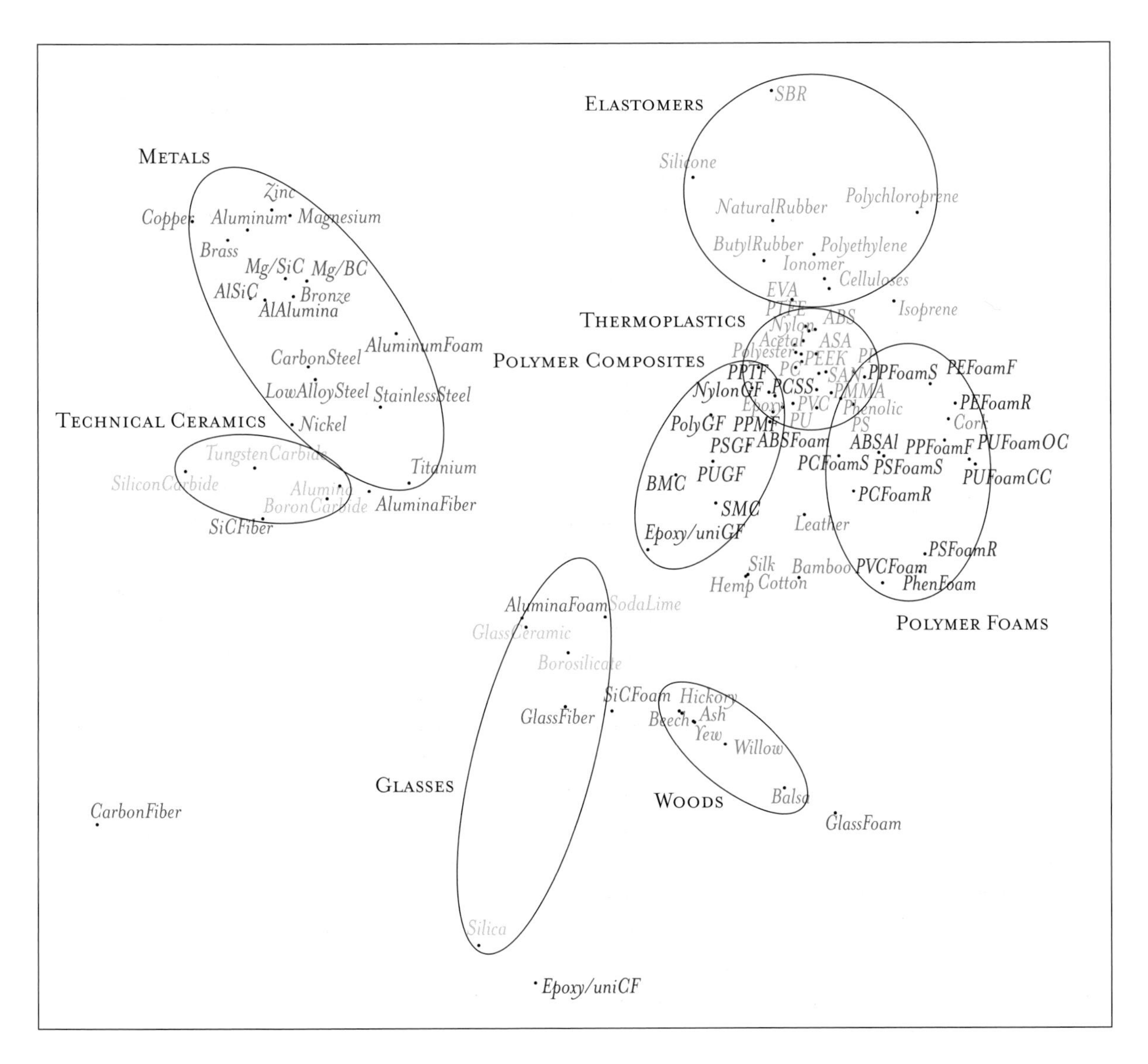

4.5 Multi-dimensional Maps: Mechanical Attributes

This map show similarity between materials based on their mechanical attributes (data from CES 4, 2002).

So, for technical attributes, the classification of 4.2 is the one we want. Classification is the basis of indexing – an essential step in cataloging materials information. We will return to indexing in Chapter 7. Here we simply note that, for technical attributes, we have a good basis, and we turn briefly to the uses of technical attributes in product design.

Using Technical Attributes

The success of a product in the marketplace depends heavily, we have argued, on its industrial design – on its appeal, visually and emotionally, to the consumer. But this presupposes that it works – that it performs its technical function well, safely, and at acceptable cost. The responsibility for this rests with technical design, and technical design relies on sound material data.

The characterization of materials for technical design – the top half of 4.1 – is well developed. Datasheets like those of 4.3 are widely available.[6] The information they contain provides the necessary inputs for technical design, using the standard methods of engineering: stress analysis, thermal analysis, optimization methods and simulation. All of these are supported by sophisticated software tools. This sounds reassuring – and it is – but one real challenge remains. You can find technical data for a material if you know the one you want. But how do you make that choice? And how do you make sure it is compatible with the other "dimensions" with which this chapter started? We return to this question in Chapter 7. But first we examine the other dimensions and the materials information needed to address them.

6. Material suppliers provide data-sheets for their products, though the content is often limited. More comprehensive data are tabulated in Handbooks such as those published by ASM International, Butterworth Heinemann and McGraw Hill, some now available in searchable CD format. Software systems such as that of Granta Design (CES 4, 2002) offer fast access to some thousands of technical data sheets in standard format, compiled from many different sources. Increasingly, technical data can be found on the Worldwide Web sites such as www.matweb.com.

The Use Dimension: Ergonomics and the Product Interface

Everyone knows the TV with the switch so cleverly integrated into the styling that it can't be found in the dark, the tap too smooth for soapy fingers to turn, the hairdryer so noisy that you can't hear the phone... well, the list is long. Each of these products works but has a poorly designed interface – it is difficult to *use*. Well-designed products not only work, they are convenient, safe and easy to interact with – they are user-friendly.

The design of a product's interface has three broad aspects: the first concerns the matching of the product to the capabilities of the human body; the second, matching to the reasoning power of the human mind; and the last, matching with the surroundings in which the human lives and works. Collectively, these are known as human factors and their study is called ergonomics, interface design, or human-factor engineering, which we now examine.

Anthropomorphics and Bio-mechanics

During World War II it became apparent that the effectiveness of service men depended strongly on the ease of use of the increasingly complex equipment on which they had to depend. This led to the study of *anthropomorphics* – the measurement of the size and shape of the human body – and of *human biomechanics* – the analysis of the movement, forces, power and stamina of which the body is capable.

Figure 4.6 lists some of these for European and North American populations. Physically, people vary in size and strength. As a yardstick, designs often use ranges that bracket 90% of the mature population of working age, and it is this range that appears in these tables. They define what might be called the Standard Design Person. In design for usability, these characteristics are treated as constraints: the product should accommodate any person with the dimensions and physical capabilities that lie within these ranges. Increasingly, in the workplace, these constraints must be met to conform with legislation intended to prevent undue fatigue or muscle strain and the accidents they cause.

But there is a story, relevant at this point, of the man who invented a shaving machine. You clamped it to your face, pressed the button, and – hey presto – you were shaved. When asked whose face it fitted he said it fitted the standard face. But

Characteristic	**Man (18-40)**	**Woman (18-40)**
Standing Height, m	1.63–1.85	1.54–1.76
Seated Height above seat, m	0.84–0.96	0.79–0.91
Seat Width, m	0.33–0.40	0.35–0.43
Forward Reach, m	0.77–0.92	0.60–0.75
Shoulder Width, m	0.42–0.50	0.38–0.46
Eye Height, Standing, m	1.52–1.74	1.44–1.65
Body Mass, kg	60–94	44–80
Head Mass, kg	1.4–1.7	1.25–1.6

4.6 Human Measurements
Average attributes of European and North American men and women of working age.

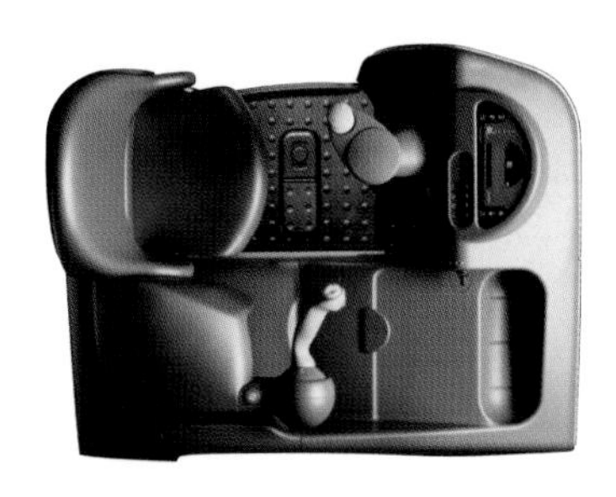

4.7 Ergonomic Design
The control center of the forklift, designed to make complex maneuvers as intuitive as possible. (Courtesy Ergonomic Systems Inc.)

what happened if someone who used the machine did not have a standard face? After they've used my machine, he said, they will.

So the idea of a standard person is a little ridiculous. For true usability, equipment must be matched to the capabilities of the individual, not to some average. Nowhere is this matching more evident than in the design and marketing of sports equipment. Here products are carefully tuned to the bio-metric and bio-mechanical characteristics of the individual athlete or sportsman, and doing so has become a very big and profitable business. These however are not the only groups that require special attention. Others with Non Standard Design Bodies (meaning that they lie outside – sometimes far outside – the ranges of the previous tables) include children, the disabled, and the elderly. Their needs, often neglected in the past, are now more widely recognized and designing to meet them is seen as a priority.

Materials have a role to play here. Heavy objects, often a source of physical strain when lifted or moved, can be made lighter by the use of light metal alloys, polymers and foam-cored sandwich-structures. Elastomers can provide grip where it is needed, polymer gels and foams allow soft tactile surfaces, and woven and non-woven fabrics conform to the body shape. Molded shapes that fit well in the hand and have smooth rounded corners that do not catch on clothing reduce the risk of accidents. Differences in size can be accommodated by using materials that allow adjustability – Velcro, demountable adhesives, elastomers.

All this is straightforward. But products interact with their operators in other ways – ways that require the communication of specific information.

Information Management

The products with which we interact today are more complex and have more functionality than at any time in the past. Much of the functionality now derives from, or is controlled by, electronics. Electrons, unlike simple mechanical things, are invisible and give few clues of what they are up to or that they are responding to the user's wishes. Thus two sorts of communication must be built-in to the overall design – the passive one, indicating function, and the active one, indicating response to an input.

The first need is met by clear icon or pictogram-based identification. Icons (4.8) are easily read and, in a world of international trade and travel, have the considerable asset that they are language independent.[7] Emoticons (4.8) – icons adapted to electronic messaging systems – are based on the specific combinations of key strokes and can be interpreted as emotions. When complete words must be printed on the surface of a product, experiments show that the best legibility is given by a *san serif* typeface such as Helvetica or Arial, the use of lower case letters except for initials, and high contrast.[8]

A person operating a device is part of a closed control-loop (4.9). Such a loop requires, first, one or more *displays* that document its current state. Visual displays are the most efficient way to transmit this information accurately, though for urgent warnings, sound is better. Second, the loop requires *control elements* or input methods. Push buttons, knobs and levers have been largely replaced by keyboards, mice, laser pointers, touch-sensitive screens and speech recognition systems. The inputs are interpreted by the device, which then reacts to them. The third requirement is an *indicator* – visual, tactile, or acoustic – that the input has been received; without it the operator has no immediate confirmation of acceptance and may mistakenly repeat or override the operation. The remaining element of the loop – that of *decision* – is provided by the operator.

Here it is materials for sensors, actuators and displays that are important. Sensors are often based on piezo-electric materials, thermal sensing materials and light sensing materials (photo diodes). Actuator materials include muscle (the old way of activating everything), bi-metallic material, shape-memory alloys, piezo-electric, electro-strictive and magneto-strictive materials and polymer gels. Thermo-chromic materials change color on heating; electro-chromic materials do so when a field is applied. Flexible materials can be shaped to give bi-stable key response, confirming keyboard and keypad inputs through

a

b

"What a great day!" *"Why must I have such a long nose?"*

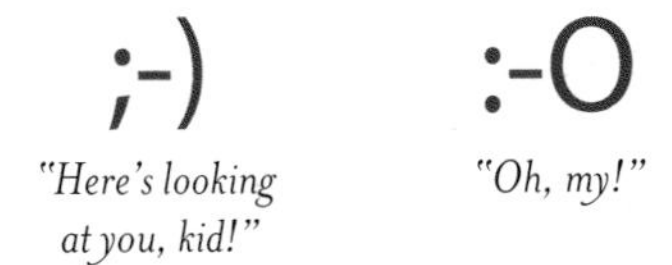

"Here's looking at you, kid!" *"Oh, my!"*

4.8 Icons and Emoticons
Visual triggers of direction in our everyday life and of emotion (in electronic text).

7. Caution is needed here – the interpretation of an icon can depend on culture. A knife-and-fork icon indicating "restaurant," could be meaningless to someone who eats with chopsticks.

8. The same is not true of text in a book. A typeface such as Times, with serifs – the thin horizontals that terminate the verticals – leads the eye more easily along and between lines.

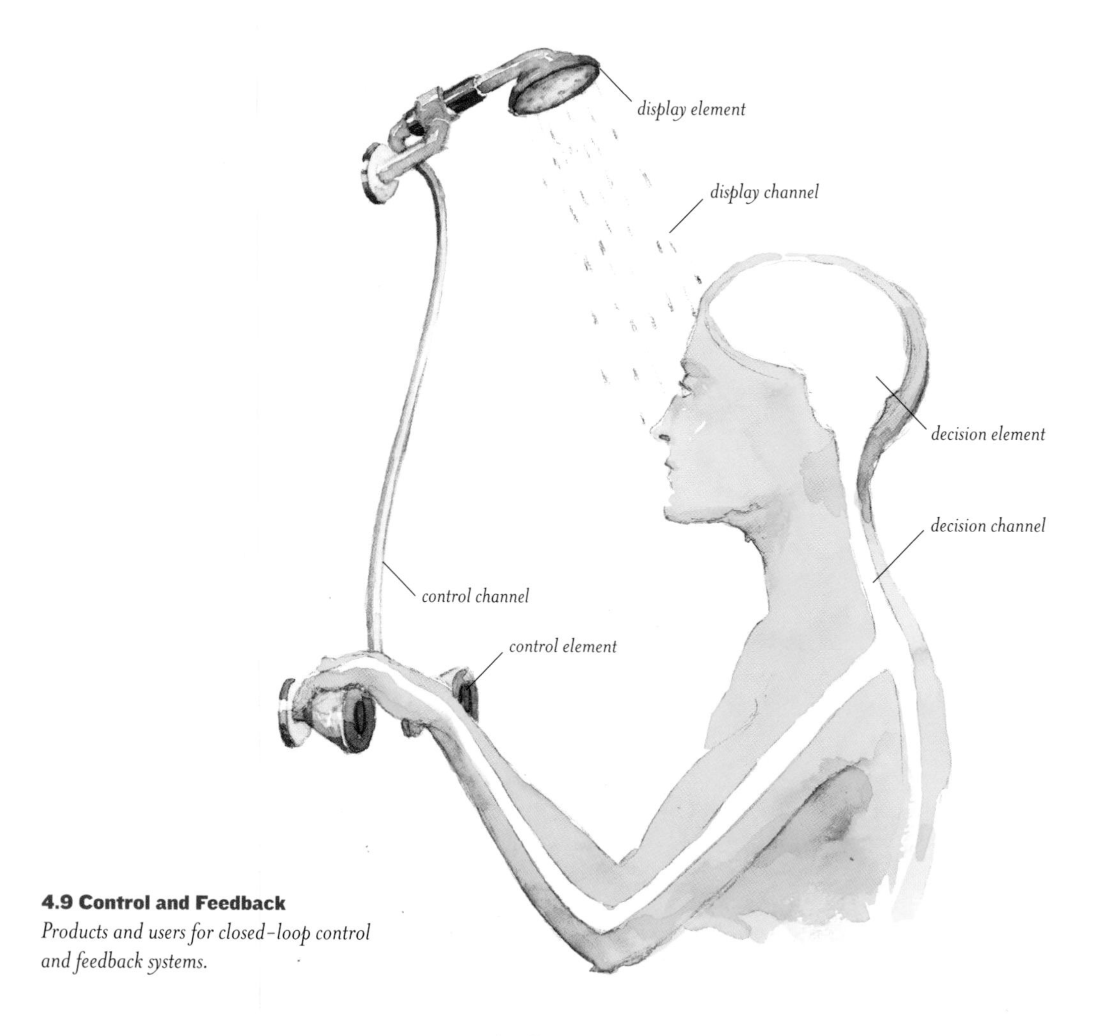

4.9 Control and Feedback
Products and users for closed-loop control and feedback systems.

touch. Surface profiling gives tactile feedback of the location of important keys. Displays are based on phosphors, liquid-crystals, light-emitting diodes and, most recently, light-emitting polymers. Surface printing techniques allow the application of icons, color coding, bar-coding and text.

Communication is a desirable way in which products interact with people. Not all interactions are so desirable.

Noise Management

Sound is caused by vibration; its pitch depends on its frequency. The (youthful) human ear responds to frequencies from about 20 to about 20,000 Hz, corresponding to wave-lengths of 17 m to 17 mm. The bottom note on a piano is 28 Hz, the top note 4186 Hz. The most important range, from the point of view of acoustic design, is roughly 500–4000 Hz. Sound pressure is measured in Pascals (Pa), but because audible sound pressure has a range of about 10^6 it is more convenient to use a logarithmic scale with units of decibels (dB). Figure 4.10 shows levels, measured in dB. Sound levels above 50 dB can impair concentration. Sustained sound levels above 90 dB cause damage to hearing.

Vibration, like noise, becomes damaging above a critical threshold. The level of vibration is characterized by the acceleration (units, m/s^2) associated with it; safe practice requires levels below 1 m/s^2 (0.1g where g is the acceleration due to gravity). The frequency, too, is important: less than 1 Hz gives seasickness; 1–100 Hz gives breathing difficulties and headaches; 10–1000 Hz causes loss of sense of touch or "white finger." Duration, too, is important – the longer the duration, the lower the level should be.

The mechanism for reducing sound and vibration levels within a given enclosed space (a room, for instance) depends where it comes from. If it is generated within the room, one seeks to *absorb* the sound. If it comes from outside, one seeks to *insulate* the space to keep the sound out. And if it is transmitted through the frame of the structure itself (deriving from a machine tool, or from traffic), one seeks to *isolate* the structure from the source of vibration.

In product design it is usually necessary to absorb sound. Soft porous materials absorb incident, airborne sound waves, converting them into heat. (Sound power, even for a very loud noise, is small, so the temperature rise is negligible.) Porous or highly flexible materials such as low density polymer and ceramic foams, plaster and fiberglass absorb well; so do woven polymers like carpets and curtains. The proportion of sound absorbed by a surface is called the sound-absorption coefficient. A material

4.10 How Loud is Loud?
Sound levels in decibels.

Sound Source	dB
Threshold of hearing	0
Office background noise	50
Road traffic	80
Discotheque	100
Pneumatic drill at 1 m	110
Jet take-off at 100 m	120

4.11 Soaking Up Sound
Sound absorption coefficients

Material	at 500 - 4000 Hz
Glazed tiles	0.01–0.02
Rough concrete	0.02–0.04
Wood	0.15–0.80
Cork tiles	0.20–0.55
Thick carpet	0.30–0.80
Expanded polystyrene	0.35–0.55
Acoustic spray plaster	0.50–0.60
Glass wool	0.50–0.99

with a coefficient of 0.8 absorbs 80% of the sound that hits it, reflecting 20% back; one with a coefficient of 0.03, absorbs only 3% of the sound, reflecting 97%. Figure 4.11 shows sound absorption coefficients for a number of materials.

Vibration is damped by isolating the source of vibration from the rest of the structure with high-damping rubber mounts, foams, elastomer grips, etc. A composite of rubber filled with cork particles is a good choice here. The low shear modulus of the rubber isolates shear waves, and the compressibility of the cork adds a high impedance to compressive waves.

4.12 Managing Heat
Thermal conductivities of selected materials.

Material	W/m K
Copper	384
Aluminum	230
Glass	1.1
Solid polymers	0.15–0.35
Pine	0.112
Balsa	0.055
Cork	0.045
Glass foam	0.050
Mineral fiber	0.046
Glass wool	0.042
Polymer foams	0.02–0.10

Thermal Management

All products that consume power generate heat, creating two classes of problem in thermal management. In the first, the challenge is to insulate parts that are sensitive to heat or that will be touched by the operator so that they don't get too hot: the hair dryer in which the casing must be insulated from the heating element. In the second, the challenge is the opposite one: that of transmitting or spreading heat, requiring materials that conduct: quality saucepans use high-conducting materials to spread heat laterally for even cooking.

Polymers, with thermal conductivities, λ, around 0.25 W/mK, are good insulators. Still air (λ = 0.025) is much better; CO_2 (λ = 0.016) and CFCs such as trichlorofluoromethane, CCl_3F (λ = 0.008) are better still. The best insulating materials are foams that trap one of these gasses in closed pores, or fibrous materials like glasswool or mineral fiber that hold air still. If the temperature of the heat source is low (<150° C) and strength is not required, polymer foams are a good choice. If the source-temperature is high, glass wool or foam, mineral fibers and ceramic foams are a better choice. And if high strength is also needed, compacted mica or porous ceramic may be best.

When conduction is the aim, the best materials are solid silver, gold or diamond (!) but these, for obvious reasons, may not be viable choices. Solid copper and solid aluminum come close and are inexpensive and easy to shape: most heat-spreading systems use them. Occasionally, materials that conduct heat well but do not conduct electricity (they are electrical insulators) are required: here the best choice is aluminum nitride, AlN, or beryllia, BeO. Figure 4.12 lists conductivities of representative conductors and insulators.

Light Management

Few products generate excessive light intensity, but many reflect light in ways that interfere with the vision of the operator. The reflectivity of a surface measures the fraction of the light intensity, incident at an angle of 60° to the surface, that is reflected back to an observer at 120° (that is, an equal angle on the other side). A black matte surface reflects less than 1%; a high gloss surface reflects 80% or more; a mirror reflects it all. Reflectivities are listed in 4.13.

The reflectivity of a surface depends on the material of which it is made or coated, and the smoothness of this surface: both color and texture are important. It is controlled by the use of low reflectivity coatings and light absorbing screens.

4.13 Mirror or Matte?
Reflectivity of materials and surfaces.

Material	%
Aluminum	60
Beryllium	49
Carbon steel	57
Invar	41
Magnesium	72
Stainless steel	70
Titanium	56
Surface Texture	**%**
Dead matte	< 1
Matte	1–10
Eggshell	15–20
Semi-gloss	40–50
Full gloss	>80
Mirror	>95

The Environmental Dimension: "Green" Design

What have you discarded lately that still worked or, if it didn't, could have been fixed? Changing trends, pushed by seductive advertising, reinforce the desire for the new and urge the replacement of still-useful objects. Industrial design carries a heavy responsibility here – it has, at certain periods, been directed towards creative obsolescence: designing products that are desirable only if new, and urging the consumer to buy the latest models, using marketing techniques that imply that acquiring them is a social and psychological necessity. And – as we said in Chapter 2 – this accelerating consumption creates problems that are an increasing source of concern.

But that is only half the picture. A well-designed product can outlive its design life by centuries, and – far from becoming unwanted – can acquire value with age. The auction houses and antique dealers of New York, London and Paris thrive on the sale of products that, often, were designed for practical purposes but are now valued more highly for their aesthetics, associations and perceived qualities. People do not throw away products for which they feel emotional attachment. So there you have it: industrial design both as villain and as hero. Let us look into this in a bit more detail because it is a complex maze, with the potential for many wrong turns.

Materials as Resources

Will we really run out of materials, or of oil and gas? The present prodigal use suggests that we might; reports have been written, starting with the famous Club of Rome report[9] in 1972, suggesting that we will. A market economy may have many faults, but it also contains feedback mechanisms of a restraining nature. As the natural resources of materials are consumed and ore-grades diminish, prices rise; the lower-grade uses become economically untenable and consumption falls. A more balanced view is that we will never "run out" of anything, but that the adjustments imposed by rising cost may be very uncomfortable to live with, giving sufficient reason to urge a policy of restraint and conservation.

9. "The Limits to Growth – 1st Report of the Club of Rome," by Meadows et al. (1972), provoked a strong reaction when first published, but proved, in some ways, to be visionary.

A more immediate concern is the impact of manufacturing on the environment. As already mentioned in Chapter 2, there are those who, observing the present level of consumption and the rate of its growth, argue that only a massive reduction can reduce the impact on the environment to an acceptable level. Can such changes be achieved without a change in lifestyle that few could accept? They might, though it is far from certain. Take the example of the automobile. Twenty five years ago the average car did 20 miles per gallon (14 litres/100 km); today you can buy a family car that does 60 mpg (4.7 litres/100 km) – a factor of 3. The average life of a tire, twenty five years ago, was about 20,000 miles; now it is nearer 50,000 – a factor of 2.5. So large changes are possible, but they are not enough. Car ownership and the average distance driven per year have both increased over the same period by a factor of roughly 6, far outweighing even these massive changes.

Balancing Material and Energy Consumption

The most obvious ways to conserve material is to make products smaller, make them last longer and recycle them when they finally reach the end of their lives. But the seemingly obvious can sometimes be deceptive. Materials and energy form part of a complex and highly interactive system, sketched in 4.14. Here primary catalysts of consumption such as population growth, increasing wealth and new technology, appear to accelerate the consumption of materials and energy and the by-products that these produce. But – to take just one example – wealth also brings education, and with it, greater awareness of the problems these by-products create, restraining consumption.

The influences have complex interactions. New technology, as an example, offers more material and energy-efficient products, yet by also offering new functionality it creates obsolescence and the desire to replace a product that has useful life left in it. Electronic products are prime examples of this: 80% are discarded while still functional. And observe, even at this simple level, the consequences of longer product life – a seemingly obvious measure. It can certainly help to conserve materials but, in an era in which new technology delivers cleaner, leaner products (particularly true of cars, electronics, and household appliances), extending the life of old products can have a negative influence on the environment. And as a final example, consider the bi-valent influence of industrial

4.14 A Complicated Interaction with the Environment
The interactive nature of the influences on consumption of materials and energy.

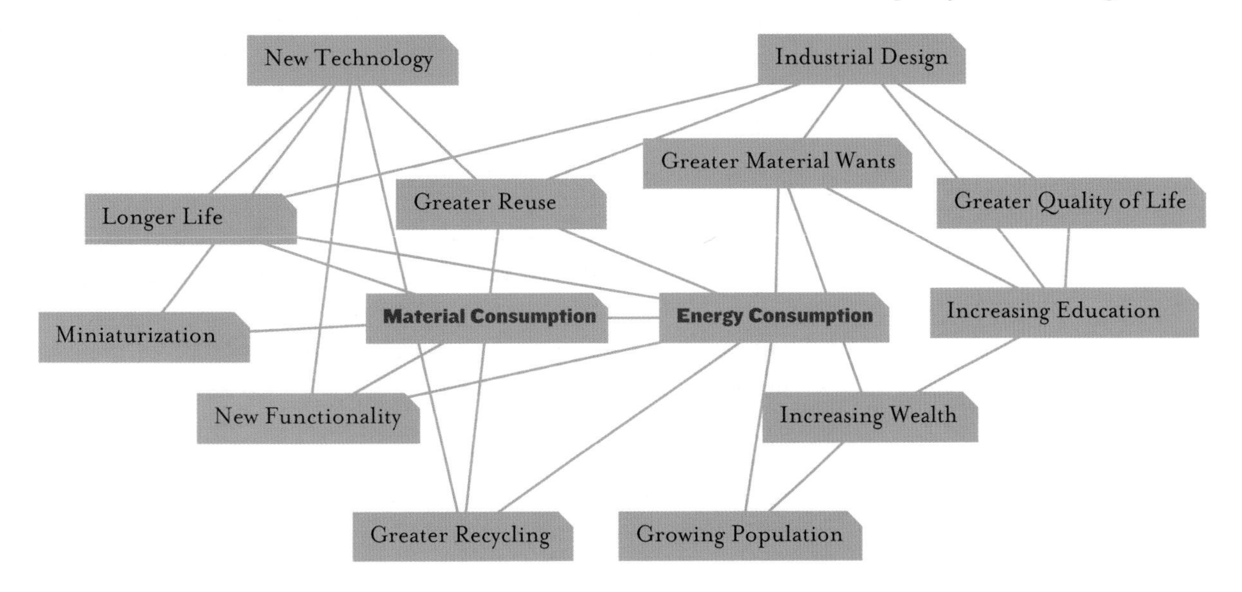

design. The lasting designs of the past are evidence of its ability to create products that are treasured and conserved. But today it is frequently used as a potent tool to stimulate consumption, creating the perception that "new" is desirable and that even slightly "old" is unappealing. How can design be used to support and enhance the environment, rather than threatening it?

Eco-Design

When your house no longer suits you, you have two choices: you can buy a new house or you can adapt the one you have got, and in adapting it you make it more personally yours. Houses allow this. Most other products do not; and an old product (unlike an old house) is often perceived to have such low value that it is simply discarded. And that highlights a design challenge: to create products that can be adapted and personalized so that they acquire, like a house, a character of their own and transmit the message "keep me, I'm part of your life." This suggests a union of technical and industrial design to create produces that can accommodate evolving technology, but at the same time are made with a quality of material, design and adaptability that creates lasting and individual character, something to pass on to your children.

Well, that is perhaps an ideal,[10] one that relies as much on cultural change as on design. To address the problem at a more practical level and to identify more clearly the role of materials, it is necessary to recognize two key features. The first is the point made by 4.14: the consumption of materials and energy is part of a highly complex and interactive system. Simple solutions ("ban the use of lead," "don't use plastics because they are energy-intensive") can have the opposite of the desired effect. The second follows from the concept of the material's lifecycle shown in 2.3. The point is that there are four significant stages. Ore and feedstock are consumed to create materials; these are manufactured into products that, almost always, consume energy throughout their lives; and these, at life's end, are discarded or recycled. Passive products are those that do not require much energy to meet their primary function (furniture, carpets and fittings, bridges, unheated buildings); for these, the material

10. For an organization with such an ideal, see www.eternally-yours.nl.

production and manufacture phases dominate the consumption of energy and materials (4.15). Here extending product life makes good sense; doubling the life of any one of them almost halves resource consumption. We have already seen that well-designed furniture, made from quality materials, can acquire value and desirability with time, and its unlimited life has no negative consequences in the system of 4.14.

By contrast, energy-consuming products (household appliances, vehicles, heated or cooled buildings) consume more resources – above all, more energy – in the use phase of their life than in all the others put together (4.16). The greatest potential for improvement then lies in examining use and disposal rather than manufacture. Extending the life of energy-consuming products, particularly those in which technology is changing rapidly, may be counter-productive, for reasons already given. Instead, the focus is on lightweight materials to reduce fuel consumption in transport systems, more efficient electronics with stand-by or "sleep" modes to cut power

4.15 The Ecoplanter
This planter, designed for the small apartment, is made of recyclable polypropylene and composts the packaging in which it was delivered. (Courtesy Cynthia Garden Design)

Production	Manufacture	Use	Disposal	
3%	91%	2%	4%	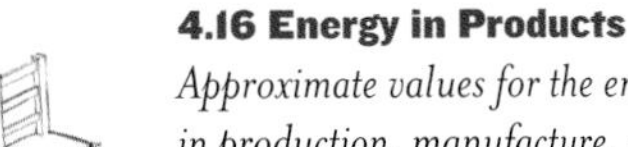
89%	4%	6%	1%	
4%	1%	94%	1%	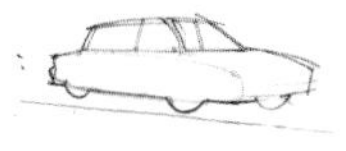
10%	3%	85%	2%	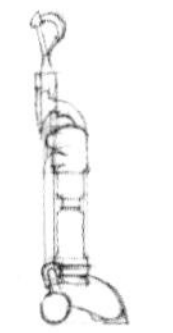

4.16 Energy in Products
Approximate values for the energy consumed in production, manufacture, use and disposal of four classes of products.

consumption when the product is inactive, and materials with better thermal insulation to reduce energy loss in refrigerators, freezers and heating systems. And if product life is short, reuse or recycling offers particularly large gains.

The Aesthetic Dimensions: the Senses

4.17 Reaching the Senses
The tactile, visual, acoustic attributes of materials.

Sense	Attribute
Touch	Warm Cold Soft Hard Flexible Stiff
Sight	Optically clear Transparent Translucent Opaque Reflective Glossy Matte Textured
Hearing	Muffled Dull Sharp Resonant Ringing Low pitched High pitched
Taste/Smell	Bitter Sweet

Aesthetic attributes are those that relate to the senses: sight, touch, sound, smell and taste (4.17). Almost everyone would agree that metals feel "cold;" that cork feels "warm;" that a wine glass, when struck, "rings;" that a pewter mug sounds "dull," even "dead." A polystyrene drinking glass can be visually indistinguishable from one made of glass, but pick it up and it feels lighter, warmer, less rigid; and tap it and it sounds quite different. The impression it leaves is so different from glass that, in an expensive restaurant, it would be completely unacceptable. These are the simplest of aesthetic distinctions, and, in an approximate way, they can be quantified. There are obvious limitations here, but within these it is worth a try, since it could allow a first-level screening of materials to meet specified aesthetic requirements. So here goes.

Touch: the Tactile Attributes

Steel is "hard;" so is glass; diamond is harder than either of them. Hard materials do not scratch easily; indeed they can be used to scratch other materials. They generally accept the high polish, resist distortion and are durable. The impression that a material is hard is directly related to the material property "Hardness," measured by materials engineers and tabulated in handbooks. Here is an example of a sensory attribute that relates directly to a technical one.

"Soft" sounds like the opposite of "hard" but, in engineering terms, it is not – there is no engineering property called "Softness." A soft material deflects when handled, it gives a little, it is squashy, but when it is released it returns to its original shape. This is elastic (or viscoelastic) behavior, and the material property that has the most influence on it is the modulus, not the hardness. Elastomers (rubbers) feel soft; so do polymer foams: both classes of material have moduli that are 100 to 10,000 lower than ordinary "hard" solids; it is this that makes them feel soft. Hard materials can be made "soft" by forming them into shapes in which they bend or twist: hard steels shaped into soft springs; glass drawn into fibers and woven into cloth. To compare the intrinsic softness of materials (as opposed to the softness acquired by shape) materials must be compared in the same shape, and then it is the modulus that is the key property. An appropriate measure of hardness and softness is dis-cussed in the Appendix of this chapter. It is used as one axis of 4.18.

A material feels "cold" to the touch if it conducts heat away from the finger quickly; it is "warm" if it does not. This has something to do with the technical attribute "thermal conductivity" but there is more to it than that – it depends also on specific heat. A measure of the perceived coldness or warmth of a material (in the sense of heat, not of color) is developed in the Appendix. It is shown as the other axis of 4.18. The figure nicely displays the tactile properties of materials. Polymer foams and low-density woods are warm and soft; so are balsa and cork. Ceramics, stone and metals are cold and hard; so is glass. Polymers and composites lie in between.

Sight: the Visual Attributes

Metals are opaque. Most ceramics, because they are polycrystalline and the crystals scatter light, are either opaque or translucent. Glasses, and single crystals of some ceramics, are transparent. Polymers have the greatest diversity of transparency, ranging from transparency of optical quality to completely opaque. Transparency is commonly described by a 4-level ranking: opaque, translucent, transparent, and water-clear or optical quality. These every-day words are easily understood; they are used in the materials profiles later in this book. Figure 4.19 ranks the transparency of common polymers. In order to spread the data in a useful way, it is plotted against cost. The cheapest polymers offering optical transparency are PET, PS and PMMA. Epoxies can be transparent but not with optical quality. Nylons are, at best, translucent. Phenolics, most

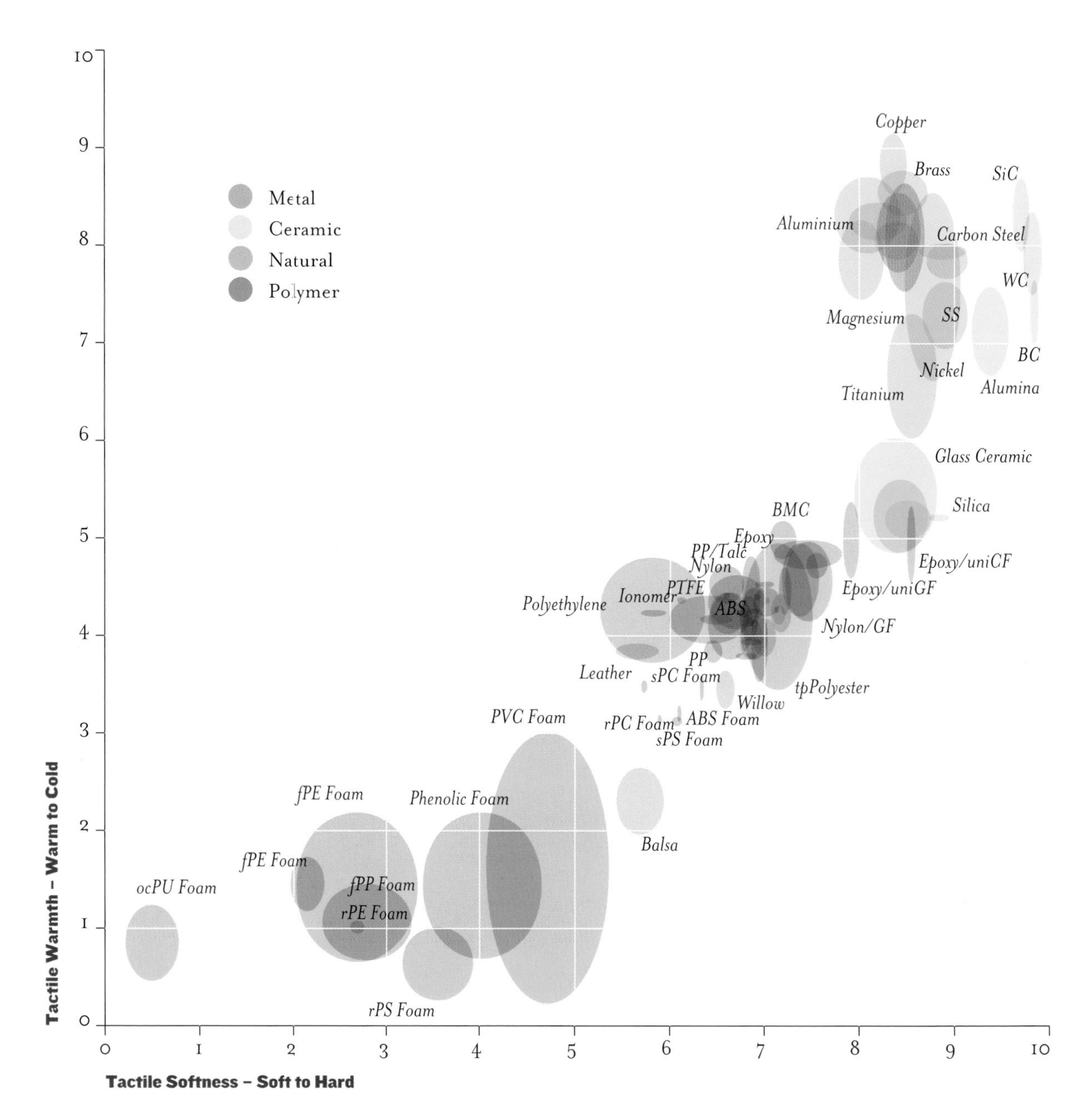

4.18 The Tactile Attributes of Materials
Some materials feel soft and warm; others, hard and cold. Here these attributes are charted, revealing a striking correlation between the two (CES 4,2002).

ABS and all carbon-filled or reinforced polymers are translucent or opaque.

Color can be quantified by analyzing spectra but this – from a design standpoint – doesn't help much. A more effective method is one of color matching, using color charts such as those

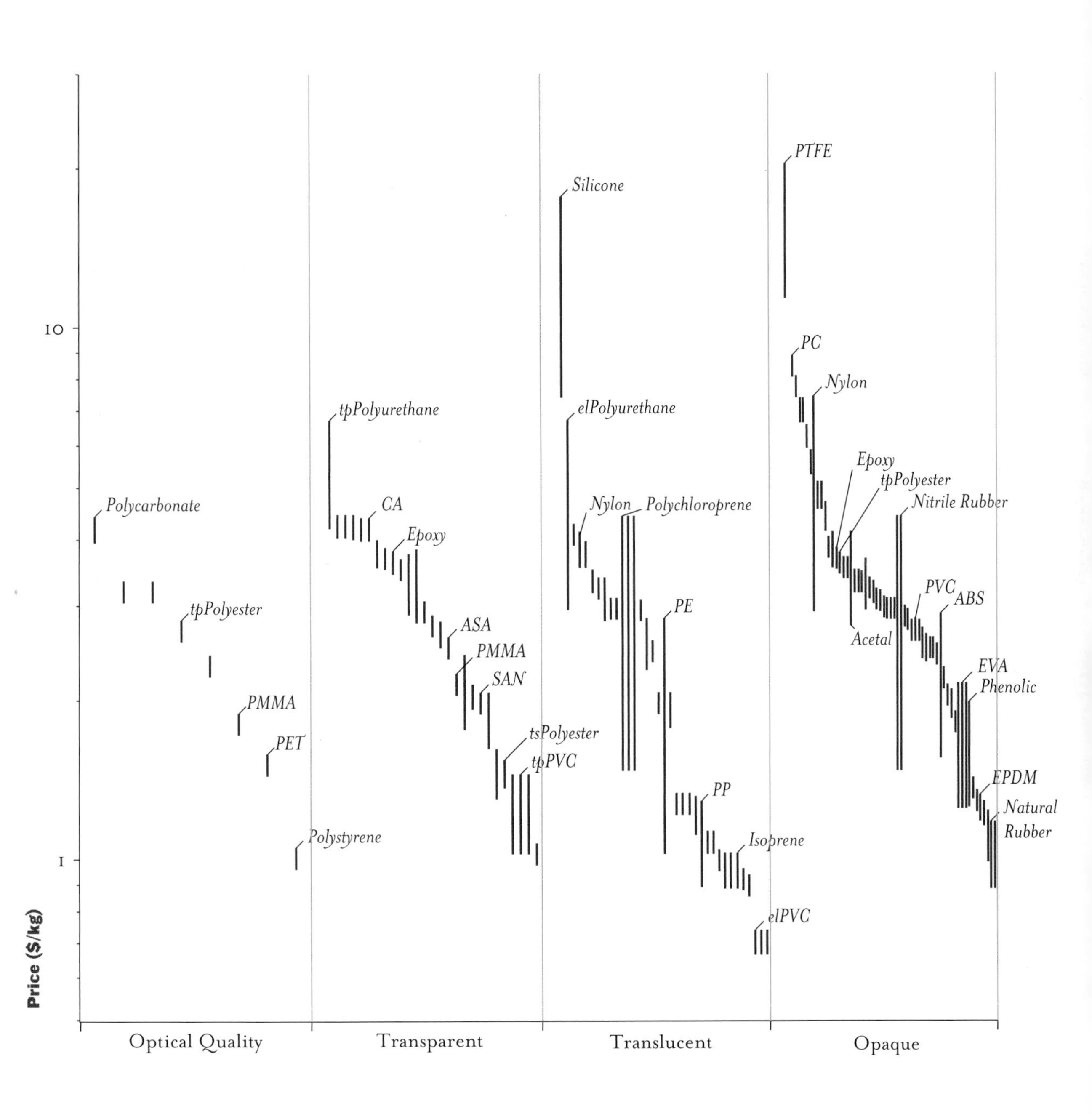

provided by Pantone;[11] once a match is found it can be described by the code that each color carries. Finally there is reflectivity, an attribute that depends partly on material and partly on surface detail. Like transparency, it is commonly described by a ranking: dead matte, eggshell, semi-gloss, gloss, mirror (4.13).

4.19 Price and Light Transmission
The optical attributes of polymers, shown in order of decreasing cost in each category (CES 4,2002).

Hearing: the Acoustic Attributes

The frequency of sound (pitch) emitted when an object is struck relates to two material properties: modulus and density. A measure of this pitch, detailed in the Appendix, is used as one axis of 4.20. Frequency is not the only aspect of acoustic

4.20 Pitch and Brightness

The acoustic attributes of materials (CES 4, 2002).

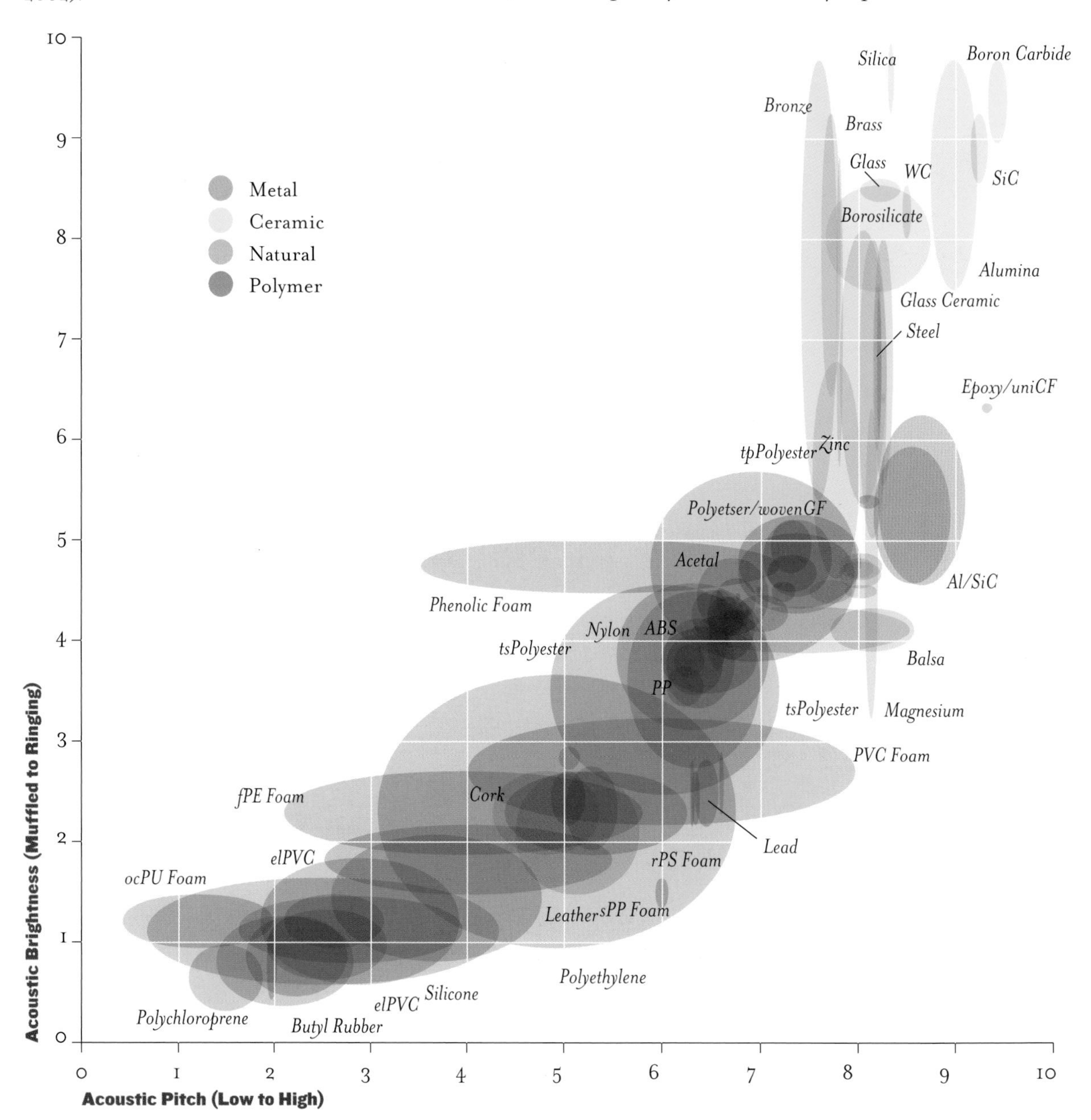

response – another has to do with damping. A highly damped material sounds dull and muffled; one with low damping rings. Acoustic brightness – the inverse of damping (the Appendix again) – is used as the other axis of 4.19. It groups materials that have similar acoustic behavior.

Bronze, glass and steel ring when struck, and the sound they emit has – on a relative scale – a high pitch; they are used to make bells; alumina, on this ranking, has the same bell-like qualities. Rubber, foams and many polymers sound dull, and, relative to metals, they vibrate at low frequencies; they are used for sound damping. Lead, too, is dull and low-pitched; it is used to clad buildings for sound insulation.

What happens if all these are combined? Figure 4.21 shows the result. It is an MDS map combing the attributes we have just described. As with the earlier MDS map, there is pronounced clustering of a kind that is broadly consistent with experience – leather is near PVC, ceramics near metals, woods near phenolics. The map should not be over-interpreted, for the reasons given already; but it does have the power of suggestion.

11. Pantone (www.pantone.com.) provide detailed advice on color selection, including color-matching charts and good descriptions of the associations and perceptions of color.

The Personal Dimension

We interact with materials through products. The interaction involves their technical and aesthetic attributes, but these are not all. A product has perceived attributes and associations – it is these, in part, that give it its personality, something designers work hard to create. But can a *material* be said to have perceived attributes or indisputable associations? A personality? At first sight, no – it only acquires these when used in a product. Like an actor, it can assume many different personalities, depending on the role it is asked to play.

And yet... think of wood. It is a natural material with a grain that gives a surface texture, pattern and feel that other materials do not have. It is tactile – it is perceived as warmer than many other materials, and seemingly softer. It is associated with characteristic sounds and smells. It has a tradition; it carries associations of craftsmanship. And it ages well, acquiring additional character with time. Things made of wood are valued more highly when they are old than when they are new. There is more to this than just aesthetics; there are the makings of a personality, to be brought out by the designer, certainly, but there none the less.

And metals... Metals seem cold, clean, precise. They ring when struck. They can reflect – particularly when polished. They are accepted and trusted: machined metal looks strong, its very nature suggests it has been engineered. The strength of

4.21 Multi-dimensional Maps: Aesthetic Attributes

An MDS map of materials by aesthetic attributes (Data from CES 4, 2002).

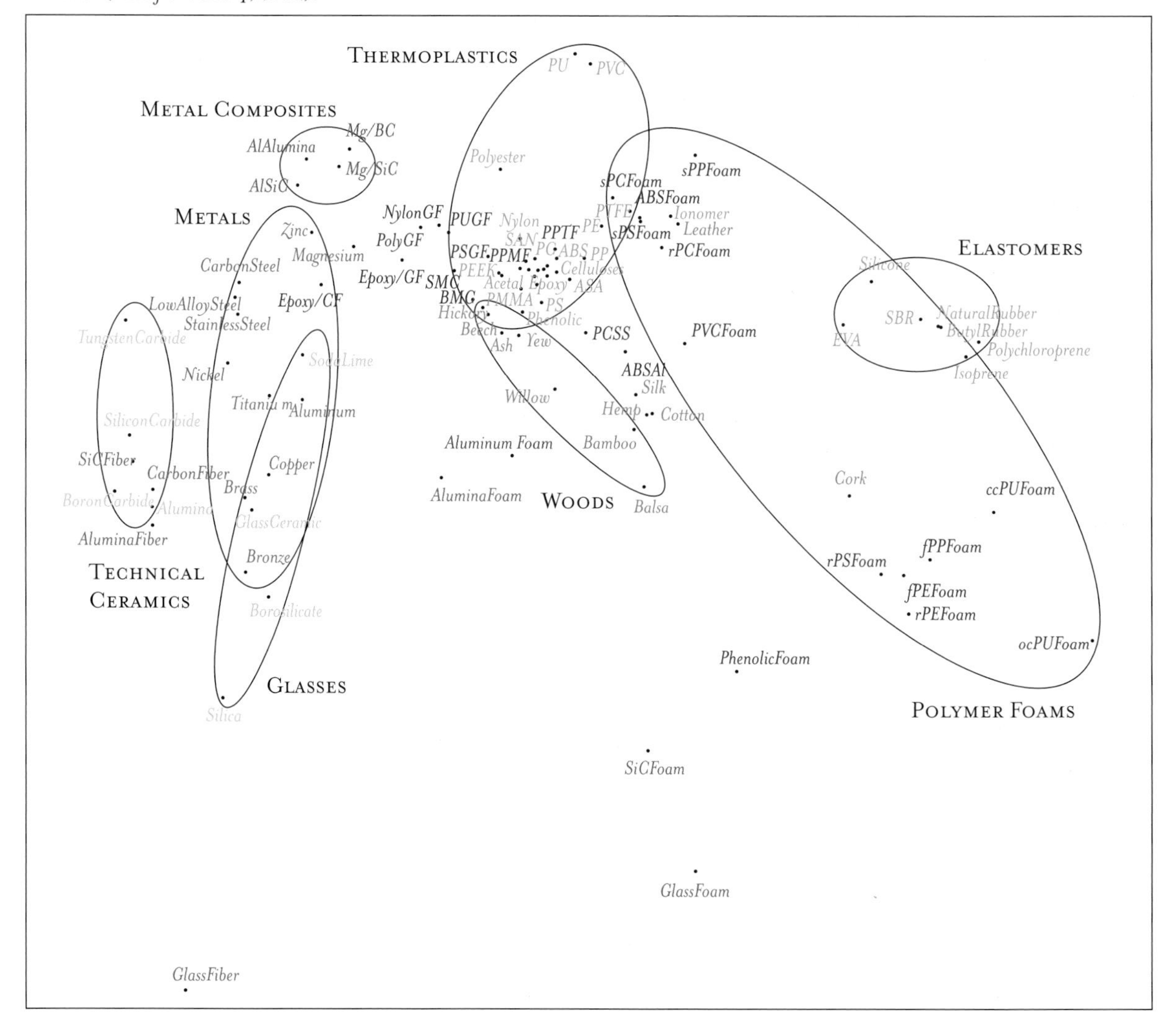

metals allows slender structures – the cathedral-like space of railway stations or the span of bridges. It can be worked into flowing forms like intricate lace or cast into solid shapes with integral detail and complexity. And – like wood – metals can age well, acquiring a patina that makes them more attractive than when newly polished – think of the bronze of sculptures, the pewter of mugs, the lead and copper of roofs.

And ceramics or glass? They have a long tradition: think of Greek pottery and Roman glass. They accept almost any color; this and their total resistance to scratching, abrasion, discoloration and corrosion gives them a certain immortality, threatened only by their brittleness. They are – or were – the materials of great craft-based industries: Venetian glass, Meissen porcelain, Wedgwood pottery, valued, sometimes, as highly as silver. And ceramic today has an additional association – that of advanced technology: kitchen stove-tops, high pressure/high temperature valves, space shuttle tiles... materials for extreme conditions.

And, finally, polymers. "A cheap, plastic imitation" used to be a common phrase – and that is a hard reputation to live down. It derives from an early use of plastics: to simulate the color and gloss of Japanese handmade pottery, much valued in Europe. Commodity polymers *are* cheap. They are easily colored and molded (that is why they are called "plastic"), making imitation easy. Unlike ceramics, their gloss is easily scratched, and their colors fade – they do not age gracefully. You can see where the reputation came from. But is it justified? No other class of material can take on as many characters as polymers: colored, they look like ceramics; printed, they can look like wood or textile; metalized, they look exactly like metal. They can be as transparent as glass or as opaque as lead, as flexible as rubber or as stiff – when reinforced – as aluminum. But despite this chameleon-like behavior they do have a certain personality: they feel warm – much warmer than metal or glass; they are adaptable – that is part of their special character; and they lend themselves, particularly, to brightly colored, lighthearted, even humorous, design.

So there is a character hidden in a material even before it has been made into an recognizable form – a sort of embedded personality, a shy one, not always visible, easily concealed or disguised, but one that, when appropriately manipulated, can contribute to good design. The attributes we discuss here are qualitative, sometimes subjective – a far remove from the sharp precision of technical attributes with which this chapter started. But – given their importance in product design – it is worth exploring them further. And here we must digress for a moment, turning from materials to products, to explore the easier question: how can the perceived attributes of products be characterized? We need a language – a way of expressing these things in words.

A Vocabulary of Perception

Research exists that explores the aesthetics and perceived attributes of products.[12] These suggest that agreement can, to some degree, be reached in describing perceptions. Attempting to put words to perceived attributes is, of course, a risky business – not everyone sees them in the same way. They are subjective; they depend on the product itself, and, importantly, on the context and the culture in which it is used. They are sometimes ambiguous, and their meaning changes with time: a product that appears aggressive or luxurious today can seem humorously retro or just plain ugly tomorrow (think of car styling). But we are bound by language – it is the most important way in which we communicate. So it is worth a try.

One way to assemble a vocabulary for perceptions is to do what any linguist, studying a new language, would do – listen to the words that native speakers use and observe the objects about which they are speaking. Newspapers, magazines, exhibition catalogs and design books carry product reviews.[13] They include pictures of the product, a profile of the designer, and, typically, a statement by the critic – the "native speaker" – of his or her perceptions of the product. A survey of these leads to the list in 4.22 – words used regularly to describe products and their attributes – each shown with an opposite to sharpen the meaning. The list has been simplified and reduced in length by replacing near-equivalent words with a single word (e.g. comical, funny ≈ humorous; durable, long-wearing ≈ lasting).

12. There are a number of such studies. They make fascinating reading. Manzini (1986) illustrates how the combination of material and form is used to obtain specific product attributes and human responses. Harni (1996) explores the sensory experiences and associations of objects using, as prototypes, 12 different styles: folklore, deluxe, kitsch, porno, toy, cartoon, sport, pseudo-eco, army, professional, space and "white," each applied to 5 products: a toaster, an iron, a hairdryer, a kitchen mixer and an electric shaver. The work demonstrates dramatically the way in materials, surface treatments and forms can be used to create perceived attributes. At a more formal level there is the field of product semiotics – the study of signs and their role in socio-cultural behavior. A product transmits signs that are interpreted differently depending on the observer and the context. Monö and Søndergaard (1997) use the terms icon, index and symbol as a basic vocabulary in describing product semiotics. An icon is prototype, an idealized abstraction of the attributes of a product (a frame representing a painting), an index is connected to what it signifies by cause (tracks in the snow indicate that someone was here) and a symbol represents a recognition of agreement between people (a crown indicates royalty). These can be used to build meaning into products.

13. Examples are the ID Magazine, the International Design Year Book (1998, 1999), the ID Magazine Annual Design Review (1998, 1999), the books by Byars (1995, 1997a, b, 1998), and the catalogs of the MOMA exhibition called "Mutant Materials" (Antionelli, 1995) and that mounted by ConneXion called "Materials and Ideas for the Future" (Arredo, 2000).

How useful is such a vocabulary? The critics use it to communicate with their readers, and given the popularity of the publications, they clearly speak a language that carries meaning for them. But is it one that non-experts can use? Experiments in which test groups are asked to assign words from the list to products that they are given (and can handle) show that, at least within a given socio-cultural group, agreement is, in a statistical sense, significant – about 80% of the group assign one or more of the same words to the same product. The motive of such studies is one of understanding perception and communicating it – the ability to attach words from the perceptions list of 4.22 (or some expansion of it) to products, so that those with similar perceived attributes can be compared. It is a useful idea and one to which we return in Chapter 7.

And – just in case this idea of perceptions still seems intellectually dubious – think of product styles (4.23). Describing a product as having elements of "Art Nouveau," or "Streamform" or "Pop" conveys, in a single word or phrase, a cluster of ideas, some concrete, some abstract. Each style is associated with certain materials – Art Nouveau with carved wood, cast bronze, and wrought iron; Streamform with steel and aluminum; Pop with plastics. Each carries associations of form, decoration and coloring. But above all is the association with an era – Art Nouveau with late 19th Century Europe, Streamform with the US in the 1930s and Pop with the US in the 1950s – and with the cultural, economic and intellectual character of that era – with its *personality*.

Thus a vocabulary to describe the perceived attributes of products is possible. If the words are put in context with images to illustrate them, they give a channel of communication – one that is useful in design, and often used (and misused) in advertising. As with any language the words and their meanings evolve with time. The difference here is that the rate of evolution is faster than in language as a whole, requiring regular rewriting of the lexicon. Like all words, these do not have universal meaning – but this we have to live with. The alternative is no communication at all.

So products have perceived attributes, and they can be

4.22 Perceptions
The perceived attributes of products, listed as pairs of opposites.

Perception		Opposite
Aggressive	·	Passive
Cheap	·	Expensive
Classic	·	Trendy
Clinical	·	Friendly
Clever	·	Silly
Common	·	Exclusive
Decorated	·	Plain
Delicate	·	Rugged
Disposable	·	Lasting
Dull	·	Sexy
Elegant	·	Clumsy
Extravagant	·	Restrained
Feminine	·	Masculine
Formal	·	Informal
Hand-made	·	Mass-produced
Honest	·	Deceptive
Humorous	·	Serious
Informal	·	Formal
Irritating	·	Loveable
Lasting	·	Disposable
Mature	·	Youthful
Nostalgic	·	Futuristic

4.23 Design Styles
A chronology of selected movements in design, sometimes linked to material innovations.

1900 and Before	
Art Nouveau	1890
Arts and Crafts	1890
Functional	1900
1900 to 1950	
Modernist	1900
Futurist	1910
Art Deco	1920
Streamform	1930
Contemporary	1945
1950 and After	
Pop	1960
Retro	1960
Classic	1970
Post-Modernist	1970
Memphis	1980

associated with a certain style. But can the same be said of materials?

Technical Perceptions

You don't need a degree in engineering to know that some materials are tougher than others (some foods, too); some more resilient (some people, too); some more flexible (like your savings account); some less abrasion resistant (like children). These words – tough, resilient, flexible, abrasion resistant – are words of everyday life deployed to describe the features of a material. These are perceptions of technical behavior, to be sure, and in this sense different from those of the list of 4.22. And because they are often based on technical attributes, they can, to some degree, be quantified or measured, allowing materials to be ranked by these familiar terms. They are used in the Reference Profiles in the later part of this book.

For a given shape, the mass of an object depends on the density, ρ, of the material of which it is made. Abrasion resistance is more complex – its closest link is that to hardness, H. These two properties are mapped in 4.24. There is a striking correlation between them: as a general rule materials that are difficult to scratch are also stiff.

Resilience and stiffness are a little more complicated. Resilience is the ability to absorb deflection without damage – a car tire is resilient; it can be driven over a curb and survive. Stiffness is resistance to deflection, measured – for a given shape – by the modulus of the material. These behaviors can also be expressed in terms of technical attributes. They are mapped in 4.25. Note that steel and other metals are stiff but not very resilient. Rubber and many polymers are resilient, but not at all stiff.

Conclusions: Expression through Material

There is a school of design thinking that holds as a central tenant that materials must be used "honestly." By this they mean that deception and disguise are unacceptable – each material must be used in ways that expose its strengths, its natural appearance, and its intrinsic qualities. The idea has its roots in the tradition of craftsmanship – the potters use of clays and glazes, the carpenters use of woods, the skills of silversmiths in crafting beautiful objects, each exploiting the unique qualities of the materials with which they work – an integrity to craft and material.

4.24 Abrasion Resistance and Weight

Materials with high hardness resist abrasion; those with high density are heavy. There is a strong correlation between the two (CES 4, 2002).

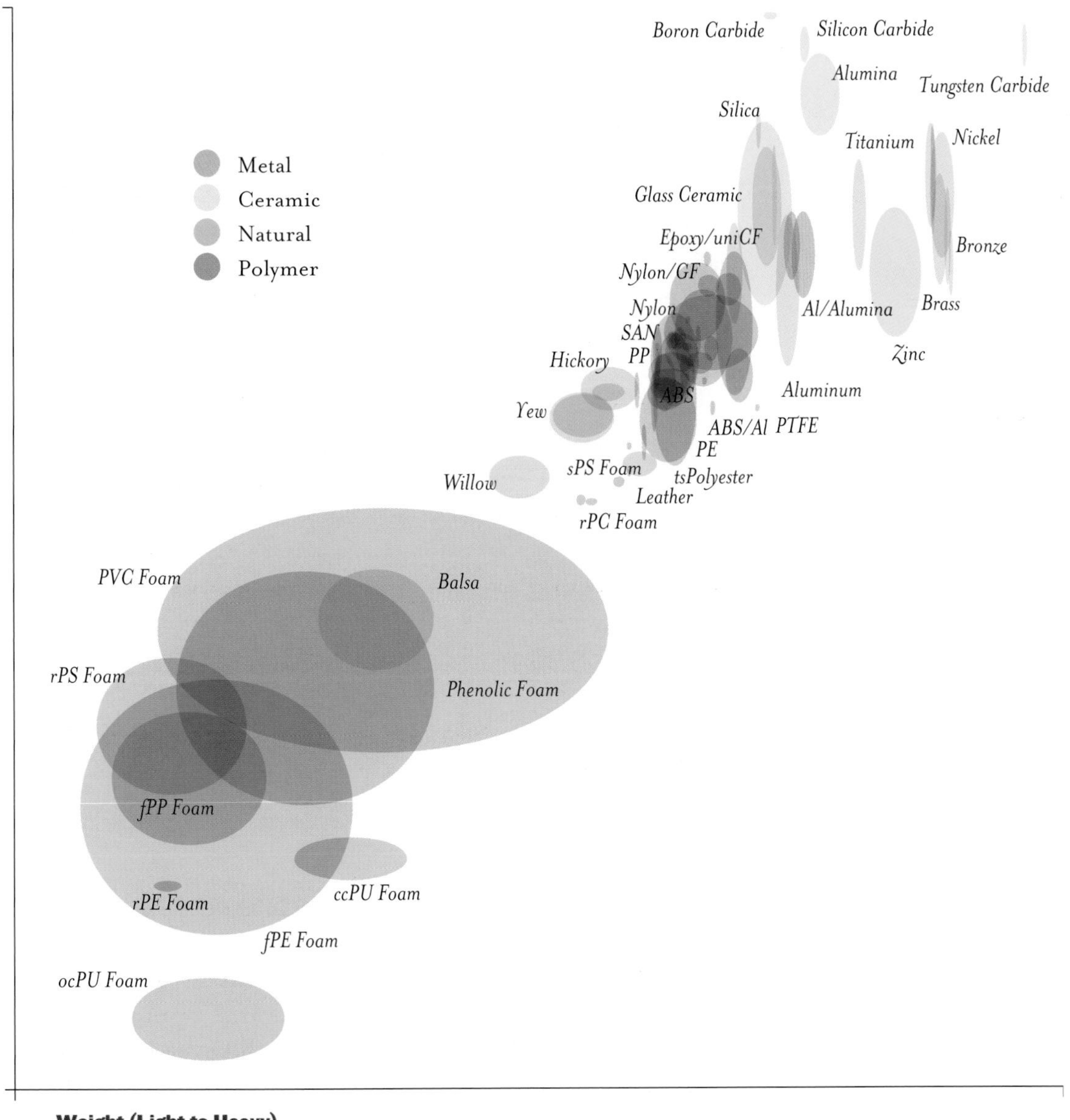

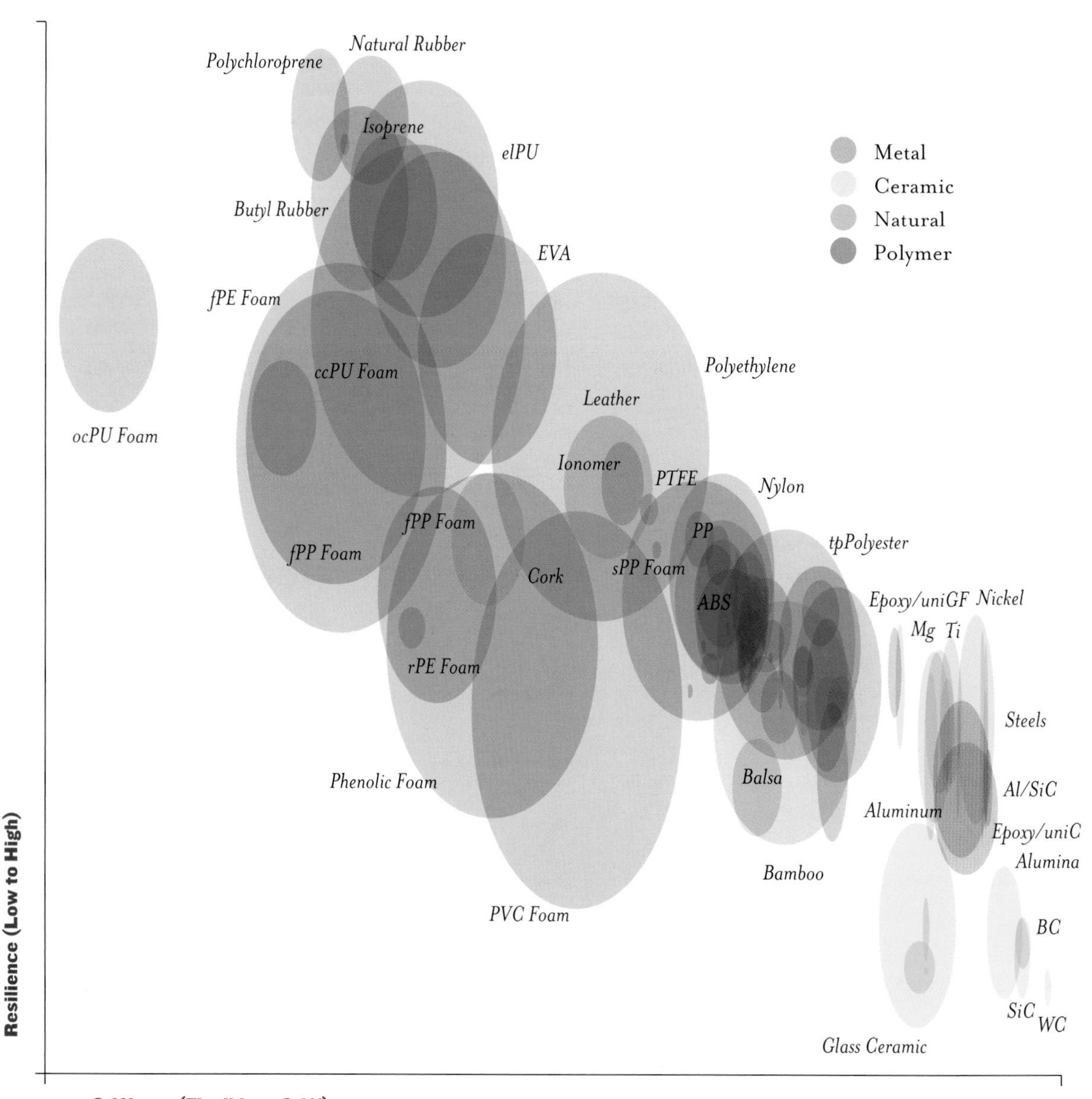

4.25 Resilience and Stiffness

Resilient materials can be flexed without breaking. Stiff materials cannot be easily bent (CES 4,2002).

This is a view to be respected. But it is not the only one. Design integrity is a quality that consumers value, but they also value other qualities: humor, sympathy, surprise, provocation, even shock. You don't have to look far to find a product that has any one of these, and often it is achieved by using materials in ways that deceive. Polymers, as we have said, are frequently used in this way – their adaptability invites it. And, of course, it is partly a question of definition – if you say that a characterizing attribute of polymers is their ability to mimic other materials, then using them in this way is honest.

So we end this chapter with the question: to what extent can the attributes of materials – technical, aesthetic, perceived – be used for expression? There are some obvious examples. Gold, silver, platinum, diamond and sapphire have associations of wealth, success, sophistication and lasting value; used in a product they give it the same associations. Polished woods suggest craftsmanship; granite – permanence; steel – strength. Metals, generally, are recognizable as metals; a product made from them acquires, as part of its personality, the character-traits of metals. But polymers can – at least in appearance – assume the character of almost any material; in particular, they can be made to look like metal, or like wood, or even like glass. Some example products are shown in 4.26: Gortex (or PTFE) in running shoes; acrylic in tail lights; nylon and aluminum in the folding bicycle.

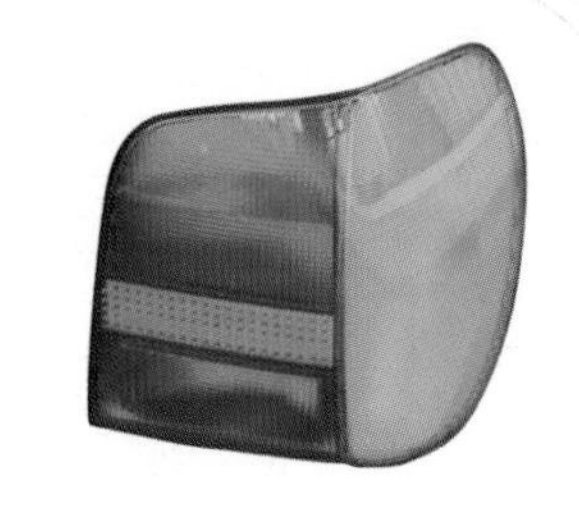

4.26 Expression through Materials
The choice of material in each of these products expresses, in part, functionality and, in part, desired aesthetics. Top, PTFE; center, acrylic; bottom, aluminum. (Bicycle image courtesy of MAS Design.)

Further Reading

Antonelli, P. (1995) "Mutant Materials in Contemporary Design," Museum of Modern Art, New York, USA. ISBN 0-87070-132-0 and 0-8109-6145-8. *(A MOMA publication to accompany their extensive 1995 review of materials in product.)*

Arredo, F. (2000) "Materials and Ideas for the Future," Material Connexion, 4, Columbus Circle, New York, NY, USA. *(The catalog of an exhibition at the Salon Internatazionale del Mobile, Milan, 2000.)*

Ashby, M.F. (1999) "Materials Selection in Mechanical Design," 2nd edition, Butterworth Heinemann, Oxford, UK. ISBN 0-7506-4357-9. *(A text that complements this book, presenting methods for selecting materials and processes to meet technical design requirements, and presenting a large number of material property charts.)*

Byars, M. (1995) "Innovations in Design and Materials: 50 Chairs;" (1997) "50 Lights"and "50 Tables;" (1998) "50 Products," RotoVision SA, Switzerland. ISBN 2-88046-264-9. *(Byars assembles pictures of contemporary products, listing designer and principal materials.)*

Cardwell, S., Cather, R. and Groák, S. (1997) "New Materials for Construction," The Arup Journal, Volume 3, p 18–20, Ove Arup Partnership, 13 Fitzroy Street, London WIP 6BQ. *(The authors devise an intriguing classification for materials, based on levels of familiarity.)*

CES 4 (2002) "The Cambridge Engineering Selector," Version 4, Granta Design, Cambridge, UK, www.granta.co.uk. *(Software for materials and process selection containing data-sheets for over 2000 materials and 150 processes.)*

Charles, J.A., Crane, F.A.A. and Furness, J.A.G. (1997) "Selection and Use of Engineering Materials," 3rd edition, Butterworth Heinemann, Oxford, UK. ISBN 0-7506-3277-1. *(A Materials-Science approach to the selection of materials – nothing on aesthetics.)*

Dieter, G.E. (1991) "Engineering Design, A Materials and Processing Approach," 2nd edition, McGraw-Hill, New York, USA. ISBN 0-07-100829-2. *(A well-balanced and respected text focusing on the place of materials and processing in technical design. Nothing significant on industrial design.)*

Dul, J. and Weerdneester, B. (1993) "Ergonomics for Beginners," Taylor and Francis, London, UK. ISBN 0-7484-0079-6. *(A brief but comprehensive and readable introduction to human-factor engineering.)*

Eternally Yours (2001), www.eternally-yours.nl. *(The Eternally Yours Foundation seeks to analyze why people surround themselves more and more with products they feel less and less attached to, and to explore the design of products that retain their appeal over long periods of time.)*

Farag, M.M. (1989) "Selection of Materials and Manufacturing Processes for Engineering Design," Prentice-Hall, Englewood Cliffs, NJ, USA. ISBN 0-13-575192-6. *(A Materials-Science approach to the selection of materials — nothing on aesthetics.)*

Flurschein, C. H. (1983) "Industrial Design in Engineering," The Design Council and Springer-Verlag, London, UK. ISBN 0-85072-123-7 and ISBN 3-540-12627-9. *(A collection of essays by a number of authors dealing with aspects of human-factor engineering and aesthetics).*

Harper, C.A. (Editor) (2001) "Handbook of Materials for Product Design," 3rd edition, McGraw Hill, New York, USA. ISBN 0-07-135406. *(A set of articles by different authors detailing properties of steels, aluminum and titanium alloys, common polymers, composites, ceramics and glasses. Heterogeneous in style and content, but a useful single-volume reference source.)*

Hawkes, B and Abinett, R. (1984) "The Engineering Design Process," Longman Scientific and Technical, Harlow, UK. ISBN 0-582-99471-3. *(An elementary introduction to the technical aspects of product design).*

ID Magazine (1998, 1999), Pearlman, C. editor, 440 Park Avenue South, Floor 14, New York, NY. *(The International Design Magazine carries reviews of contemporary and experimental products and graphics.)*

International Design Yearbook (1998), Sapper, R. editor, and (1999), Morrison, J. editor, Laurence King, London, UK. *(An annual review of innovative product designs.)*

Koodi Book (1996) "The Koodi Code, 12 Styles, 5 Products," The University of Art and Design Helsinki, Finland. ISBN 951-9384-97-9. *(An extraordinary little book illustrating a design exercise in which 5 products — a toaster, an iron, a hairdryer, a kitchen mixer and an electric shaver — were redesigned in 12 styles — folklore, deluxe, kitsch, porno, toy, cartoon, sport, pseudo-eco, army, professional, space and "white" — as part of the Industrial Design Program at Helsinki.)*

Kruskal, J.B. and Wish, M. (1978) "Multidimensional Scaling," Sage Publication No 11, Sage Publications Inc. 275 South Beverly Drive, Beverly Hills, CA 90212, USA. ISBN 0-8039-0940-3. *(A short, clear treatise on the methods and application of multi-dimensional scaling — a method for visualizing the hidden structure of a database — particularly one in which the data are imprecise.)*

Lewis, G. (1990) "Selection of Engineering Materials," Prentice-Hall, Englewood Cliffs, NJ, USA. ISBN 0-13-802190-2 *(A text on material selection for technical design, based largely on case studies.)*

Manzini, E. (1989) "The Material of Invention," The Design Council, London, UK. ISBN 0-85072-247-0 *(Intriguing descriptions of the role of material in design and in inventions. The translation from Italian to English provides interesting – and often inspiring – commentary and vocabulary that is rarely used in traditional writings about materials.)*

Meadows, D.H., Meadows, D.L., Randers, J. and Behrens, W.W. (1972) "The Limits to Growth – 1st Report of the Club of Rome," Universe Books, New York, NY, USA. ISBN 0-87663-165-0. *(A pivotal publication, alerting the world to the possibility of resource depletion, undermined by the questionable quality of the data used for the analysis, but despite that, the catalyst for subsequent studies and for views that are now more widely accepted.)*

Norman, D.A. (1988) "The Design of Everyday Things," Doubleday, New York, USA. ISBN 0-385-26774-6. *(A book that provides insight into the design of products with particular emphasis on ergonomics and ease of use.)*

Pantone (2001) www.pantone.com. *(Pantone provide detailed advice on color selection, including color-matching charts and good descriptions of the associations and perceptions of color.)*

Appendix: Modeling Aesthetic Attributes and Features

Softness (to the touch)

Hardness is resistance to indentation and scratching. It is directly related to the material property of hardness H. Softness has to do with stiffness – or, rather, to lack of stiffness. The stiffness of a material in a given shape is proportional to its modulus E, another material property. It is convenient to have a single measure that allows materials to be ranked along a single axis. One that works here is the measure

$$S = EH$$

If S is small, the material feels soft; as S increases it feels harder.

Warmth (to the touch)

A material feels "cold" to touch if it conducts heat away from the finger quickly; it is "warm" if it does not. Heat flows from the finger into the surface such that, after time t a depth x of material has been warmed significantly while its remoter part has not. Transient heat-flow problems of this class all have solutions with the feature that

$$x = \sqrt{at}$$

where a is the thermal diffusivity of the material

$$a \approx \frac{\lambda}{\rho C_p}$$

here, λ is the thermal conductivity, C_p is the specific heat and ρ is the density. The quantity of heat that has left each unit area of finger in time t is

$$Q = x\rho C_p = \sqrt{\rho\lambda C_p} \cdot \sqrt{t}$$

If Q is small the material feels warm; if large, it feels cold. Thus $\sqrt{\rho\lambda C_p}$ is a measure of the "coldness" of the material. Softness and Warmth are used as the axes of 4.18.

Pitch (of sound)

Sound frequency (pitch) when an object is struck relates to the modulus, E, and density, ρ, of the material of which it is made. We use the quantity

$$P = \sqrt{E/\rho}$$

as a relative measure of natural vibration frequency, and thus pitch. If P is small the material's pitch is low; as P increases the material's pitch becomes higher.

Brightness (of sound)

Sound attenuation (damping or muffling) in a material depends on its loss coefficient, η. We use the quantity

$$L = \frac{1}{\eta}$$

as a measure for ranking materials by acoustic brightness. If L is small the material sounds muffled; as L increases the material rings more. Pitch and Brightness are used as axes of 4.20.

Abrasion Resistant

Resistance to abrasion depends on many properties, but one – hardness, H – is of particular importance. One material is scratched by another that is harder, and it, in turn, can scratch those that are less hard. So we use hardness H as a relative measure of abrasion resistance.

Stiffness

Stiffness, for a given shape, is directly related to the material property "Modulus," E. Abrasion resistance and Stiffness are used as the axes of 4.24.

Resilience

Resilience is the ability to accept large deflection without damage. Deflection is limited either by yield (permanent deformation) or by fracture, that is, by the more limiting of the two quantities

$$\frac{\sigma_y}{E} \text{ and } \frac{K_{IC}}{E}$$

where σ_y is the yield strength and K_{IC} is the fracture toughness. We use the quantity

$$R = \frac{\sqrt{\sigma_y K_{IC}}}{E}$$

as a measure of resilience.

Chapter 5

Other Stuff... Shaping, Joining and Surfaces

If you want to make something out of material you need a process. Processes create shape, they allow parts to be joined, and they impart textures, finishes or coatings that protect and decorate. Processes must be matched to materials – the processes that can shape or join polymers differ from those that can do the same to ceramics or glasses or metals, and even within this family, the process must be matched to the polymer type. Here we review the technical attributes of processes and explore the features they possess that contribute to the industrial design of a product.

The Engineering Dimension: Technical Attributes

Classifying processes is not as easy as materials. For materials, an unambiguous classification could be built on the scientific understanding of the bonding between atoms or molecules. But processes are devised by man, not dictated by nature – science does not help much here. The best approach is to classify according to the purposes for which the process will be used, though ambiguities are unavoidable. One adapted to their use in technical design is shown in 5.1. At the highest level, processes are segregated into the three families suggested above, which we will abbreviate as *shaping, joining* and *surface*. Each of these is developed in the way shown in 5.1, partly expanded to indicate classes and members.

For shaping, the first question is "how can I shape this material?" The classes of shaping are organized, as far as this is possible, by the families of materials to which they can be applied and the underlying similarities between the processes themselves. Each has many members (5.1, top row). Technical data for each member define its attributes – the details of the shapes it can make and the restrictions it imposes on size, precision and cost, as indicated in the last column of the figure.

The choice of process is determined by the materials to be shaped, the shape itself and the economics of the process. When the material is a thermoplastic and the shape is complex,

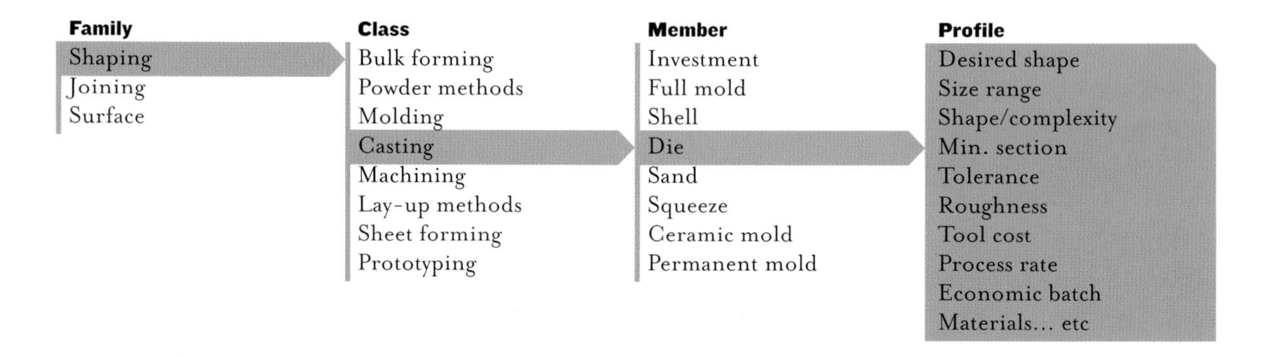

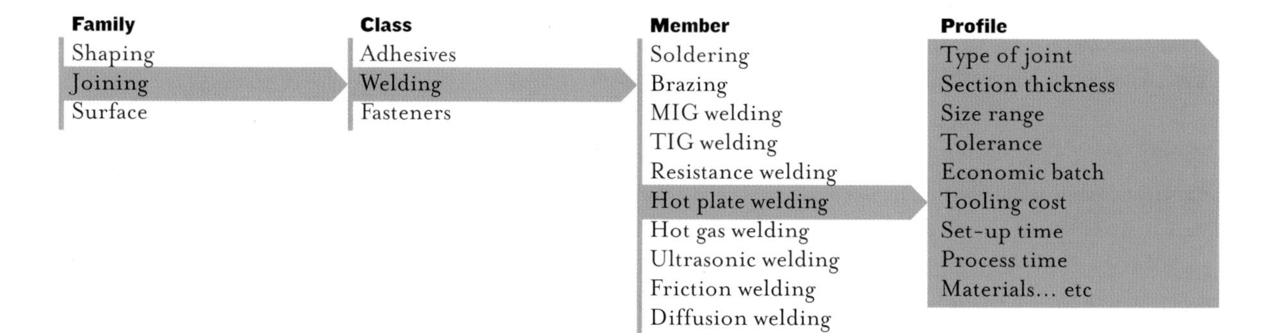

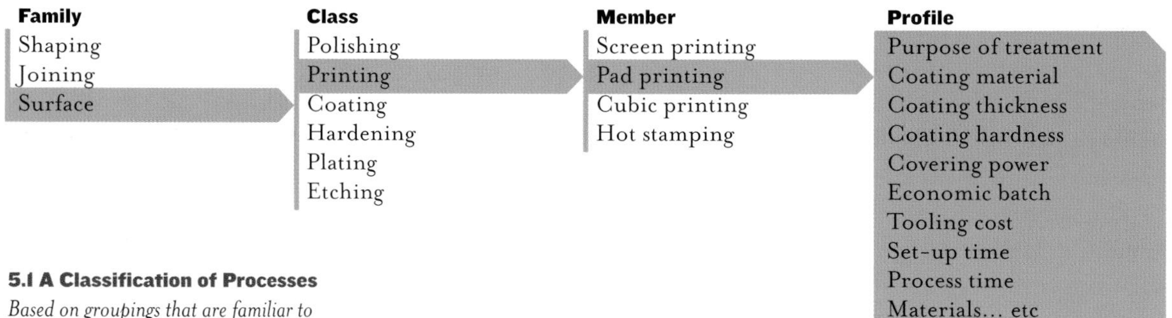

5.1 A Classification of Processes
Based on groupings that are familiar to engineers. The final column shows a list of possible attributes for a specific process. Top, Classification of Shaping Processes; center, Classification of Joining Processes; bottom, Classification of Surface Processes

with doubly-curved surfaces, economics favor injection molding. Steel molds are very expensive, so for very small batch sizes (around 10) an epoxy mold is used; for 100, a metal-filled epoxy mold just survives; for 100,000, tool steel is necessary, and above that only molds of Stellite (an alloy based on cobalt, tungsten and chromium) suffice.

Metals compete directly with polymers in filling these roles. Die-casting of aluminum and magnesium alloys allows economic production of complex shapes with double curvature. Pressure die-casting is faster and so cheaper for larger batch sizes than gravity die-casting. As with injection molding, dies are expensive. Aluminum has a small solubility for iron, limiting the die-life of steel molds to about 100,000 parts. Magnesium has no solubility for iron, giving an almost unlimited die-life.

High labor costs and low material costs have created a society that discards many products after a single, often trivial use. Such things must obviously be made by very low cost processes. Yogurt pots, butter containers, soft-drink bottles and such are made by pressing or blow molding. The dies are cheap, the process is fast, and decorative labeling can often be molded in at the same time. Sheet pressing of metals, too, can be fast, cheap and can be decorated in a continuous process (think of soda cans).

Joining processes are classified in a similar way (central tree of 5.1). The first distinction, giving the classes, is the way the joint is achieved – by adhesives, by welding or by fasteners. Beyond the classes lie the members. The last column suggests the nature of the technical data that characterizes them.

The last tree (5.1, bottom row) shows a classification for surface processes. Almost all manufactured parts are given a surface treatment of some sort. The reasons are diverse: to make the surface harder or more scratch resistant, to protect against corrosion or wear, or to enhance the visual and tactile qualities of the finished product. There are many different ways of doing each of these, shown in the second level of 5.1: chemical, mechanical, painting, printing, and so forth, each with many members. The final column again suggests the nature of technical data for a member.

Technical data for the members of 5.1 can be found in texts on processing,[1] in software[2] and in suppliers' datasheets.[3] Profiles for 65 of the most important of these are given in the second part of this book.

1. Some texts give a broad survey of processes, others specialize in one of the classes of 5.1. Lesko (1999) and Swift and Booker (1997) give good introductions; but the bible is Bralla (1998), comprehensive, but not a book you'd read for pleasure. Of the more specialized texts, Houldcroft (1990) gives a good survey of welding processes; Wise (1999) gives more specialized information for the joining of polymers; and Grainger and Blunt (1998) detail processes for surface coatings for technical applications. We have been unable to find any text that deals with surface processes with predominantly aesthetic function – though the later part of this book goes some way to correcting that.

2. The ASM offer their 20 volume Metals Handbook, and the Desk Editions of both the Metals Handbook and the Engineered Materials Handbook, both on CD and the Internet, allowing rapid searching for text about all types of processes. The Cambridge Materials Selector, CES 4 (2002), contains data sheets for about 120 processes of all types, combining it with an advanced search engine.

3. An example can be found in Poeton (1999), "The Poeton Guide to Surface Engineering," www.poeton.co.uk/pands/surface/.

The Other Dimensions

Processes influence the ergonomics, aesthetics and perception of products and of the materials of which they are made. Shaping gives form. The shapes of which a process is capable are clearly important here: extrusion, for instance, allows only prismatic shapes, whereas die casting of metals and injection molding of polymers allows shapes of great complexity. Joining – an essential step in the assembly of a product – can also create contrasts and other expressive features: some processes give near-invisible joints, others allow them to be prominent. Surface processes, particularly, influence aesthetics and perceptions through color, reflectivity, texture and feel, and they are also important for ergonomic reasons, creating the visual and tactile communication with the user. We explore them by examining expression through processing.

Expression through Shaping

Creating form is one of the earliest forms of human expression: carved stone and molded pottery figures, beaten ornaments and cast jewelry pre-date any documented ability to write or draw. The sculpture, pottery and architecture of today are evolutionary descendants of these pre-historical antecedents, and still use many of their processes: molding (initially with clay, now with polymers), carving (now a subset of machining) and lost wax (now investment) casting. Shaping, here, is a channel for self-expression.

Figure 5.2 shows ways in which form and materials have been used to create different aesthetics and perceptions. The lamp uses ABS and acrylic formed by thermal forming and injection molding; in both its folded and its unfolded state it is perceived as humorous, friendly, appealingly dotty yet completely functional. The tape dispenser expresses a sense of humor and surprise in both its biomorphic form and bizarre color.

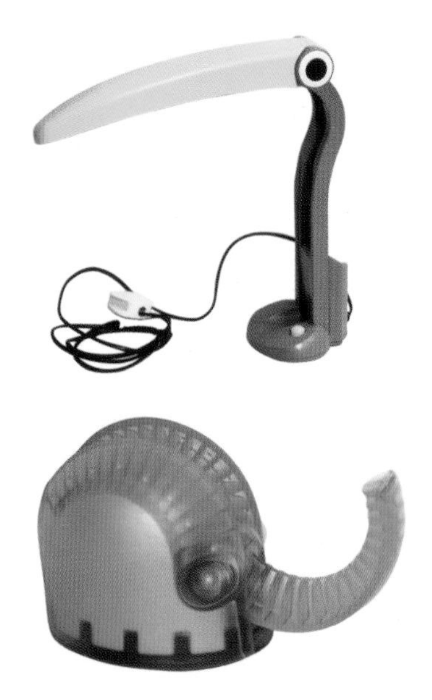

5.2 Expression through Shaping
In both the bird-lamp and the elephant tape-dispenser, humor is expresses through shape.

Expression through Joining

Joining reaches an art form in book binding, in the dovetailing of woods, and in the decorative seaming of garments. In product design, too, joints can be used as a mode of expression (5.3). The fuel cap of the Audi TT machined from stainless steel and attached by eight Allen screws, is an expression of precision technology that implies the same about the rest of the car. The prominent welds on the frame of a mountain bike express the robustness of the design. Deliberately highlighted joints are used as decorative motifs, often to emphasize the function of the product. Co-molding of two contrasting polymers to form the toothbrush both decorates and highlights the different functions of the two (one structural, the other to give grip).

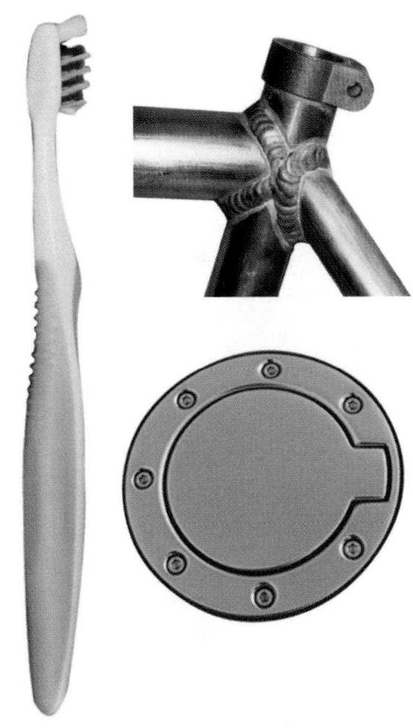

5.3 Expression through Joining
The uses of Allen screws to mount the machined steel filler cap of the Audi TT and the prominent welds of the mountain bike express the engineered robustness of both products. The contrasting co-molded handle of the toothbrush emphasizes its structural and ergonomic functions.

Expression through Surfaces

One could describe the paintings and prints of the world's great museums as surface treatment elevated to the status of high-art. Both techniques have been adapted to the more lowly task of product design. These, and a range of others, some very ingenious, allow expression in a number of ways.

The late 20th and early 21st century is addicted to flawless perfection.[4] Makers of earth-moving equipment have long known that – if their products are to sell – they must deliver them with the same high finish that is required for a passenger car. And this, despite the fact that the first thing a purchaser does is to lower the equipment into a hole half full of mud to start digging. It is because the perfection of the finish expresses the perfection of the equipment as a whole; a poor finish implies, however mistakenly, poor quality throughout.

So surface processes can serve to attract, as with the digger. It can suggest, sometimes with the aim of deceiving; metalized plastic is an example. It can surprise, adding novelty – a jug kettle with a thermo-chromic surface coating that changes color as the water heats up. It can entertain: holographic surface films can suggest something lurking inside the article to which it is applied. It can add function: non-slip coatings add an ergonomic function, and technologies for printing electronic circuitry onto products give them sensing and information-processing functions.

Figure 5.4 illustrates expression through surface treatment.

4. Perfection and Imperfection. Surface perfection is violated by the slightest defect – it has no hope of ageing gracefully. Better, to make visual imperfection a part of the personality of the product – something that gives it individuality. It is this, in part, that makes natural materials – wood, leather and stone – attractive.

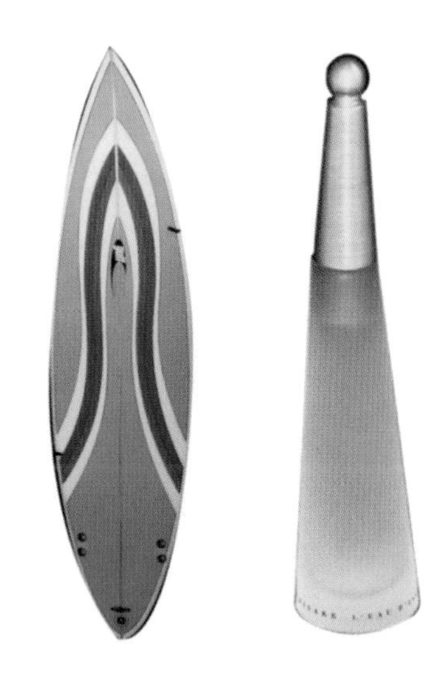

5.4 Expression through Surfaces
Surface processes give textures, text, colors and effects that together create the image of a product in the mind and hand of the user.

Screen printing is used to distinguish surfboards, board style and personal style. Bead blasting of a glass surface gives a frosty sense of elegance. The metal oxide coated sunglasses express high-tech fashion through brilliant, reflected color.

Conclusions

The aesthetics of a product are created by the materials of which it is made and the processes used to shape, join and finish them. Ideas are expressed and perceptions and associations created by the ways in which these are used. The technical requirements of the product impose certain constraints on shape, but within these there is still room for expressing quality, or humor, or delicacy, or sophistication. Joining, too, can be used to suggest craftsmanship, or robustness, or to differentiate parts of the product that have different purposes. Above all, surface treatments modify color and reflectivity, texture and feel, and can add pattern, symbol or text to instruct, amuse or deceive.

The later part of this book contains profiles for some 65 processes, chosen because of their importance in product design, that bring out these features.

Further Reading

Bralla, J.G. (1998) Handbook of Product Design for Manufacture, 2nd edition, McGraw Hill, New York, USA. ISBN 0-07-007139-X. *(The bible – a massive compilation of information about manufacturing processes, authored by experts from different fields, and compiled by Bralla. A handbook, not a teaching text.)*

ASM (2001) "ASM Metals Handbook Desk Edition" and "Engineered Materials Desk Edition," both on CD and offered on-line, American Society for Metals, Metals Park, OH 44073, USA. *(Both volumes contain extensive technical information about processes for shaping, joining and surface treatment. The searchable nature of software allows rapid retrieval of information.)*

Canning, W. (1978) "The Canning Handbook on Electroplating," W. Canning Limited, Gt Hampton St., Birmingham B18 6AS, UK. *(Just what it sounds like – everything you need to know about electroplating.)*

CES 4 (2002) "The Cambridge Engineering Selector," Version 4, Granta Design, Cambridge, UK www.grantadesign.com. *(Software with technical attributes for some 120 processes.)*

DeGarmo, E.P., Black, J.T. and Kohser, R.A. (1984) "Materials and Processes in Manufacturing," Macmillan Publishing Company, USA. ISBN 0-471-29769-0. *(A comprehensive text focusing on manufacturing processes, with a brief introduction to materials. The perspective is that of metal processing; the processing of polymers, ceramics and glasses gets much briefer treatment.)*

Grainger, S. and Blunt, J. (1998) "Engineering Coatings, Design and Application," 2nd edition, Abington Publishing, Woodhead Publishing Ltd, Abington Hall, Abington, Cambridge, UK. ISBN 1-85573-008-1. *(A monograph aimed at technical engineers detailing processes for enhancing the wear and corrosion resistance of surfaces.)*

Houldcroft, P. (1990) "Which Process?," Abington Publishing, Woodhead Publishing Ltd, Abington Hall, Abington, Cambridge, UK. ISBN 1-085573-008-1. *(Brief profiles of 28 welding and other processes for joining metals, largely based on the expertise of TWI International (formerly the Welding Institute) in the UK.)*

Hussey, R. and Wilson, J. (1996) "Structural Adhesives: Directory and Data Book," Chapman & Hall, London. *(A comprehensive compilation of data for structural adhesives from many different suppliers.)*

Kalpakjian, S. (1984) "Manufacturing Processes for Engineering Materials," Addison Wesley, Reading, Mass, USA. ISBN 0-201-11690-1. *(A comprehensive and widely used text on manufacuring processes for all classes of materials.)*

Lesko, J. (1999) Materials and Manufacturing Guide: Industrial Design, John Wiley and Sons, New York, NY, USA. ISBN 0-471-29769-0. *(Brief descriptions, drawings and photographs of materials and manufacturing processes, with useful matrices of characteristics, written by a consultant with many years of experience of industrial design).*

Mayer, R.M. (1993) "Design with Reinforced Plastics," The Design Council and Bourne Press Ltd. London, UK. ISBN 0-85072-294-2. *(A text aimed at engineers who wish to design with fiber-reinforced plastics; a useful source of information for shaping processes for polymer composites.)*

Pantone (2001) www.pantone.com. *(When it comes to defining, describing and exploring the associations of color, start with Pantone.)*

Poeton (1999) "The Poeton Guide to Surface Engineering," www.poeton.co.uk/pands/surface/. *(A Web site with useful information on surface treatment for "engineering" properties.)*

Roobol, N.R. (1997) "Industrial Painting, Principles and Practice," 2nd edition, Hanser Gardner Publications, Cincinnati, OH, USA. ISBN 1-56990-215-1. *(A comprehensive guide to painting and resin-coating.)*

Swift, K. and Booker, J.D. (1998) "Process Selection: From Design to Manufacture," John Wiley & Sons, UK. ISBN 0-340-69249-9. *(Data-sheets in a standard format for 48 processes for shaping and joining metals. Nothing on other materials, or on treatments for visual effect.)*

Wise, R.J. (1999) "Thermal Welding of Polymers," Abington Publishing, Woodhead Publishing Ltd, Abington Hall, Abington, Cambridge, UK. *(A compilation of information on processes for welding polymers, largely based on the expertise of TWI International (formerly the Welding Institute) in the UK.)*

Chapter 6

Form Follows Material

Materials exert a profound influence on the form of products. Nowhere is this more visible than in architecture. The Parthanon, the Eiffel Tower, the Golden Gate Bridge, all great symbols of their age, are unique expressions of what is possible with a particular material. An Eiffel Tower made from stone is as inconceivable as a Parthanon made from wrought iron or a Golden Gate Bridge made of reinforced concrete. The material has constrained each design, but within these constraints the designer has created a form that subsequent generations see as structural art.

Materials and Architecture

The most direct links between material and form arise through the forces that materials can carry (6.1). To bring out the point, look at the images of 6.2, each pair illustrating the forms that are made possible – even forced onto the design – by the nature of the material of the primary structures. Stone and cemented brick (6.2a,b) are strong in compression, weak in bending and tension, requiring designs that transmit loads by compression: closely spaced columns (a) or compressive arches (b). Wood is strong in tension, compression and bending, allowing triangulated space-frames or trusses (c) and cantilevered structures (d). Early cast iron (e and f), like stone, is brittle in tension or bending, forcing designs that follow the column/arch principle to transmit loads by compression, but – because of the ease with which it can cast to delicate shapes, it allows a new lightness to

6.1 Bending, Tension and Compression

Three ways of supporting a heavy object. Its mass exerts a load on the support. If a material that can only support compression, such as stone, is to be used, the form must resemble that on the right. If a material is strong in tension, but cannot support compression or bending, as is the case with strong wires or fibres, then a suspended form like that in the centre is required. If a material that can support bending, like wood or structural steel, is to be used, the form on the left becomes possible.

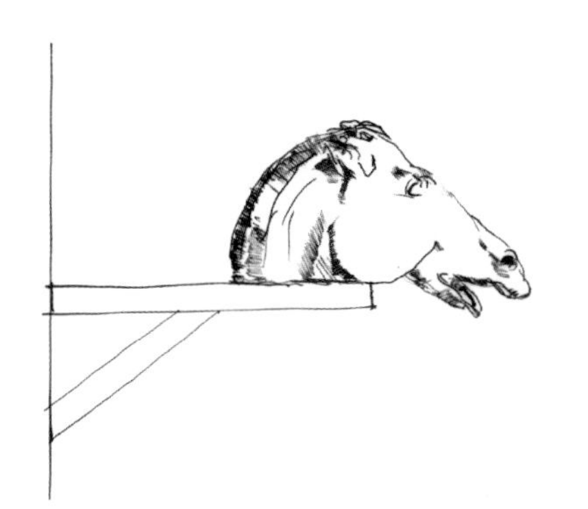

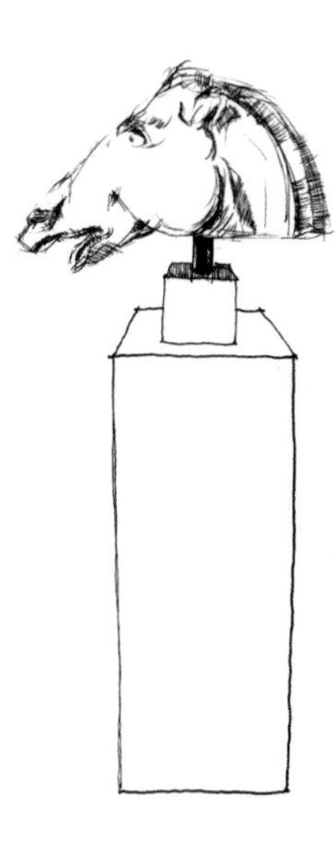

6.2 Headline for All Images

(a, b) The forms of these structures are dictated by the brittleness of the stone, making it strong in compression but weaker, by a factor of about 20, in tension and bending. Brick, like stone, is strong only in compression.

(c, d) Here the forms reflect the high strength-to-weight ratio of wood, and its ability to carry both tensile and compressive loads almost equally well.

(e, f) Here the forms reflect iron can be cast to intricate shapes, but also its relative weakness in tension and bending.

(g, h) Steel in the form of I-sections carries compression and bending efficiently; drawn to high-tensile cable, it is very efficient in tension. In the explicit design of these structures the orientation of the struts reveals the directions in which the forces act.

(i, j) Reinforced, and pre-stressed, concrete allow great freedom of shape and delicacy of form. The materials, a composite of steel and concrete, carry tensile, bending and compressive forces equally well.

(k, l) Left: high tensile steel cables with truss-like compression members supporting a glass-reinforced PTFE room membrane; Right: a rubberised nylon membrane supported by internal pressure. The forms are a direct consequence of the choice of material.

appear in the structure. Wrought iron (g) and steel (h) overcome the brittleness problem of stone and cast iron, allowing slender members that can carry tension or bending, stimulating further development of the space-frame concept. Reinforced concrete, too, has tensile as well as compressive strength, and the ease with which the reinforcement can be shaped allows fluid, shell-like structures (i, j). Fabrics and high tensile steel cables carry tensile loads well but cannot carry bending or compression (they buckle) leading to another set of forms (k, l).

In all these examples, materials have been formed into elements capable of carrying certain forces; the elements are integrated into forms designed to convert the loads on the structure into forces that are compatible with the material of which it is made. Most of the buildings in 6.2 were devised before the science of stress analysis had developed, but even without it, the designer was capable of visualizing the forces and the relationship between force, material, and form. In every case, the form of the building has been powerfully influenced by the nature of the material of which its structure is made. Form, you might say, follows material.

Materials and Chairs

Chairs must meet certain obvious mechanical constraints, but they are not severe, leaving room for diversity both in choice of material and in form. The form is not particularly limited by the function – if it were, all chairs would look the same, and a glance at 6.3 shows that this is clearly not the case. But once the material is chosen the form is constrained.

A chair, you might think, has two basic components: a frame to carry the mechanical loads, and a seat to adapt the frame to the shape of the person who will sit in it. The chair of 6.3a has these features: the frame is tubular steel and the seat is molded, corrugated polypropylene. In 6.3b, the use of a steel wire mesh, that is formed, soldered, nickel-plated and coated with epoxy combines the frame and the seat in one. In 6.3c, Philippe

a

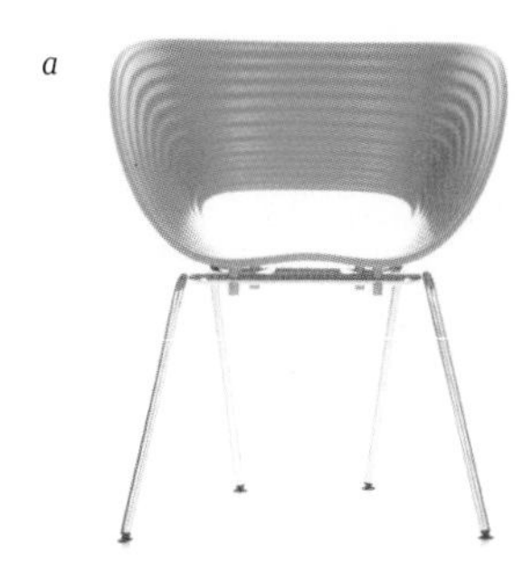

6.3 Chairs
The Chair is one of the most common forms manipulated by industrial designers or engineers; some unusual forms and materials are shown here. (Courtesy Vitra Museum and Hans Hansen.)

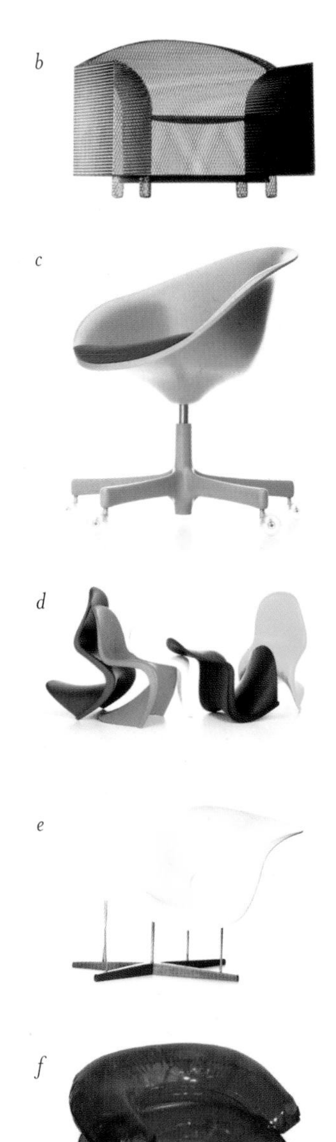

6.3 Chairs, *continued*

Starck has manipulated polypropylene to create a molded seat that is soft and playful. In 6.3d, the use of glass fiber-reinforced polypropylene allows the combination of the frame and seat in one. Charles and Ray Eames use fiberglass to create two shells, joined and bonded to a chrome-plated tubular steel base (6.3e). The last, 6.3f, is made of inflated PVC and has something of the bulbously pneumatic look of the inflated building of 6.1. Here the frame has disappeared; the seat is also the support. As with buildings, the materials of the chair have profoundly influenced its form.

Thus there is more than one way to approach the selection of materials. One is through technical function. The other is through form and features, seeking materials capable of providing them. Here we explore the idea of form and features.

Identifying a Feature List

The act of design transforms a *need* into a *product*; it starts with a featureless abstraction and ends with a concrete reality. First some definitions. A *need* is an abstract idea. The need for "illumination" carries the idea of light, but says nothing whatever about its form, its feel, how it will be achieved or how it will be perceived. A *concept* is one way in which the need might be met – a candle, an incandescent lamp, a fluorescent tube, a laser – and for each of these there is a set of possible sub-concepts. A *product* is a realization of one of these; it is a concrete object, with forms and features that can be seen and touched. It is made by processes, using materials that themselves have visual and tactile behaviors. A *feature* is an aspect of the design that contributes to its functionality, usability, or personality.[1] Features are of many types. There are topological features, defining the configuration of the product. There are geometric and dimensional features, some determined by technical requirements (strength, stability, efficiency etc), others by the needs of the user (ergonomic features), still others to give the product certain visual and tactile qualities. Beyond these, there are perceived features – those that create the associations of the product and the meaning it conveys. *Solutions* are combinations of features that embody the concept, making it real, and which do so in ways that, to some degree, meet the intentions of the designer. The solution that is chosen for manufacture becomes the product.

The starting point for developing form is the formulation of the desired features of a product. The features define the constraints to which the form and materials must comply. Formulating and applying constraints is central to any act of selection. Solutions that do not meet the constraints, applied successively, are rejected, until a manageably small subset remains. But while designing includes the act of reducing the set of possible solutions, it importantly also contains also the act of expanding them. The designer, experimenting with ways in which the features are used, evolved and combined, visualizes new solutions. Thus moving from concept to product involves both narrowing the number of solutions by screening out those that fail to meet the constraints and expanding it by creating new solutions that are in turn screened. The feature list exerts the constraints and at the same time provides the ingredients of inspiration.

Features are identified, and solutions synthesized and screened in ways that we shall group under the headings: Identification, Visualization and Materialization (6.4). The sequential nature of a book – read in the order in which it is written – suggests that this is a linear progression, but that is not intended here. Using a text-based analogy: design, and particularly the creation of form, is not a linear procedure like filling out a tax form ("if this, then that..."), it is more like a multi-dimensional crossword puzzle ("does this fit with these... or perhaps these are not yet quite right... what if we tried this instead...").

Identification

Identification of concept is the first step in developing a feature list for an initially abstract idea. We use the need for "seating" as an example. The idea is initially abstract: it has as yet no form. There are many possible concepts for seating – basic ways of pro-

1. Functionality is a measure of how well a product meets its technical expectations. Personality measures the degree to which it provides emotional delight and satisfaction in ownership. Thus a desk lamp could – by providing light of the right intensity at the right place – have high functionality, yet – by having a form, color and feel that clashed with the rest of the décor – have low personality.

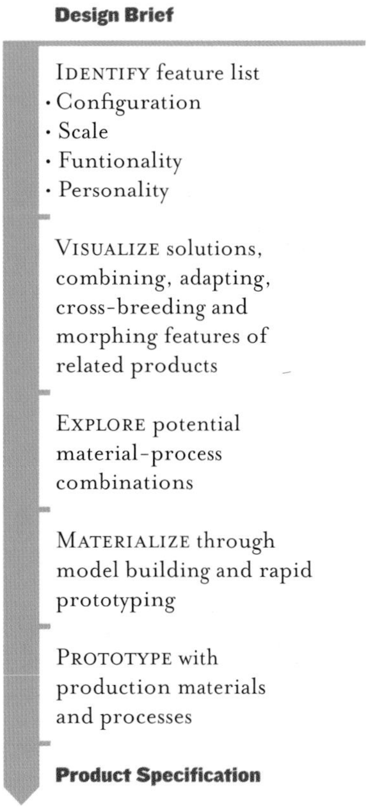

Design Brief

IDENTIFY feature list
- Configuration
- Scale
- Funtionality
- Personality

VISUALIZE solutions, combining, adapting, cross-breeding and morphing features of related products

EXPLORE potential material-process combinations

MATERIALIZE through model building and rapid prototyping

PROTOTYPE with production materials and processes

Product Specification

6.4 Identification, Visualization and Materialization
The steps in developing product form. The list implies a sequential process, but it is far from that.

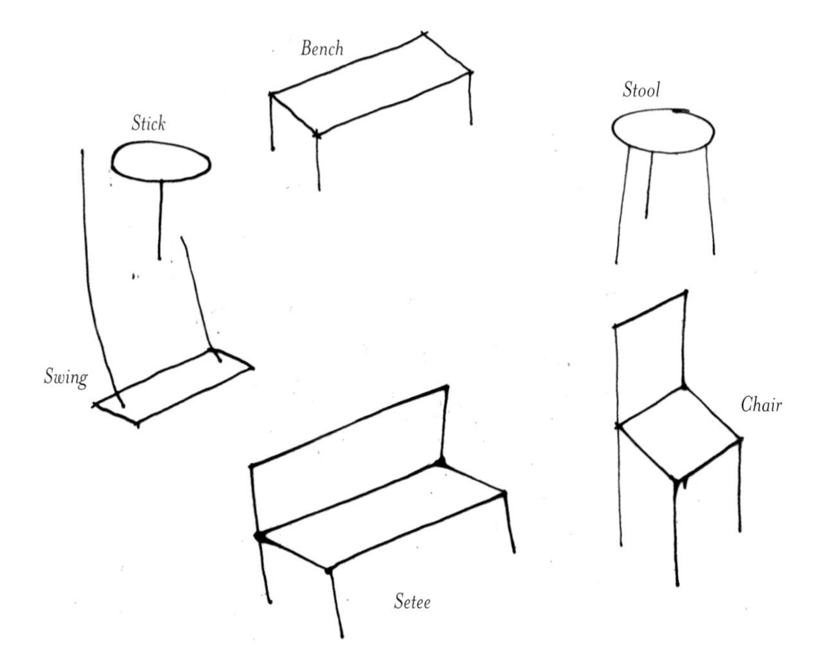

6.5 Concepts for Seating
The idea of "seating" can involve many concepts: some with 4 legs, some with none; some for one person, some for many.

viding support: the bench, the stool, the chair, the settee, the swing, the shooting stick, and many more, as suggested by 6.5.

We choose the concept "chair." This is a concept that has been used before and although we wish to have the maximum freedom in its design, the choice of chair implies that it will have certain generic features – features that are accepted characteristics of a chair. The chair stick-diagram of 6.5 illustrates these. They are of two types. The first has to do with configuration: a chair, normally, is designed to support one person; it has a horizontal seat, supported above the floor by legs or some equivalent structure, and it provides support for the back. This combination of generic features already differentiates it from all the other concepts of 6.5. The second type of feature has to do with scale. The chair is to support an adult; if it is to do this, the depth and width of the seat, and its height above the floor are constrained to lie within certain broad ranges. Thus the choice of concept has identified two sets of features, one constraining configuration, the other constraining scale (6.6). The creativity lies in choosing features and combining them into solutions that meet these constraints and, at the

6.6 Features of a Concept "Chair"
A chair has cartain generic features that constrain the form.

Configuration
Support one person
Horizontal seat
Back support

Scale
$0.3 < \text{Width} < 0.4$ m
$0.7 < \text{Height} < 0.9$ m
$0.3 < \text{Depth} < 0.4$ m

same time, optimally achieve the designer's *intentions*. That is a word that needs explanation. It comes next.

The features that characterize a chair for an airport differ for those of a chair for a nursery; each is designed for its intended use. The phrase "Design for X" nicely captures the way in which the features of the chair are selected with one or more broad underlying intentions in mind – here, design for public use, or design for children. This use of *intentions* to guide choice of features goes further than simple market objectives: design can be adapted to meet economic objectives, environmental objectives, and, of course, performance objectives. Figure 6.7 suggests the categories.

Consider, as an example, the design of an office chair (6.8), defining an intention – "design for the office." Some features of office chairs are required by legislation: an office chair, in parts of Europe , must have five, not four, points where it touches the floor. Others are identified by examining existing solutions to concepts for office furniture, seeking those that appear well-adapted to other intentions of the design. Thus the prioritization of features in "design for light weight office equipment" is different from that in "design for recyclable office equipment." The intentions provide filters through which features are screened. Indexing by intention allows the exploration of features of other products designed with the same intentions in mind (6.9).

A successful product is well adapted for the circumstances under which it will be used. The office chair will be used for many hours a day, and not always by the same person. It must offer comfort and be adjustable in height and in the angle of the seat and back. For freedom of movement it must rotate. It must be strong and stable – industrial injuries lead to damaging compensation. It must be durable, immune to coffee stains and scuffing (6.10).

And finally there is its personality. Personality describes the associations and meaning that a product has for those who own or use it. Personality relies most heavily on the visual and tactile: a sense of order, proportion and internal coherency, and on shape, color and texture. And it includes the sense of

6.7 Intentions in Design

Intentions describe the broad visions that guide the design. A single product may have more than one dominant intention.

Market

Design for public use
Design for women
Design for the elderly...

Economics

Design for minimum cost
Design for assembly
Design for mass production...

Sustainable

Design for the environment
Design for recycling
Design for biodegradability...

Performance

Design for maximum insulation
Design for minimum mass
Design for minimum volume...

6.8 Visualizing an Office Chair

An office chair has certain required features: by legislation, in Europe, it must have five – not four – points where it touches the floor.

6.9 Visualizing an Office Environment
The environment in which the chair will be used suggests visual features that relate it to its surroundings and purpose, and to the aspirations of the user.

compatibility with the lifestyle and aspirations of the consumer. Features that create and transmit these visual messages become another part of the feature list from which the final solution ultimately emerges.

The easiest of these to describe in words are those of color, texture, feel and form (see 4.17 in Chapter 4). Perceptions are harder to express precisely, and are usually described by loosely defined words or metaphors like "rugged," "feminine" or "classic" that suggest rather than define the perception, but still convey a certain set of features created by the choice of form and material. Figure 4.22 listed such words, drawn from reviews in design magazines and exhibition catalogs. Some that might apply to office chairs are listed in 6.11.

6.10 Feature List for Functionality
Technical and ergonomic features are required for products to function properly.

Comfort
Range of motion
Adjustable
Strong
Stable
Easy to clean
Abrasion resistant

Visualization

Strategies for creating personality are found by examining the environment in which the product will be used and the ways in which other products in this environment succeed in creating a similar personality. A common technique is to assemble images of products or environments, color and texture samples, and samples of materials themselves that exhibit some aspect of the personality that is sought. These are assembled into a collage or mood board allowing them to be moved, rearranged and overlapped to suggest alternative combinations, as described in Chapter 3.

Building the feature list, as we have said, has the effect of narrowing and focusing the constraints on product form. But at the same time the list provides a tool for inspiration. The freedom to rearrange features can suggest new solutions, some of which may come closer to achieving the desired personality than any existing solution. It is a kind of visual "reasoning," often supported by hand sketching, which, in this context, can be thought of as "data processing;" the "data" are the individual visual features.

The feature list thus far defines both a set of constraints and a set of inputs to the creative synthesis of new solutions. But what can really be made? Actuality is limited by materials and by the processes used to shape, join and surface them. They are, so to speak, the free variables of the design; their choice makes possible a range of solutions.

6.11 Feature List for Personality

Aesthetic and perceived features are required for each product's personality.

Aesthetics

Black
Metallic
Soft
Organic

Perceptions

High tech
Masculine
Expensive

Materialization

Figure 6.12 illustrates the way in which the choice of material and process can allow different groups of features. The matrix lists four materials and six processes.[2] The numbered boxes

2. *Profiles for most of these appear at the end of this book.*

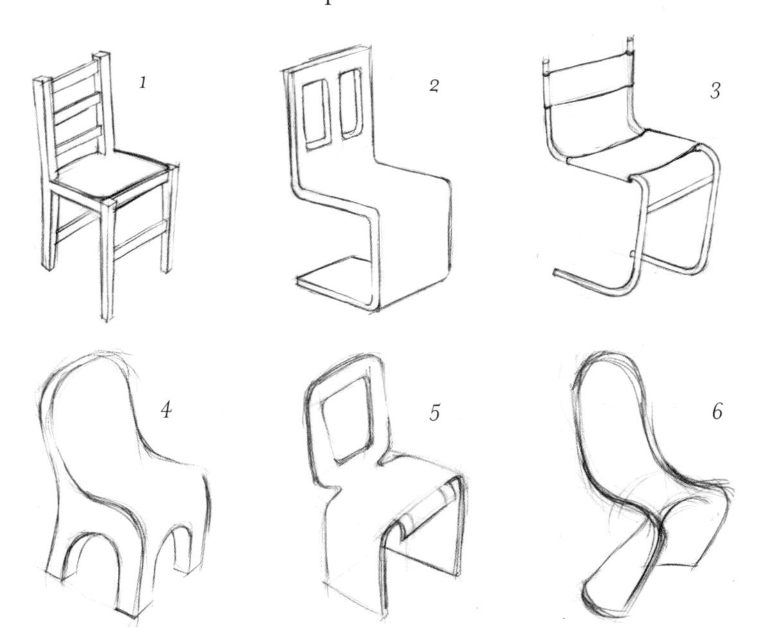

6.12 A Material/Process Matrix for Chairs

Form is partly determined by the choice of material and process.

	Machining	*Sheet forming*	*Drawing*	*Rotation moulding*	*Lay-up methods*	*Thermoforming*
Solid wood	1					
Plywood					5	
ABS				4		6
Stainless steel	2	2	3			

mark combinations that are explored in the sketches. Each combination leads to forms that have characteristic features: lamination of wood (chair 5) is well adapted to structures with flat planes or single curvatures; rotational molding (chair 4) gives closed hollow structures with doubly-curved surfaces. The choice also influences the aesthetics of the product – color, texture and feel. Wood (chairs 1 and 5) is naturally textured, it is light in weight and it feels warm. Stainless steel (chair 2) is metallic and reflective, heavy and cold. The degree to which these combinations match the desired feature list is the metric of success. It is here that technical and industrial design reasonings merge. Materialization – creating an object that can be seen and touched – is partly conditioned by choice of material and process, partly by the necessity that the product carried design loads safely, and this is established by standard methods of technical design. The environmental impact of the product, too, is influenced by the initial choice of material and process. So too is the energy it will consume during its lifetime and the length of this life.

Before leaving this topic we look briefly at two more examples of material and form: the bicycle, and the bottle opener.

The Bicycle: Materials and Form

Bicycles are made in a great diversity of materials and forms (6.13). The requirements here are primarily mechanical – strength and stiffness – to be achieved at low weight and acceptable cost. The triangulated space frames convert the forces generated by the rider into tensile, compressive and torsional loads in its members, like the architectural space frames of 6.1c and g; the materials are chosen and shaped to carry these loads. Early bicycles had a solid hickory frame joined by adhesives – they were light but not very durable. Solid steel is 10 times denser than wood, but shaped into a hollow tube and joined by brazing it is as light as wood for the same stiffness and strength (form 1 of 6.13). Most metals can be drawn to tube, so that metal bicycles frequently have this form.

But joining, for some materials, is difficult, slow and expensive. Might it not be possible to make a frame with fewer parts,

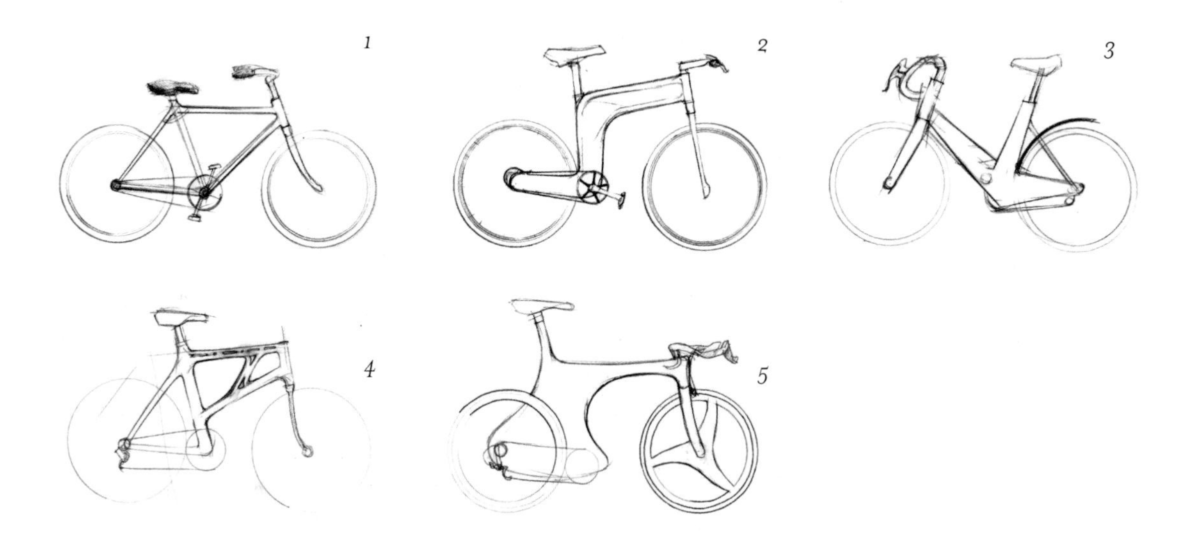

or even as a single integral unit? Form 2 shows a frame made by pressing and adhesively bonding two aluminum sheets to give a light, hollow shell. Form 3 uses injection molding to integrate parts into a single unit; here both the frame and wheels are single nylon moldings. Made in large numbers, moldings can be cheap, and nylon, unlike steel, does not rust. There is, however, a problem: nylon, even when filled, is more that 50 times less stiff and 10 times less strong than high-strength steel, requiring that the cross-sections have to be larger. The chunky, heavy appearance of a polymer bicycle is the result; it looks clumsy – an unfortunate personality for a bicycle. If not injection molding, why not die-casting? Magnesium is light and easy to die cast. But tubes can't be die cast; an I-section of equivalent stiffness can. Form 4 is the result – a form dictated by the choice of process and material. The lightest, stiffest material-choice for a bicycle is CFRP (an epoxy resin reinforced with carbon fibers). Early CFRP bicycles used tubes, emulating the steel bike, but joining them was difficult, and the joints added weight; the form did not make good use of the material. The properties of CFRP are better utilized in a monocoque construction (a shell-like skin), as in form 5.

In each of these examples the form reflects the properties of the material and the ways in which it can be processed.

	Extrusion, Drawing	*Sheet forming*	*Die-casting*	*Injection molding*	*Lay-up methods*
Steel	1				
Aluminum		2	4		
Magnesium			4		
Nylon				3	
CFRP					5

6.13 A Material/Process Matrix for Bicycles

The matrix suggests possible combinations of material and process; the empty spaces represent opportunities or limitations.

Bottle Openers: Materials and Form

A bottle opener performs a simple mechanical function: it captures the rim of the cap with a tooth and – when leverage is applied to the handle – lifts it off. That's one concept (there are others) and it implies certain constraints on configuration and scale. It can be embodied in many ways, a few of which are sketched in 6.14. That at (3) is stamped from flat sheet steel. Those at (2) and (4) start with an extruded section, requiring materials that are easy to extrude. That at (1) is made by stamping and folding of steel sheet, with a polymer handle molded on.

Matrices[3] like those of 6.12, 6.13 and 6.14 are a useful way to explore materialization, and have the ability to suggest ways in which material-process combinations might lead to new solutions.

3. *The suggestion originates with Haudrum (1994).*

Before concluding this chapter it is interesting to switch perspective briefly – to think not of the designer designing, but of the observer analyzing the origins of a design. What can be inferred about a product from its features? Let us consider the connection between material and form in other fields of research.

6.14 A Material/Process Matrix for Bottle Openers

Again, the matrix can also suggest possible combinations.

	Stamping, Press forming	*Injection moulding*	*Extrusion*	*Stamping*
Steel	1		4	3
Thermoplastic		1		
Aluminum			2	

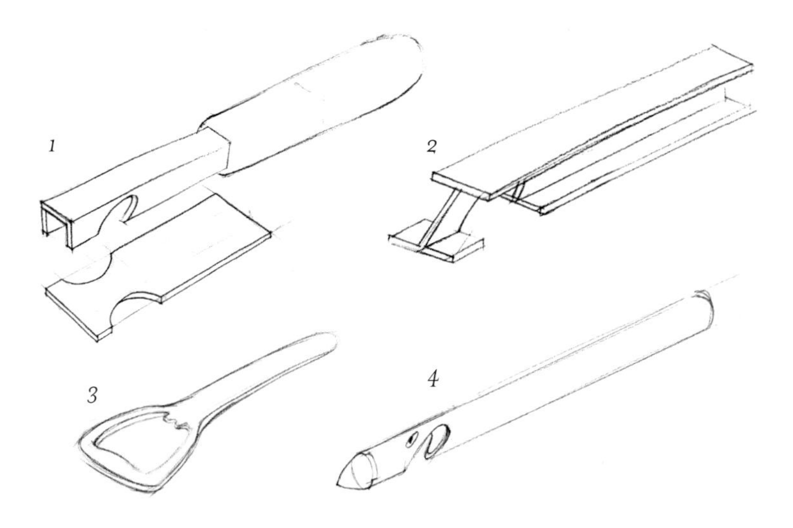

The Inverse Process: Product Archeology

An archeologist, unearthing an incised and patterned object like that of 6.15, seeks to infer from its features the needs and desires that inspired its making. Its diameter and slightly concave shape imply the concept: a plate or dish. Its shallowness and absence of rim are features suggesting its functionality: used to hold solids, not liquids. The geometry of the design associates it with other, previously discovered, artefacts with different forms and imagery but with the same style: the proportions, the geometry of the patterning, the material and the way it was worked. This in turn associates the plate with a period, a society and a culture – that of the Sumerian civilization – about which much is already known. The material and the process – silver, beaten and incised – and the quality of the workmanship begin to give meaning to the dish: silver and the skills to work it have never been resources to which the average man could aspire. The perception of the plate, then as now, was of a rare and precious object, the possession of an individual or organization of wealth, importance and power – the last reinforced by the image of the lion.

6.15 Sumarian Plates
A Sumarian silver dish with beaten and incised pattern and decoration.

This kind of archeological method is known as "Typology." Its basis is that of clustering objects into groups ("types") that share the same attributes of material, form and surface decoration. Behind it lies the premise that objects from a given time and place share a recognizable style – what today would be called a "brand identity" – and that changes in style are gradual. Three broad groups of attributes are used in Typological Archeology, and they are revealing. This is the way one well-known text[4] describes them:

4. *Fagan (1999) gives a readable account of archeological methods.*

- *Formal attributes* are features such as shape, measurable dimensions and components.
- *Stylistic attributes* include decorations, pattern, color and surface finish.

· *Technological attributes* are those describing the material and process used to make the object.

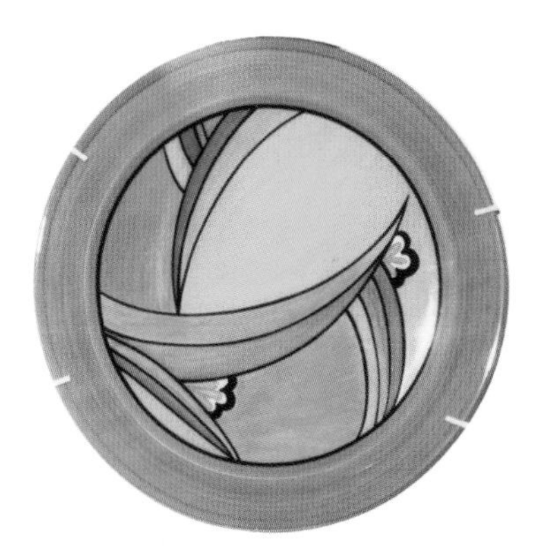

6.16 Art Deco (Clarice Cliff) Plates
The material and form, and the color and geometry of the decoration, typify the Art Deco movement.

The parallel with the feature-lists developed in this chapter, based on current design thinking, is striking.

Let us try it on a 20th century design. The object shown in 6.16 is the work of a design-house – Clarice Cliff – and epitomizes a style – Art Deco – that gripped both Europe and the US in the 1920s. Art-Deco styling invokes semi-abstract geometry involving horizontals, zigzag lines and dynamic streamlining. Geometric ornamentation is based on hexagons, octagons, ovals, circles, and triangles. Decoration is borrowed from classicism; pre-Colombian, Egyptian and African art using motifs such as the fan, the lotus-blossom, the rising sun, and the lightning strike. Art Deco was the first style movement to make expressive use of polymers. Bakelite (a phenolic resin), the first commercial plastic that could accept a wide range of colors and textures, lent itself to the adventurous curved and terraced shapes of the Art Deco movement – shapes that would be hard to make from wood. At one extreme, that of the portable radio, the use of plastic, chrome, steel and aluminum allowed a range of cheap mass-produced products. At the other, that of state rooms of great ocean liners and grand hotels, the use of exotic hardwoods, leather, ivory, glass and bronze created a décor for the very rich. The associations of Art Deco are those of cosmopolitan luxury, industrial progress and power, faith in technology, and futurism.

Some styles, such as those listed earlier (4.23), have such significance that they have been studied in depth; the cluster of attributes – form, pattern, decoration, material and process – that characterizes each is well documented. But many other "styles" exist that have arisen naturally in attempting to meet a need. Thus there is a style associated with military vehicles that has its origins in purely practical needs, with the obvious associations of rugged durability, toughness and aggression. Car designers, particularly designers of off-road vehicles, draw on the form, color and material of military vehicles to create, in their products, the same personality. There is a style associated with electronic equipment in aircraft and ships, again purely utilitarian in origin, which carries associations of cutting-edge technical performance, precision and reliability. Designers of high fidelity equipment for the connoisseur use the visual prompts of this style – understated rectangular form, matte-black finish, switches and controls that mimic those of the professional equipment – to suggest uncompromising technical leadership. There is a style that characterizes aerospace that goes beyond simple streamlining, extending to color and material (polished or matte aluminum), shapes of windows and doors (rounded) and method of joining (riveting). Designers of household appliances use these and other characterizing attributes of aerospace to suggest associations with the advanced nature of that industry.

This digression into archeology has a point. It is that the methods of Typology allow products to be indexed in a number of ways: by configuration, by scale, by intention, by functionality, by personality and by the materials and processes used to make them. A catalog, of products indexed in this way (or a software-based equivalent) becomes a tool for industrial design, a kind of computerized collage or mood board, and yet another way of suggesting materials and processes. This idea is developed further in Chapter 7.

Conclusions: What Influences Form?

One of the most quoted and influential phrases of design history is "form follows function." Its originator, Louis Sullivan, had a simple point he wished to make[5] but its brevity, its alliteration, its practicality – its *neatness* – gives it a compelling quality, seeming to say that the best form for a product is that which best fulfills its function. This reasoning has an obvious appeal to engineers, who's job it is to make things function, and there are many instances in which the pursuit of function has created a thing of beauty: the Golden Gate bridge in San Francisco, the original Volkswagen Beetle, the Concorde aircraft (though the Boeing 747, it could be argued, is nearer to

5. Sullivan was an architect, a designer of skyscrapers. His phrase laid down a basic principle of modernism, and with it a misunderstanding that has proved as hard-wearing as concrete. It has become an alibi for products with brutish lack of imagination. Sullivan himself explained what he really meant, namely that building banks that look like Greek temples is ridiculous – after all, the clerks behind the counter do not wear togas. A bank that serves its purpose should look like a bank; a museum like a museum; a residential building like a residential building.

the "form follows function" principle) – all these were designed with technical rather than aesthetic challenges in mind. Perceptions like these have given the phrase the status of a religious dogma.

But can it really be true? Is the best design – the only design – that which best fulfills its function? Form, as just shown, is also influenced by materials; in the case of architecture, the influence is strong; in product design, though sometimes less obvious, it is certainly present. A case could be made for the dictum "form follows material" – it doesn't sound as good but it may be closer to reality. It seems to be a general rule that good design uses materials in ways that make the most efficient, and often visible, use of their properties and the way they can be shaped. Rules, of course, have their exceptions: incongruous use of materials can convey the surreal (the furry watch of 2.6) or the ridiculous (a knife blade made of rubber) or transfer an association from one object to another (chocolate Eiffel Towers). All have their small place. But at the center of the stage are forms that use materials elegantly, efficiently and economically.

And function and material are not the only directors of form. Consumers buy things they like; and with many alternatives of nearly equal technical merit, the consumer is influenced by trends in taste, by advertising and by the things that other people buy. Form, inevitably, follows fashion. There are still other drivers of form. Form follows delight. The near-satiated society of 2000+ is attracted by humor (the Snoopy hair dryer, the banana telephone), by perfection (the immaculate body paint of a new car), by luxury (deep, soft leather-upholstered furniture) and by whimsical novelty (any Philippe Starck design). Product form is influenced by many things, but material and process are among the strongest.

Further Reading

Bahn, P. (1996) "Archaeology – A very short history," Oxford University Press, Oxford, UK. ISBN 0-19-285379-1. *(A lively, critical and entertaining survey of archeological methods and schools of thought.)*

Billington, D.P. (1985) "The Tower and the Bridge, New Art of Structural Engineering," Princeton University Press, Princeton, NJ, USA. ISBN 0-69102393-X. *(Professor Billington, a civil engineer, argues that great structures – the Brooklyn Bridge, the skyscrapers of Chicago, the concrete roof-shells of Pier Luigi Nervi not only overcome technical challenges, but also achieve the status of art. Much is biographical; the achievements of individual structural architects, but the commentary in between is forceful and enlightening.)*

Cowan, H.J. and Smith, P.R. (1988) "The Science and Technology of Building Materials," Van Nostrand-Reinhold, New York, USA. ISBN 0-442-21799-4. *(A good introduction to materials in architecture, drawing distinctions between structural materials, "externals" or materials for the outside, and "internals" or materials for control of sound, heat and light within the building. There are chapters on stone, steel, wood, concrete glass and plastics, with guidelines for their selection.)*

Duncan, A. (1988) "Art Deco," Thames and Hudson, London, UK. ISBN 0-500-20230-3. *(A comprehensive review of the history and characteristics of the Art Deco design movement.)*

Fagan, B.M. (1999) "Archaeology – A Brief Introduction," 7th edition, Prentice Hall International, London, UK. ISBN 0-321-04705-2. *(An introduction to the methods and techniques of modern archeology).*

Haudrum, J. (1994) "Creating the Basis for Process Selection in the Design Stage," PhD thesis, The Institute of Manufacturing Engineering, Technical University of Denmark. ISBN 87-89867-13-0. *(An exploration of process selection, viewed from a design theory perspective.)*

Jakobsen, K. (1989) "The Interrelation between Product Shape, Material and Production Method," ICED '89, Harrowgate, August 1993, pp.775 – 784. *(Jakobsen identifies the interdependence of function, shape, material and process, and that different sequences for selection of the four aspects can be followed.)*

Lenau, T. (2001) "Material and Process Selection using Product Examples," presented at the Euromat 2001 Conference, Rimini, Italy 10-14 June 2001. *(An analysis of the material-information needs of industrial designers.)*

Muller, W. (2001) "Order and Meaning in Design," LEMMA publishers, Utrecht, The Netherlands. ISBN 90-5189-629-8. *(Prof. Muller teaches Design in the School of Industrial Design of the Delft Technical University. The book describes the process he has developed for understanding and teaching Industrial Design. It contains many useful insights, some of which we have used in this chapter.)*

Chapter 7

A Structure for Material Selection

To create a tool for material selection we need to structure the information that has been presented so far. This tool – as described in Chapter 3 – should (a) capture and store material, process and product information, organizing it in a way that allows rapid retrieval, (b) present the same information in a creative format, and (c) allow browsing, retrieval, and combination of, "bits" of information about materials, processes and the products they create. The first step is that of classification.

Classification and Indexing

Classification is the first step in bringing order into any scientific endeavor. The founders of biology, zoology and geology were those who created the classification systems. Classification segregates an initially disordered population into groups that in some way have significant similarities. These groups can be further subdivided by seeking finer levels of similarities within each. The success of the great classification systems relies on the choice of relevant attributes and the judgment of what is "significantly similar." Mendeléeff's classification of the elements into the periodic table, Darwin's classification of the natural life of the Galapagos islands, Linnaeus's classification of plants – all these not only brought order to their subjects but also suggested the existence of missing members of the populations – undiscovered elements and species. Not all classifications survive – genetic mapping is even now reconstructing the classifications of biology. But this in no way diminishes the importance of the original classification; it was an essential step in reaching the new structure.

Classification has a key role in design. Design involves choice, and choice from enormous range of ideas and data – among them, the choice of materials and processes. Classification is closely linked to indexing, a central activity for both information retrieval and selection. But to be efficient, the classification and indexing must be adapted to the nature of the population of objects that are to be classified and the purpose of the search. Consider the following example.

An Information Structure for a Bookshop

Imagine that you run a small bookshop. You have access to the standard books-in-print database. It lists certain attributes of books: the author, the publisher, title, ISB number, date of publication, price and so on. A customer wanting a book and knowing any one of these pieces of information can be helped. The retrieval is objective and systematic: anyone entering the same request will get exactly the same response (7.1a). We will refer to this as "simple indexing."

But suppose the customer knows none of this and simply wants "a historical romance." The customer can still be helped if you – the bookshop owner – know the character of your books or have a database that contains such information (7.1b). Here the attributes describe the nature of the books at a more abstract level: biography, documentary or fiction; romantic, tragic or humorous; contemporary or historical, and so on. Now, a search on "romantic and historical" will return books that meet the customer's wishes. The method is subjective, and is non-systematic – it is based on perceptions, not on simple facts, and views on book character will sometimes differ. But if a broad consensus exists (as with books it certainly does), selection by character is productive: it is fast, flexible and copes with imprecisely defined inputs. Although it is not exact, it is a big step forward; the alternative is the response "if you don't know the ISB number I can't help you." We will refer to it as "deep indexing."

And there is more. Figure 7.1c, shows selection of a different sort. Here, the selection is no longer a simple search using the relevant index; it depends on analogy. It is a response to the customer who likes books by author X, but having read them all now wants books with a similar character, written by other authors. The books by author X have one or more abstract characteristics, and these same characteristics are linked back to other authors. And here the precise words used to describe book character are less important, they are, as it were, "dummy variables," creating associations between one author and another.

Taking this one step further (7.1d), the coupled pair of databases allows "browsing." Browsing is curiosity-based

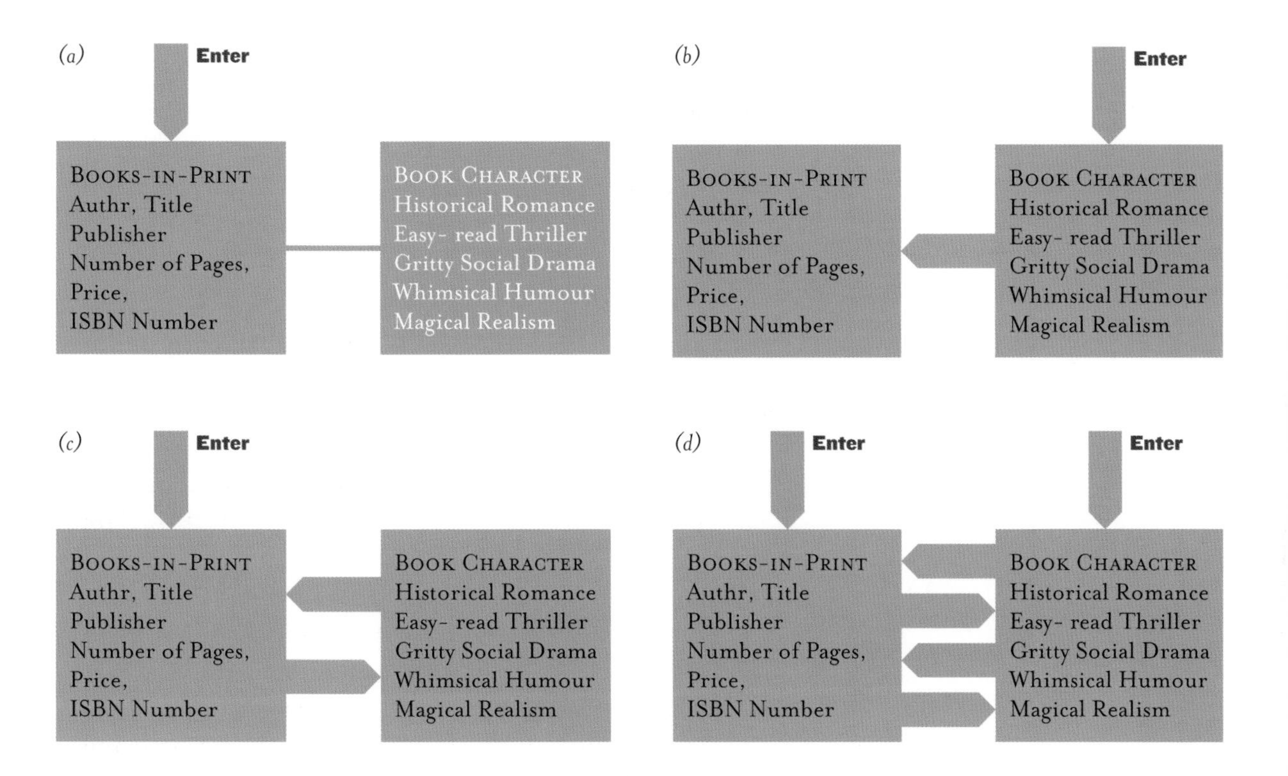

7.1 Indexing and Selection
(a) Simple indexing by objective attributes, using "the object has attributes..." (b) Deep indexing by associations or perceptions, using "the object has features..." (c) Selection by synthesis, using analogies – "the object is like..." (d) Selection by inspiration, by browsing...

searching, the way to satisfy a need that is felt but imprecisely defined. It is like wandering through the bookshop pulling books from the shelves and examining their character until a satisfying combination is found. Browsing allows unplanned suggestion, serendipity: you leave the bookshop with books that – when you entered – you didn't know you wanted. And that is a sort of inspiration.

This may seem a long way from the central subject of this book, but there are useful messages in it. Here are two. First the choice of attributes, and the classification and indexing that follows from it, must be adapted to the purpose for which the information will be used. And second: multiple indexing allows a richer diversity of selection methods, among which the most striking distinction is that between those based on analysis – "find an object that has something" – and those using analogy – "find an object that is like something."

So how are classifications constructed?

· *Empirically* – using verifiable, universal facts – as social scientists might classify the population by age, sex, married status, number of children, profession, income, and the like.

· *Visually* – by seeking clusters in "density-maps" of attributes – as in classifying crime statistics by plotting occurrences on a map of a country and grouping those areas with a given area-density of such occurrences. The methods of attribute-mapping and of multi-dimensional scaling, described in Chapter 4, reveals "similarities" within populations of materials, processes or products.

· Using underlying *scientific knowledge* – as in the classification of materials based on the nature of their inter-atomic bonds and crystal structures, or of the classification of animals based on their DNA.

· By more abstract qualities of *perception* or *association* – as you might classify friends as sensible, emotional, tolerant or short-tempered; or as we classified books as thrillers, or romantic novels, or social dramas.

The science-based classification, reinforced by property maps, worked well for the engineering attributes of materials (Chapter 4). The empirical method worked adequately for processes (Chapter 5). But for products we need something more abstract.

An Information Structure for Product Design

Figure 7.2 shows a schematic of an information structure for product design. It draws together many of the ideas developed in earlier chapters. It is made up of six linked data-tables. The central table – Products – contains factual data for product attributes: the product name, manufacturer, model number,

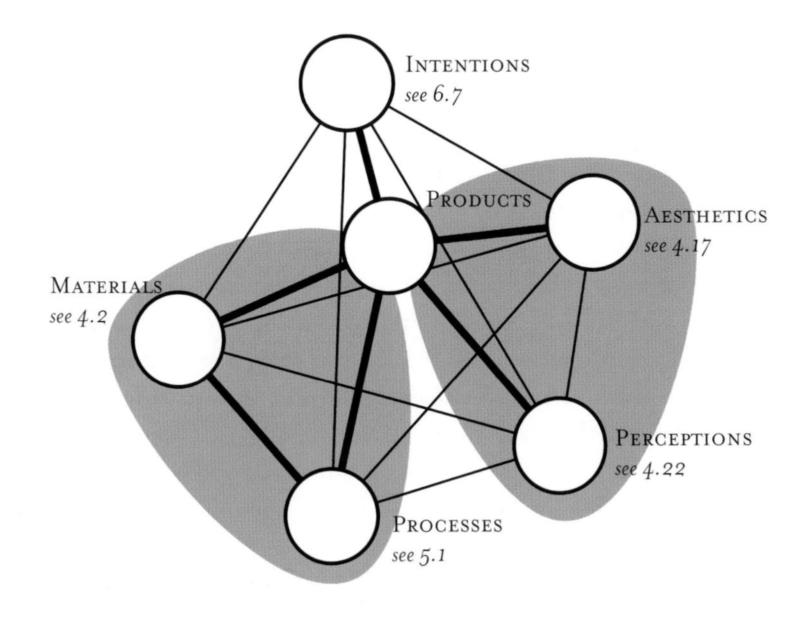

7.2 An Information Structure for Product Design
Each circle and its links is explained in the text. The numbers beside each circle refer to a table in the text.

price, and details of performance. Products can be retrieved by searching on these attributes, just as books can be retrieved by title or ISB number. It allows simple indexing.

A deeper level of indexing is made possible by linking each product to the materials and processes used in its manufacture (left-hand side of 7.2). It is now possible to search for all products made, say, of ABS, or of stainless steel, or those decorated, perhaps, by pad-printing. And it opens up new ways of selecting materials by analogy and browsing, to which we return in a moment. But first: the tables on the right of 7.2.

The first contains information about product aesthetics: the visual, tactile, acoustic and – if relevant – olfactory attributes listed in 4.17. It allows products to be indexed by these features, and indirectly, through products, it links these features to the materials and processes able to provide them. The second contains information about perceptions – the attributes listed in 4.22. It is like the "book character" table of 7.1. Its content and the links from it to products are matters of judgement: they depend on culture, taste and fashion, and they evolve with time. But, as argued earlier, the fact that they are imprecise does not mean that they are useless; they allow products to be indexed in new ways and give new paths for retrieving information about materials and processes based on the ways in which they are used.

One table remains – that at the top of 7.2. In designing a product the designer is guided by certain broad priorities that condition, to a greater or lesser degree, all decisions and choices. We encountered these in Chapter 6, where they were called Intentions (6.7). The word "intention" describes what the product is *meant* to be – the priorities in the mind of the designer. Indexing by intentions allows a degree of abstraction, grouping together products that may differ radically in function but are designed with a similar set of priorities. The intention "Design for the elderly" implies a vision of a product with functionality, appeal and price that are adapted for old people, influencing choice of configuration, form, size, color and texture, and of the materials and processes used to achieve them. "Design for mass-production" implies a vision of a product that can be made quickly and cheaply by automated equipment, directing choice of material and process for shaping, joining and finishing. "Design for recycling" implies a choice of materials and of ways to join and finish them that will enable disassembly and their return into the production cycle. If the intentions are known or can be inferred, products can be linked to them.

It is a complex picture. An analogy will help. Think of the product as a person – anthropomorphize it. Then the left-hand side of 7.2 – the materials and process bit – becomes the flesh and bones, the things you can hold and feel – the physiology of the product. The right-hand side – the aesthetics and perceptions bit – becomes the appeal and character, the product psychology. Finding an analogy for intentions is harder – perhaps it is the characteristics with which, by parentage, the person is endowed. Add them all up and they create the *product personality*.[1]

1. Perhaps to be thought of as the "product DNA," meaning the particular combination of features that distinguish it from all other products.

This information structure allows simple indexing of a product by its attributes: name, maker, serial number and so on. More helpfully, it can be indexed in a deeper way by material, process, intention, aesthetics and perception. The deeper indexing allows a whole set of selection methods, just as the indexing of books did.

An example will illustrate some of these. The concept: a

child's car seat. The starting point is a design brief, setting out certain requirements and implying a feature list. The seat must hold and protect the child when subjected to decelerations of up to 10g; it must be strong enough to withstand the inertial forces that this demands yet not damage the child while doing so. Its dimensions must allow it to pass through a car door easily. It must be light, yet able to withstand mishandling and impact, and it must be easy to clean. Second, the intentions. There are several: "design for children" is the obvious one, but "design for mass production" and "design for low cost" probably enter too. Focus on the first: design for children. If you think of products for children, you think of robustness and tolerance of misuse, of high standards of safety, non-toxic materials, simple, bold forms and bright colors. What can be learnt from other products that were designed with this intention?

An examination of products which are designed for children can give ideas that might help with the car seat. ABS, PP and nylon are used in many products for children. They are strong, tough, corrosion resistant and, of course, they are non-toxic and carry FDA approval. All can be brightly colored and – being thermoplastics – are easily molded. This information doesn't design the child's car seat, but it does suggest options that are worth exploring and ways of creating the features we want.

The seat must meet certain technical requirements. It must be light and strong, requiring materials with high strength-to-weight characteristics. It must withstand impact yet deflect when the child is thrown against it, requiring materials that are tough and resilient. The material profiles later in this book characterize toughness and resilience: among polymers, ABS, PP and nylons are both, as are fiber-reinforced polyesters and epoxies. The material must be capable of being molded or pressed to a complex shell-like form with the necessary fixing points for straps, harnesses and padding. If the ability to be colored and molded at low cost are priorities, ABS and PP move to the top of the list. If lightweight and strength are more important, then fiber-reinforced polymers move up – but they may have to be painted or coated, and children have ways of removing paints and coatings that defeat all known ways of applying them.

Successful product design makes use of all this information, and does it in more than one way. We explore these next, focusing on the selection of materials and processes.

Material Selection for Product Design

Life is full of difficult choices. Everyone has their own way of dealing with these, some effective, some not. Studies of problem-solving distinguish two distinct reasoning processes, already touched on in Chapter 3: *deductive* and *inductive* reasoning. They are the basis of selection by analysis, by synthesis and by similarity, which we now examine in more detail, using the data structure of 7.2.

The act of selection involves converting a set of inputs – the design requirements – into a set of outputs – a list of viable materials and processes (7.3). There are several methods for doing this.

Design Requirements	Selection Methods	Possible materials processes
· Technical · Economic · Sustainability · Aesthetic · Perceptions · Intentions	1) Analysis 2) Synthesis 3) Similarity 4) Inspiration Databases of materials and products.	

7.3 Methods of Selection
1) by analysis; 2) by synthesis; 3) by similarity and 4) by inspiration.

Selection by Analysis[2]

Technical engineers are trained in analysis – it is one of the tools of the trade. The inputs here are the technical requirements. The analysis proceeds in four steps.

- *Translation* of the requirements, often expressed initially in non-technical terms, into a statement of the objectives and constraints the design must meet.

- *Analysis* of the component for which a material is sought,

2. *A brief summary of the analytical method for selecting materials is given in the Appendix to this chapter. Full explanations are given in texts on material selection such as Dieter (1991), Charles et al. (1997) or Ashby (1999); the last of these gives many worked examples.*

identifying performance metrics and expressing these as equations that measure performance.

· *Identification*, from these equations, of the material properties that determine performance.

· *Screening* of a database of materials and their properties, eliminating those that fail to meet the constraints, and *ranking* those that remain by their ability to maximize the performance metrics.

Inputs	**Path to Material Selection**
Impact resistant	· Fracture toughness, $K_{ic} > 20$ MPa.m$^{1/2}$
Corrosion resistant	· No materials that corrode in water
Available as rod	· None that cannot be extruded or drawn
Light and withstand design load	· Rank by ratio of strength to density

7.4 Technical Analysis for a Bicycle Spoke
Constraints, listed in the top three lines, allow screening, eliminating materials incapable of providing the function. The objective of minimizing mass allows those that survive to be ranked.

Thus – to take a simple example – the choice of a material for the spoke of a lightweight bicycle wheel might proceed in the way shown in 7.4. The first three design requirements impose limits on material attributes, screening out unsuitable candidates. Those that remain are ranked by the fourth requirement – that of strength at minimum mass (here mass is the metric of performance). This must be balanced against cost, using trade-off methods. The method of analysis is summarized in 7.5 and the Appendix to this chapter; it will not be followed further here.

The method of analysis has great strengths. It is systematic. It is based on a deep ("fundamental") understanding of the underlying phenomena. And it is robust – provided the inputs are precisely defined and the rules on which the modeling is based are sound. This last provision, however, is a serious one. It limits the approach to a subset of well-specified problems and well-established rules. In engineering, as in all other aspects of life, is sometimes necessary to base decisions on imprecisely specified inputs and imperfectly formulated rules. Then the

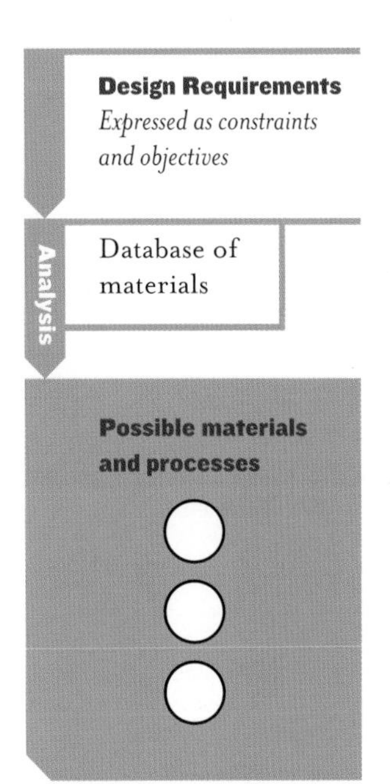

7.5 Selection by Analysis
Requires the selection of a material from a database of materials and material attributes. Here the white circles represent solutions that meet all the constraints and maximize the objective.

analytical method breaks down; methods of a different sort are needed.

Selection by Synthesis

Synthesis has its foundations in previous experience and analogy. Here the inputs are design requirements expressed as a set of features describing intentions, aesthetics and perceptions. The path to material selection exploits knowledge of other solved problems ("product cases") that have one or more features in common with the new problem, allowing new, potential solutions to be synthesized and tested for their ability to meet the design requirements.

7.6 Desired Perceptions for an Insulin Pump
Perceptions can be created by combining features from other products that are perceived in a particular way.

Inputs	Path to Material Selection
Trendy	· Transparent, brightly colored PC · Translucent PP and PE.
Humorous	· Soft over-molded elastomer surfaces · Colored, simple shapes · Injection molded forms with references to animal and human forms

Again, a specific example will help. A manufacturer of medical equipment perceives a need to adapt their insulin pump, at present perceived as clinical and intimidating, to make it more acceptable – friendly, trendy, even humorous. This leads to a list of desired perceptions listed in 7.6. A starting point is a search for other products that are perceived in this way. A search on "trendy" and "humorous" gives a number of products that are characterized by bright colors, the use of transparent polycarbonate and translucent polypropylene and polyethylene, soft over-molded elastomers, simple rounded shapes sometimes with references to human or animal forms, often made by injection molding. These suggest changes in design that the company may wish to pursue.

At first sight the analogy-based methods of synthesis (7.7) appear to have one major drawback: because they depend on past experience and designs, they cannot suggest radically new solutions, whereas analysis, requiring no such inputs, can.

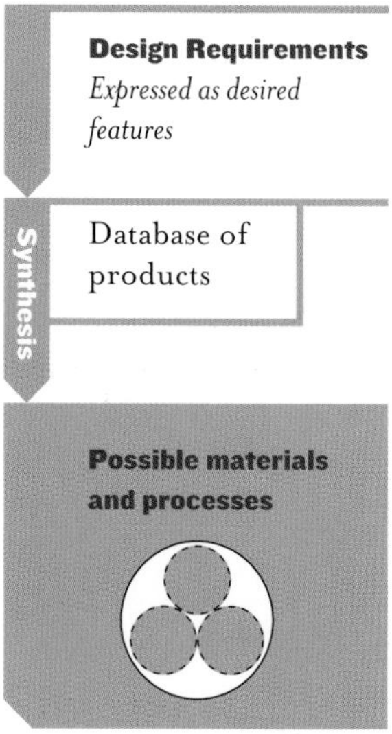

Design Requirements
Expressed as desired features

Synthesis

Database of products

Possible materials and processes

7.7 Selection by Synthesis
New solutions are generated (and materials for them selected) by examining existing products with features like those required for the new design. Here the white circle represents a synthesis of features from the three solutions (colored circles) it contains.

This, however, may be too harsh a judgement. The method encourages cross-pollination: developments in one field can be adapted for use in another where it was previously unknown; it is innovation of a different sort – what might be called *technology coupling*.

Selection by Similarity

There are many reasons why a designer may wish to consider similar materials: the need for substitution, breaking pre-conceptions or simple exploration. What is the best way to find them? Substitutes are sought when an established material ceases to be available or fails, for some reason, to meet a changed design requirement – because of new environmental legislation, for example. Analysis – the modeling of function, objective and constraints from scratch – may be possible, but in doing this we are throwing away valuable information. Established materials have a respected status because, for reasons that are not always fully understood, they have the right complex mix of attributes to meet the requirements of the design. The attribute profile of the incumbent is, as it were, a portrait of an ideal. A good portrait can capture much that is subtle and hard to analyze.

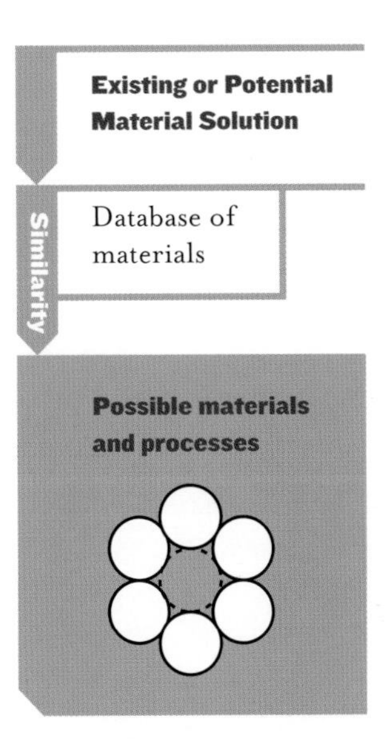

7.8 Selection by Similarity
Finding substitutes by matching attributes profiles. Here the white circles represent materials that have values of critical attributes that match, as nearly as possible, those of the established material (central circle).

The substitute, ideally, should match the incumbent in all important respects except, of course, the reason for wishing to replace it. So one way to approach such problems is that of "capture-edit-search." First, capture the attribute profile of the incumbent – easy, with present day data sources. If we use this as a template for selection we will probably retrieve only the material we started with. So the next step – edit – involves relaxing the constraints on non-critical attributes, admitting a wider range of candidates and tightening the constraint on the one attribute that was causing the problem. The last step – search – retrieves substitutes that offer the essential qualities of the incumbent, without its weaknesses (7.8).

A search for similar materials can serve as a method of expanding the set of possible material solutions. Designers often have pre-conceptions: that polycarbonate, ABS or a blend of the two are the best choices for product bodies and shells; that polyethylene, polypropylene or PET can solve most packaging problems; or that aluminum or steel are the answer for most "engineering" products. Preconceptions may be a shortcut to a quick solution but they constrain innovation. A search for similar materials can break pre-conceptions and thereby introduce creativity or novelty in a design solution.

Selection by Inspiration

Designers get most of their ideas from other designers (past as well as present) and from their environment. It is possible to search for these ideas in a systematic way – that is the basis of selection by synthesis. It relies on a mental index (or indexed books or computer files) to search for features and the way they are created. But many good ideas are triggered by accident – by an unplanned encounter. The encounter is "inspiring," meaning that it provokes creative thinking. The scientific method is of no help here; inspiration of this sort comes by immersion; by exploring ideas almost at random, like delving in a treasure chest (7.9). The word "almost" here is important; if inspiration is to be useful, the treasury of ideas needs to have a little structure. Inspiration can be fired by interaction with materials, through a material collection or – less good – through images of them. It can be fired by interaction with products, by browsing in stores that market good design, or by browsing in books – indeed, one role of the ID magazines and coffee-table books is simply that of presenting the reader with exciting images in more or less random order. Another, made possible by web-based technology, is a digital viewer of images of materials, linked to limited technical information and supplier contact. One we have made in the form of a screen saver can be downloaded from www.materialselection.com.

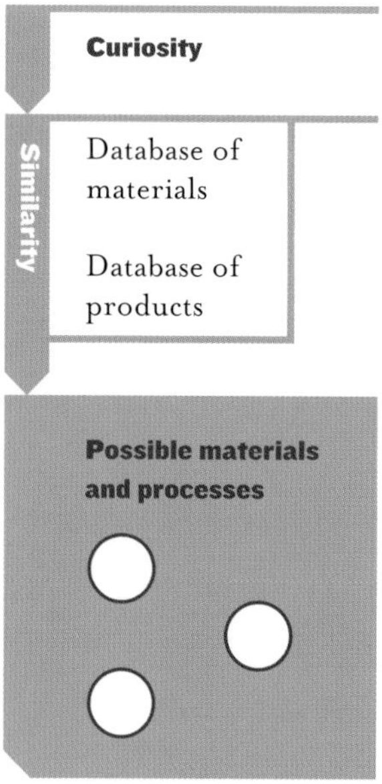

Curiosity

Similarity

Database of materials

Database of products

Possible materials and processes

7.9 Selection by Inspiration
Inspiration can be sparked by materials or processes, particularly if new or used in unusual ways. Here the white circles represent solutions inspired by a random walk through a collection or catalog of materials and products.

Combining the Methods

Figure 7.10 shows a way of thinking about all these methods. It is a development of the bubble model in Chapter 3, specialized to the selection of materials and processes. In it, the selection methods we have just described are seen as large bubbles; their job is to generate possible solutions for the design brief. Each generates a population of little bubbles – the solutions that have survived or emerged from the technique, each containing information about a material or a material combination. The selection methods – the larger bubbles – draw on two major resources: a database of materials and processes, and a database of products, stored as "cases": indexed examples of the use of materials in products.

The choice of path from design brief to product specification depends on the nature of the selection problem. If well-specified, the method of analysis efficiently screens the population

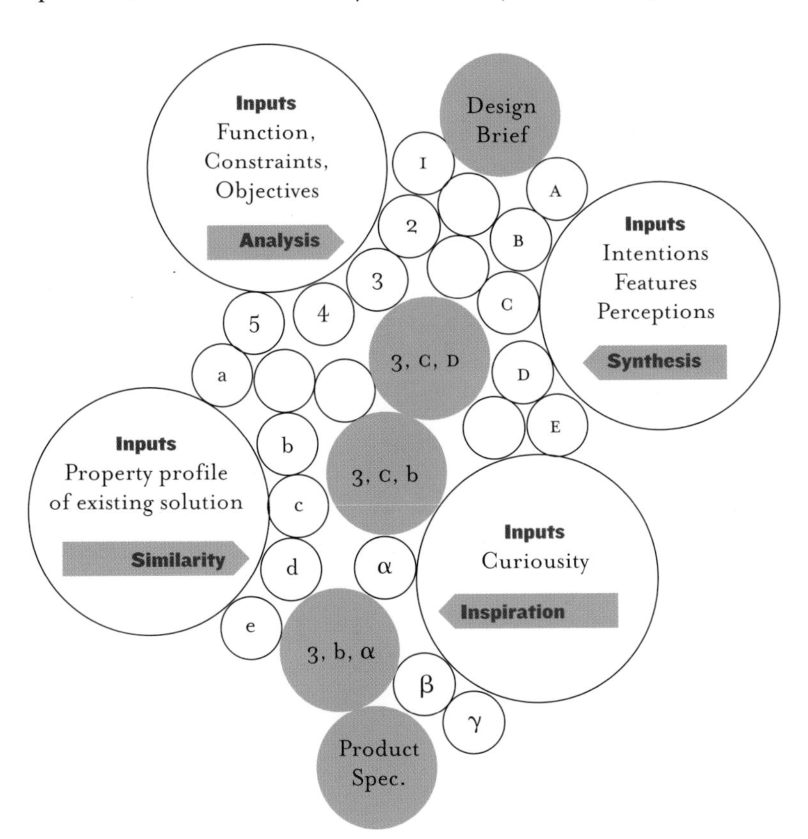

7.10 The Path of Material Selection
Selection is seen as a journey from the design brief to the product specification. The traveler is assisted, in varying degrees, by the tools of analysis, synthesis, similarity and inspiration (large circles), generating successively refined solutions as the journey proceeds.

contained in the database of materials, killing off members that cannot meet the requirements and ranking those that can by a measure of their performance, giving the possible solution "bubbles" 1, 2, 3... The database of products then provides supporting information – products in which these selected materials have been used.

If the requirements cannot be easily quantified, analysis is disabled. The method of synthesis can be used to generate possible solutions such as A, B, C... by exploring products that have the desired features, extracting material and process information about them. Here the database of products is the principal resource; that of materials simply provides the data profile once a possible material solution has been found. When the specification is even less precise, possible solutions a, b, c... can still be generated by using the method of similarity – that of seeking materials with profiles like that of the one already used – indeed, in reassessing material use in any product its helpful to know what the alternatives are. Here the databases of materials and products both act as resources. Finally a possible solution can emerge from a process of browsing – simply exploring the use of materials in products, chosen at random, using the findings as a source of inspiration (α, β, γ...).

Any one of these methods can be used in isolation, but the most effective path exploits the most useful features of each. Thus the choice of a material for a lightweight space frame is a well-specified problem, lending itself to analysis and generating possible solutions. But if one of these is unfamiliar, the designer may be reluctant to use it, fearing unexpected problems. Then it is helpful to turn to synthesis, seeking products that are space frames to see what they are made of, or products made of the unfamiliar material. The result of this can be thought of as a merging of solution bubbles from analysis with those from synthesis, generating new, hybrid, solutions. Solutions from the other methods can be combined in a similar way.

As a model for the material selection process, this bubble diagram, though complex and fuzzy, is a more accurate representation than that of the linear model of 3.3. Within one method – analysis for instance – a linear path may be found but it is rare this leads to an unambiguous choice. Almost always, a more devious path like the one just described is the best. The case studies that follow in Chapter 8 will illustrate this.

Summary and Conclusions

Creative solutions to design problems can be reached in more than one way; the more flexible the approach, the greater the likelihood of creativity. The essential requirements are:

- An *information structure* that allows both simple and deep indexing; here we have cross-indexed by Intentions, Products, Materials, Processes, Aesthetics and Perceptions.[3]

- *Selection methods* that can cope both with precisely formulated design requirements and rules, and with less precisely specified features that can include the technical, aesthetic and perceived dimensions.

Four complementary methods are described and illustrated here.

- *Selection by analysis* (deductive reasoning), using precisely specified inputs and the well-established design methods of modern engineering, drawing on databases of materials and their attributes.

- *Selection by synthesis* (inductive reasoning), drawing on past experience, retrieved by seeking a match between the desired features, intentions, perception or aesthetics and those of documented design solutions, stored in a database of product "cases."

- *Selection by similarity*, seeking materials with selected attributes that match those of an existing material, without knowing why these have the values they do, merely that they are relevant for the success of the design.

3. This structure may be stored in the mind of the designer — the "experience" of a lifetime of designing — but this is of little help to someone new to the field. For them a computer-based equivalent, storing information about products, materials and processes, indexed in the way described in this chapter, can help.

- *Selection by inspiration* – looking for ideas by randomly viewing images of products or materials (or, by visiting stores, viewing the products and materials themselves), until one or more are found that suggest solutions to the present challenge.

Traditional engineering education emphasizes analytical methods, often to the exclusion of ideas about analogy and synthesis. The justification is that analysis is an exceedingly powerful tool, and one that with modern computer support can be applied with ever greater speed and rigor; further, it can be taught as a set of formal procedures. But there is evidence that many of the most creative ideas in technical fields – science and engineering among them – arise not from analysis but from inductive thinking: from successive guessed solutions (informed guesses, of course, drawing on past experience), testing one after another until a solution, created by pulling together strands from many past solutions, is found that matches the requirements.

Each of the methods, then, has its strengths and weaknesses. The one that is best suited for one problem may not suit another – all are needed. Often, the best solution is found by combining all three; this combination, as illustrated here, is a much more powerful tool than any one used alone.

Further Reading

Ashby, M.F. (2000) "Multi-objective optimisation in Material Selection," Acta Mat 48, pp.359 – 367.

Ashby, M.F. (1999) "Materials Selection in Mechanical Design," 2nd edition, Butterworth Heinemann, Oxford, UK. ISBN 0-7506-4357-9. *(A text that complements this book, presenting methods for selecting materials and processes to meet technical design requirements, and presenting a large number of material property charts.)*

Bailey, K.D. (1995) "Typologies and Taxonomies, an introduction to classification techniques," Sage Publishers, Thousand Oaks, CA and London, UK. ISBN 0-803-95259-7. *(A monograph on classification in the social sciences, with a forceful defense of its applications there and elsewhere.)*

Bakewell, K.G.B. (1968) "Classification for information retrieval," Clive Bingley, London, UK. ISBN 85157-014-3. *(A collection of six lectures on differ-ent types of classification, interesting for the contribution on faceted classification, the precursor of relational database structures).*

Charles, J.A., Crane, F.A.A. and Furness, J.A.G. (1997) "Selection and Use of Engineering Materials," 3rd edition, Butterworth Heinemann, Oxford, UK. ISBN 0-7506-3277-1. *(A materials science approach to the selection of materials – nothing on aesthetics.)*

Dieter, G.E. (1991) "Engineering Design, A Materials and Processing Approach," 2nd edition, McGraw-Hill, New York, USA. ISBN 0-07-100829-2. *(A well-balanced and respected text focusing on the place of materials and processing in technical design. Nothing significant on industrial design.)*

Everitt, B. (1974) "Cluster Analysis," Heinemann Educational Books Ltd, London, UK. ISBN 0-435-82298-5. *(A comprehensible introduction to cluster analysis.)*

Haberlandt, K. (1997) "Cognitive Psychology," 2nd edition, Allyn and Bacon, Boston, Mass, USA. ISBN 0-205-26416-6. *(An introduction to the psychology of choice and decision-making with insights that are useful here.)*

Kolodner, J. L. (1993) "Case-Based Reasoning," Morgan Kaufmann Publishers Inc. San Mateo, CA, USA. ISBN 1-55806-237-2. *(Case-based reasoning is a technique for storing information about past situations in ways that allow useful information retrieval to solve present problems; the methods are well-presented in this book.)*

Appendix: Selection by Analysis

We here paraphrase the analytical approach to materials selection. We give only the briefest outline.[4]

4. The method is developed fully, with many examples, in the companion volume (Ashby 1999).

Function, Objectives, Constraints and Control Variables

Any engineering component has one or more functions: to support load, to contain a pressure, to transmit heat, and so forth. In designing the component, the designer has an objective: to make it as cheap as possible, perhaps, or as light, or as safe, or some combination of these. This must be achieved subject to constraints: that certain dimensions are fixed, that the component must carry a given load or pressure without failure, that it can function safely in a certain range of temperature, and in a given environment, and many more. Function, objectives and constraints (7A.1) define the boundary conditions for selecting a material and – in the case of load-bearing components – a shape for its cross-section.

7A.1 Details of Analysis

Function
What does a component do?

Objective
What is it to be maximized or minimized?

Constraints
What non-negotiable conditions must be met?

Free Variables
Which control variables are we free to adjust?

The performance of the component, measured by performance metrics, P, depends on control variables. The control variables include the dimensions of the component, the mechanical, thermal and electrical loads it must carry, and the properties of the material from which it is made. Performance is described in terms of the control variables by one or more objective functions. An objective function is an equation describing a performance metric, P, expressed such that performance is inversely related to its value, requiring that a minimum be sought for P. Thus,

$$P = f[(\text{Loads, F}), (\text{Geometry, G}), (\text{Material, M})]$$

or;

$$P = f[F, G, M]$$

where "f" means "a function of." Optimum design is the selection of the material and geometry which minimize a given performance metric, P. Multi-objective optimization is a procedure for simultaneously optimizing several interdependent performance metrics, P, Q, R.

This, in the abstract, sounds horribly obscure. A simple example will demystify the method. Consider the selection problem mentioned in the text: that of materials for the spoke of a racing bicycle wheel. The design requirements specify its length, L, and the load, F, it must carry. The *function* is that of a tie; the *objective* is to minimize the mass; the *constraints* are the specified load, F, the length, L and the requirement that the load be carried without failure. The *free variables* (those that, so far, are unspecified) are the cross-sectional area, A and the choice of material. The inputs to the model, which we now describe, are summarized in 7A.2.

The measure of performance is the mass, m – it is this that we wish to minimize. The performance equation describes this objective:

$$m = AL\rho$$

where ρ is the density of the material of which the rod is made. The length L is given but the section A is free; reducing it reduces the mass. But if it is reduced too far it will no longer carry the load F and this constrains the section, requiring that

$$\frac{F}{A} > \frac{\sigma_y}{s}$$

where σ_y is the yield strength of the material of the rod, and s is a safety factor. Using this to replace A in the equation for m gives

$$m > sFL\left[\frac{\rho}{\sigma_y}\right]$$

Everything here is specified except the term in the square brackets – and it depends only on the choice of material. The mass is minimized (and performance maximized) by choosing materials with the smallest value of ρ/σ_y, which is called a *material index.*

7A.2 Details of Analysis for a Bicycle Spoke

Function
Tie rod

Objective
Minimize mass

Constraints
Specified length L
Specified load F, no failure

Free variables
Cross-section A
Choice of material

Material Indices

Each combination of function, objective and constraint leads to a performance metric containing a group of material properties or material index; the index is characteristic of the combination. Table 7A.3 lists some of these; their derivation is detailed in the companion text. The point of interest here is that materials with extreme values of certain indices are well suited to meet certain sets of design requirements. The spoke is an example: materials with low values of ρ/σ_y are good choices.

7A.3 Material Indices

When you want to optimize material choice to meet purely technical requirements, indices act as a guide.

Function, Objective and Constraint	**Minimize index**
Tie, *minimum weight, stiffness prescribed* (Cable support of a light-weight tensile structure)	ρ/E
Beam, *minimum weight, stiffness prescribed* (Aircraft wing spar, golf club shaft)	$\rho/E^{1/2}$
Beam, *minimum weight, strength prescribed* (Suspension arm of automobile)	$\rho/\sigma_y^{2/3}$
Panel, *minimum weight, stiffness prescribed* (Automobile door panel)	$\rho/E^{1/3}$
Panel, *minimum weight, strength prescribed* (Table top)	$\rho/\sigma_y^{1/2}$
Column, *minimum weight, buckling load prescribed* (Push-rod of aircraft hydraulic system)	$\rho/E^{1/2}$
Spring, *minimum weight for given energy storage* (Return springs in space applications)	$E\rho/\sigma_y^2$
Natural hinge, *maximum distortion, flexure without failure* (Inexpensive hinges)	E/σ_y

(ρ = density; E = Young's modulus; σ_y = elastic limit)

Indices allow comparisons. If the spoke is at present made from a reference material M_O and has a mass m_O, a competing material M_I with mass m_I will be lighter than M_O, for the same load-carrying capacity, by the factor

$$\frac{m_I}{m_O} = \frac{\rho_I/\sigma_{y_I}}{\rho_o/\sigma_{y_O}}$$

(where the subscript "o" refers to the reference and "I" to the new materials) giving a direct measure of the gain in performance achieved from switching from M_O to M_I.

Software systems exist that allow selection of materials to meet technical constraints. The output of one of these is shown in 7A.4. The axes of this diagram are yield strength σ_y and density ρ; each little bubble describes a material. The diagonal line isolates materials with low values of ρ/σ_y, the starting point for

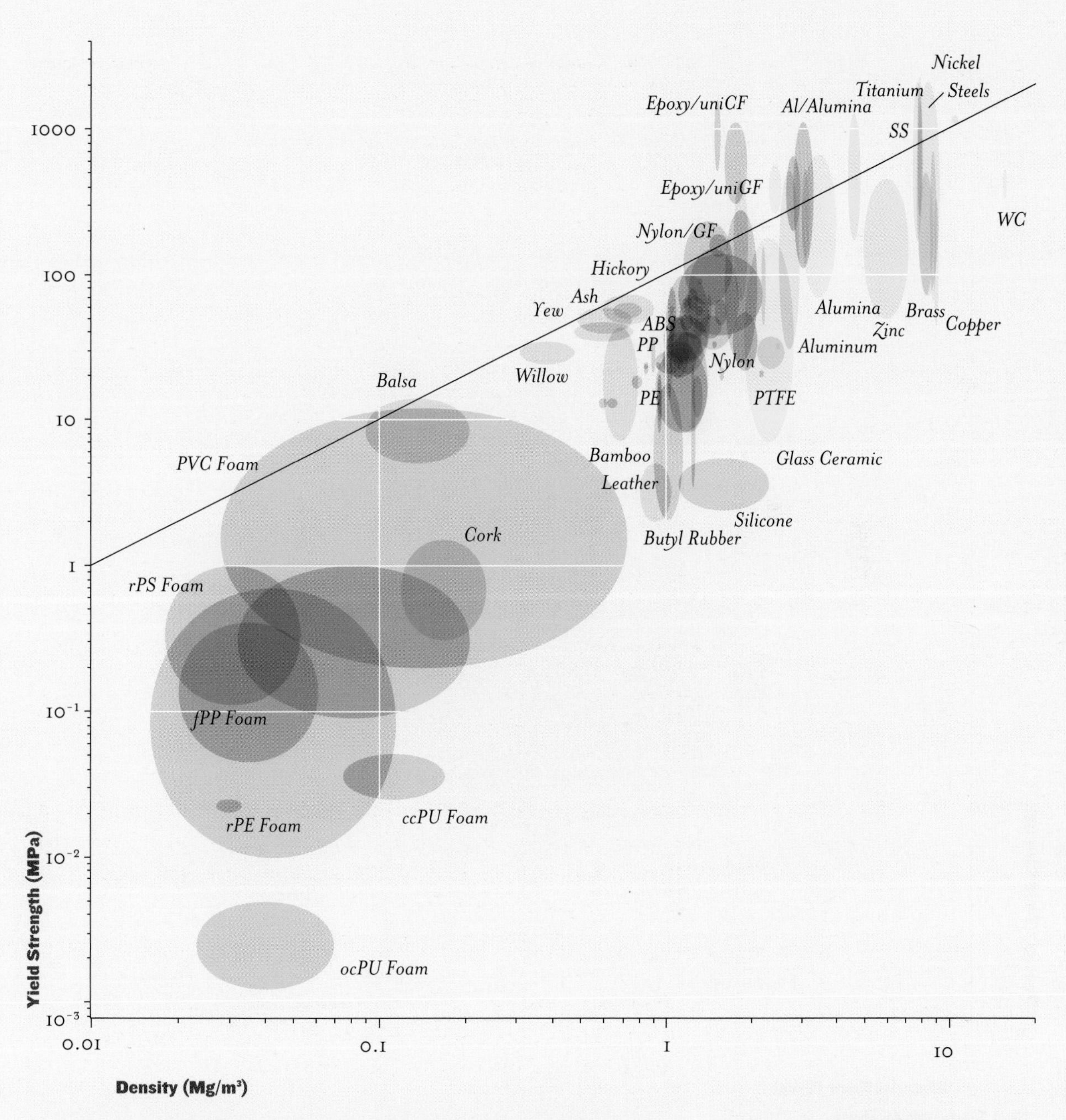

choosing materials for the spoke. To this can be added requirements of corrosion resistance, impact resistance, stiffness, and so forth, narrowing the choice. One more step is necessary: some of the candidates in 7A.4 are more expensive than others. How can weight be balanced against cost?

7A.4 Optimising Selection
A strength-density chart. Materials with low values of ρ/σ_y lie above the diagonal line, at the top left (CES 4, 2002).

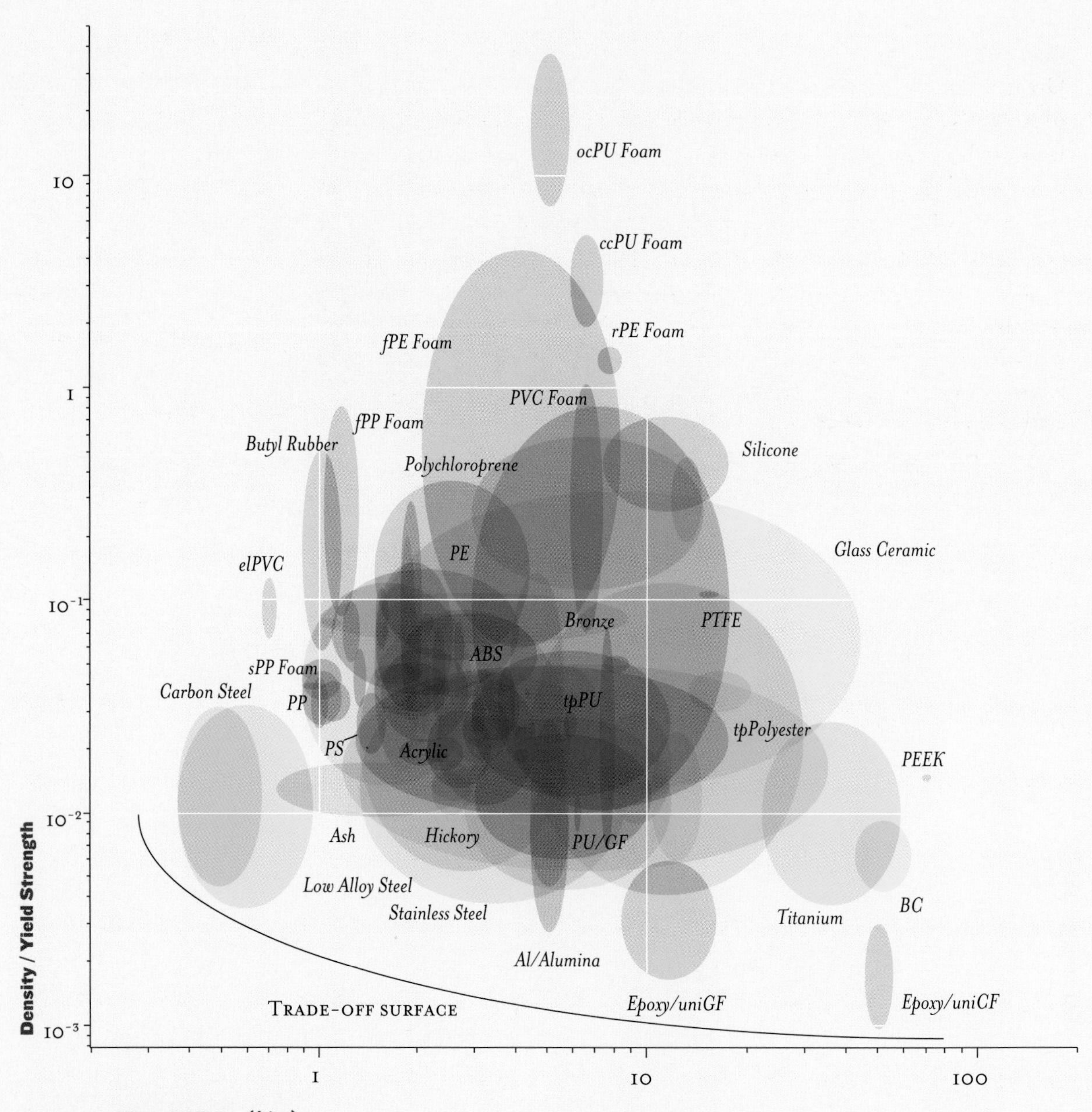

7A.5 Trade-Off Analysis

Here a compromise is sought between minimizing weight and cost. The materials at or near the trade-off surface are the candidates; the final decision depends on how much each attribute is valued. (CES 4, 2002)

Multi-Objective Optimization

Greater discrimination is possible by plotting trade-off diagrams of which 7A.5 is a simple example. Here the material group ρ/σ_y is plotted on the vertical axis, and material cost/kg on the horizontal one. The plot allows materials to be compared on the basis both of performance and of cost. The most attractive candidates lie on or near the trade-off surface (solid line); these offer both low mass and low cost. It shows that carbon steels, low alloy steels, aluminum alloys, and stainless steels are all good choices. Carbon fiber-reinforced polymers (CFRP) offers a lower mass than these, but at a considerable increase in material cost. Titanium alloys are also more expensive, and have greater mass – they are not a good choice.

Chapter 8

Case Studies in Materials and Design

The preceding chapters have explored the attributes of materials and processes, and the ways in which this information can be classified and retrieved. Here we turn from exploration to example: how is this information used? The case studies developed in this chapter are necessarily brief. They are designed to illustrate each of the broad methods of selection: analysis, synthesis, similarity, and inspiration.

The Structure

Each material selection challenge has its own special features – is, in some way, unique – so there is no single prescription for how to tackle it. It is helpful, none the less, to have some structure. The structure follows that introduced in Chapter 7: the starting point is the design brief; the endpoint, a product specification. Between these lie the steps of *identification*, *selection* and *implications*.

Identification

The first requirement in material selection is a broad understanding of the context. What does the product or the component do? Of what system is it a part? In what environment will it operate? What forces must it carry? What are the consumer's pre-conceptions and expectations? The answers to these questions set the scene within which a solution must be sought. The first priority is to identify the design requirements. Some can be stated explicitly. Constraints, it will be remembered, are conditions that *must* be met: technical, environmental, aesthetic. Objectives are a statement of what should be maximized or minimized: minimizing cost is an objective; minimizing mass is another; maximizing safety might be a third. Many requirements cannot be expressed in this way; they are described instead as features, intentions or perceptions. The nature of these inputs guides the choice of method for selection.

Material Selection

When the objectives and constraints can be expressed as well-defined limits on material properties or as material indices, systematic selection by analysis is possible. When constraints are qualitative, solutions are synthesized by exploring other products with similar features, identifying the materials and processes used to make them. When alternatives are sought for an existing material and little further is known, the method of similarity – seeking material with attributes that match those of the target material – is a way forward. And many good ideas surface when browsing. Combining the methods gives more information, clearer insight, and more confidence in the solutions.

Implications

We now have a set of selected possibilities. The next step is to scrutinize the more intimate aspects of their character; to search for details of how they are best shaped, joined and surfaced; to establish what they are likely to cost, how long they will last and how they respond to particular environments. How well does the list of possible solutions meet the original design brief? Does its use carry other implications – environmental impact, perhaps, or the need for the introduction of a new manufacturing process?

The case studies follow this general pattern. They omit much detail, and simply aim to show how the methods combine and reinforce each other (like the bubble diagram of 7.10), sometimes drawing more from one, sometimes from another, but blending to achieve more than is possible by any one alone.

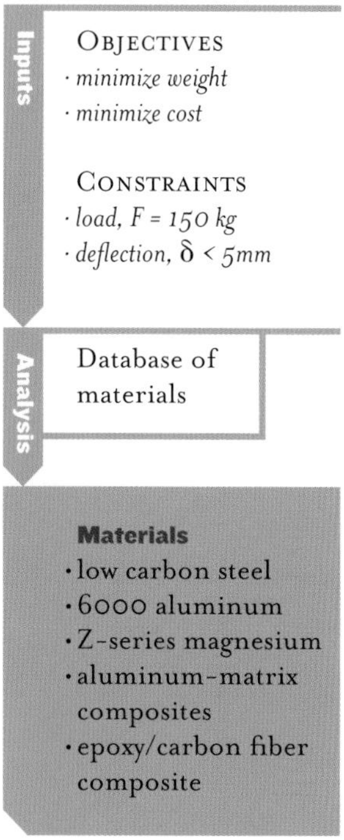

Inputs

Objectives
- *minimize weight*
- *minimize cost*

Constraints
- *load, F = 150 kg*
- *deflection, δ < 5mm*

Analysis

Database of materials

Materials
- low carbon steel
- 6000 aluminum
- Z-series magnesium
- aluminum-matrix composites
- epoxy/carbon fiber composite

8.1 Selection by Analysis
For office furniture, selection by analysis requires the optimization of a set of objectives given a set of constraints.

Office Furniture

A manufacturer of steel-framed office furniture wishes to launch a new line of tables that are lighter and easier to move. An office table, designed for mass production, has a frame supporting a work-surface. It must be light and cheap, yet support normal working loads safely – and this includes the load of a large person sitting on it – without failing or deflecting significantly (8.1). The current design has a frame made from square-section mild steel tubing, welded at the joints. There is a belief on the part of the client that a potential exists for using extrusions of a lighter metal. This throws up questions of material, joining and shape for stiffness at low weight and ease of assembly. It is a problem that lends itself both to analysis and to synthesis; combining the two gives even greater insight.

8.2 The Trade-off

Reading a compromise between low weight and low cost (CES 4, 2002).

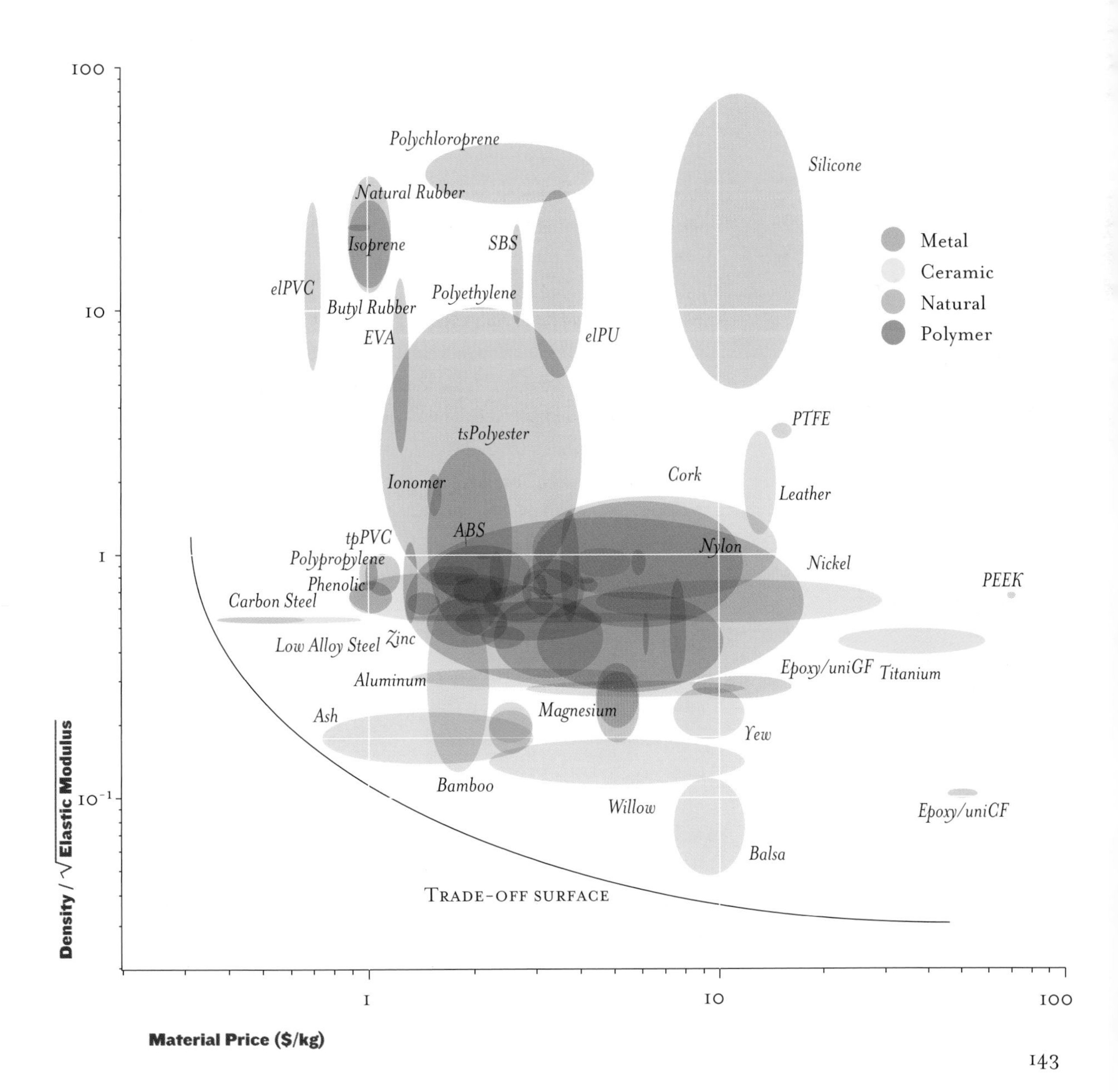

Material Selection – Analysis

A light, stiff frame, loaded in bending (as this frame is) is best made of a material with a low value of $\rho/E^{1/2}$ (see the table of material indices 7A.3) where ρ is the material density and E its modulus. This group of properties characterizes the weight of the frame. Cost, obviously, is a consideration. The material index $\rho/E^{1/2}$ is plotted against material price/kg in 8.2, on to which a trade-off surface has been sketched. Materials on or near this surface offer the best combination of low weight and low cost. The incumbent material – carbon steel – is marked. It is low in cost but it is also relatively heavy. Aluminum alloys, easily extruded to square section tube, are more expensive, but considerable lighter. Magnesium alloys allow a further reduction in weight, but at a considerable increase in price. Lighter still are GFRP and, particularly, CFRP – but CFRP is expensive.

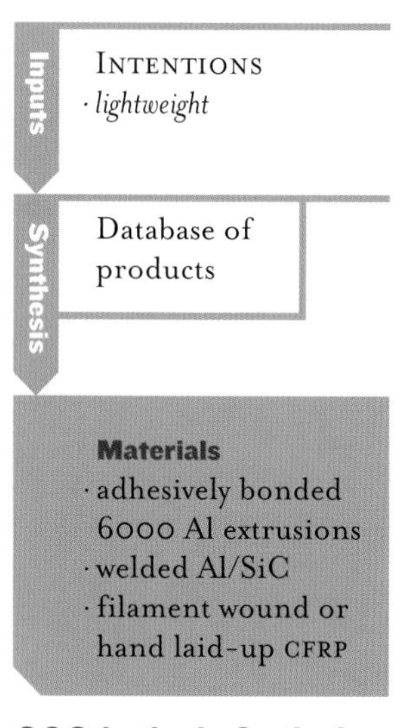

8.3 Selection by Synthesis
Here we seek products that contain eight weight space frames, enquiring for the material and process by which they are made.

Material Selection – Synthesis

What can be learnt from synthesis (8.3)? The intention here is lightweight design. Where else do you find lightweight space frames? Car manufacturers, today, strive for light weight, and achieve it by building the car round a light, stiff space frame. Design for backpacking has, as one of its primary priorities, low weight: backpacks, tents, mountain climbing and rescue equipment all have lightweight frames, here seeking strength rather than stiffness. And weight carries heavy penalties in other sports: the design of the frames of bicycles and of snowshoes are guided by minimizing mass.

A survey of materials and processes used in these products can suggest options for the desk. Both Audi and Lotus use extruded 6000 series aluminum, welded or adhesively bonded and riveted, to create light space frames for cars. Backpacks and snowshoes, too, use 6000 series aluminum tubing, bent, welded and finished by powder coating. Bicycle frames have been made of many different materials; several lightweight models use tubing made of the 6000/7000 series aluminum or of aluminum matrix composites. High performance backpacks and bicycles are made of carbon fiber-reinforced polyester or

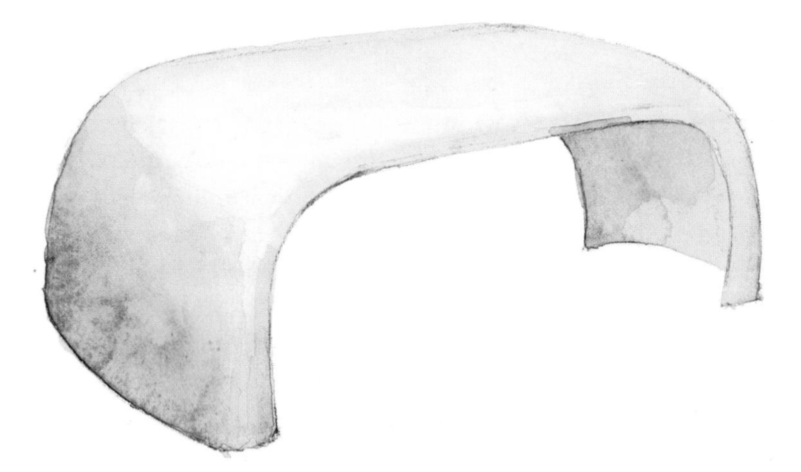

8.4 Mobile Office Furniture
With carbon fiber/epoxy composites, a stiff, light table can be formed.

epoxy, either as tubing (but this is inefficient) or as a shell made, usually, by hand lay-up of woven prepreg. None use magnesium.

The first three of these are mass-produced products that command substantial market; they boost confidence that 6000 series aluminum holds no unpleasant surprises, and point to ways in which they can be joined (welding and adhesives) and finished (powder coated). This would be the safe choice – one already used in other, similar, designs. The small additional weight saving offered by the aluminum matrix composites does not justify its cost.

But could there be a more radical solution? CFRP bicycles use lay-up methods to create elegant, doubly-curved shapes; chairs exist that do the same. A high-end desk could be made in a similar way (8.4), molding-in attachment points for drawers and other fixtures as they are on the bicycle. That level of sophistication may not be justified here. But the lightness and stiffness of polymer-matrix composites is an attraction. Pultrusion is a relatively low-cost way of making high stiffness hollow sections of composites. These can be adhesively bonded using cast aluminum nodes at the corners, a technique used in some CFRP bicycles, to form a frame. There is potential here for major weight savings, coupled with novel visual appearance (black CFRP pultrusions with polished aluminum connections) that might be manipulated to give perceptions of advance precision engineering.

CD Cases

We take many everyday products for granted. They work reasonably well, they are easy to use, and they cost next to nothing – think of disposable cups, document folders and paperclips. Because we take them for granted, we don't value them – and, to some degree, this lack-of-value transfers to their function or contents. And if, for any reason, they *don't* work well, they become a major irritation. CD cases are an example.

Most music CDs are marketed in the standard "jewel" case. Jewel cases crack easily, and when cracked, they look awful and don't close properly. The standard case is made from polystyrene, chosen, apparently, for its low cost and water-clear transparency. The challenge is to make a case that does not crack and is – if possible – just as transparent.

8.5 Resilience and Transparency
The resistance of materials to breaking when bent is plotted here, relative to that of polystyrene, PS. All the materials shown here can be injection molded and are optically clear or transparent (CES 4, 2002).

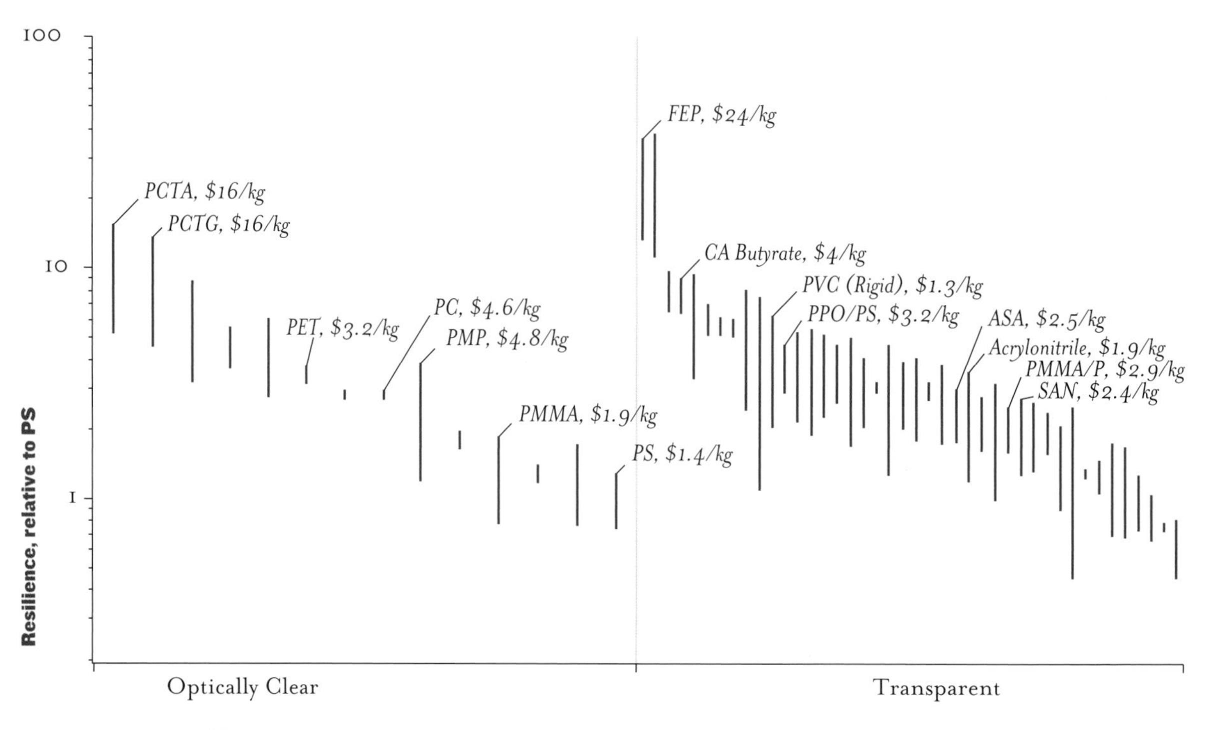

Material Selection – Analysis

A material cracks when the force put on it makes small imperfections (always present) suddenly propagate. Sharp changes of section and thin projections (as at the hinges of a CD case) concentrate stress and make things worse. The resistance of a hinge to failure by cracking is measured in more than one way. Here the measure we need is the *resilience* – it measures the amount the material can be bent before it breaks.[1] Figure 8.5 shows this measure normalized by that for the incumbent – polystyrene, PS – meaning that PS lies at the value 1; the scale then shows how much better or worse the other candidates are. Each bar describes a material, labeled with its name and approximate cost. All meet the constraints that they are optically clear or transparent, and can be injection molded. Those with the price below $4/kg (three times that of PS) are listed in order of increasing cost in the output box of 8.6.

This analysis highlights both the attraction and weakness of the incumbent PS: it is optically clear, moldable and cheap, but it is also the least resilient of all the clear polymers. A number of materials offer greater resistance to cracking, though – with the exception of PET and PC – it is with some loss of optical clarity. Among these PET seems to be a good compromise, offering three times the cracking resistance with a modest increase in price. Among transparent (less clear) polymers, SAN is also cheap (and is sometimes used for CD cases) but it, too, is relatively brittle. PVC is an interesting possibility: it is cheap, has high resilience, and could allow a design with a "living hinge" or integral joint, replacing the troublesome pin-joint of the jewel case. Its lack of clarity means that CD labeling will have to be printed or adhesively bonded onto the outside of the case or the case designed so that the entire label on the CD itself is not covered. The proposed solution: a PVC living hinge design, with cut-outs to expose the printed side of the CD, and with adhesively applied labeling if needed (8.7).

Inspiration could be sought in other ways, provoking other lines of thinking. Product reviewers in design magazines use such words as "cool," "trendy," "futuristic" when they feel that the product they are reviewing justifies them. How about a CD case

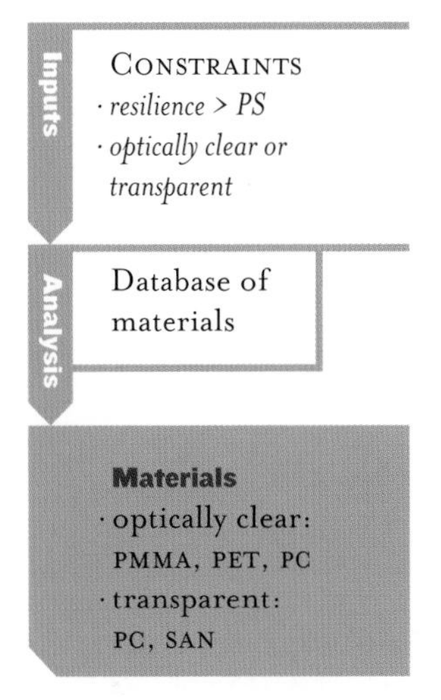

8.6 Selection by Analysis
For a CD case, the constraints on analysis are resilience and transparency.

1. *The details of this analysis can be found in the appendix to Chapter 4 and Ashby (1999); the property used here for resilience is K_{ic}/E where K_{ic} is the fracture toughness and E is the elastic modulus.*

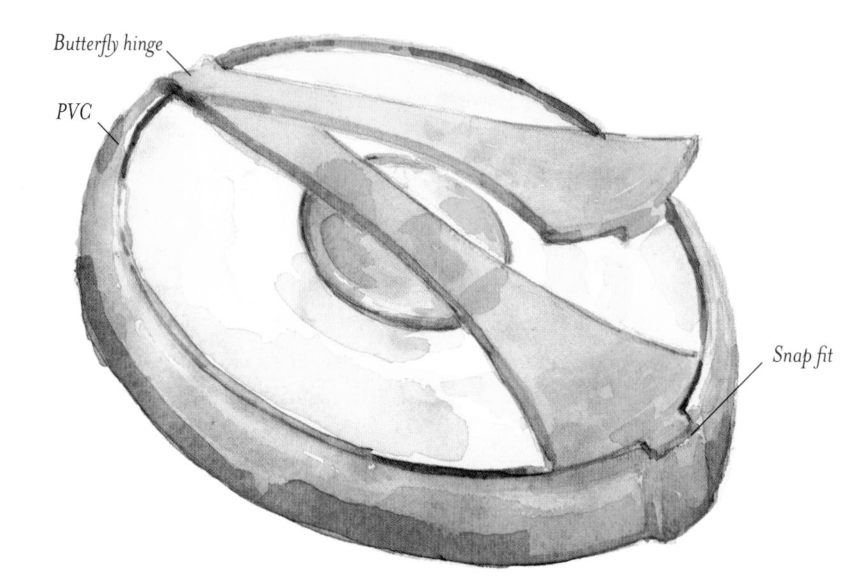

8.7 A CD Case

This case (perhaps of PVC) has an open design allowing a line of sight to the label on the CD. The closure has a "butterfly" living hinge, giving positive closure and opening, and allowing single part molding.

that evokes these perceptions? A search in recent issues of design magazines (like ID magazine) turns up products about which these words were used and identifies the materials used to make them: handbags – translucent polypropylene in irridescent colors; lighting – opalized translucent polycarbonate (rather expensive for a CD case); furniture – beaten aluminum sheet joined by riveting (a bit much for a CD); backpacks – nylon fabric reinforced with polymer tubing (but too floppy to protect the CD). All at first sight rather unpromising. But not so fast – there are ideas here. Some grades of polypropylene are nearly transparent, certainly translucent, and they can be colored and are easy to mold – those qualities of the handbags could be captured here. Acrylic (PMMA), too, can be colored and molded, and it can be metallized – the robust metallic look of the riveted aluminum could be simulated in this way (as it is in the cases of many small cameras). The appeal of the nylon fabric lay in its texture – that, too, could be simulated by printing. Browsing in this way liberates the imagination. allowing jumps of concept not possible by analysis alone.

Violin Bows

When an established material that fits its function well can no longer be used, a need arises for a substitute. This might be because the incumbent material has become too expensive; or that its availability is for some reason restricted; or that it was environmentally damaging; or that it was being exploited at a rate that was unsustainable. Any one of these can disallow its use, killing an otherwise near-perfect application of a material. The following case study is an example.[2]

Violin bows (8.8) are, by long tradition, made from a single species of tropical hardwood: pernambuco. But pernambuco is slow growing and with limited habitat. Violins, requiring bows, are made in large numbers. As a result, pernambuco is dangerously over-exploited; there is a real possibility it may become commercially extinct.[3] Is there an alternative?

2. Wegst, U.G.K. (1996) PhD Thesis, Engineering Department, Cambridge University, Cambridge, UK.

3. A similar predicament threatens the use of Honduras mahogany for electric guitars.

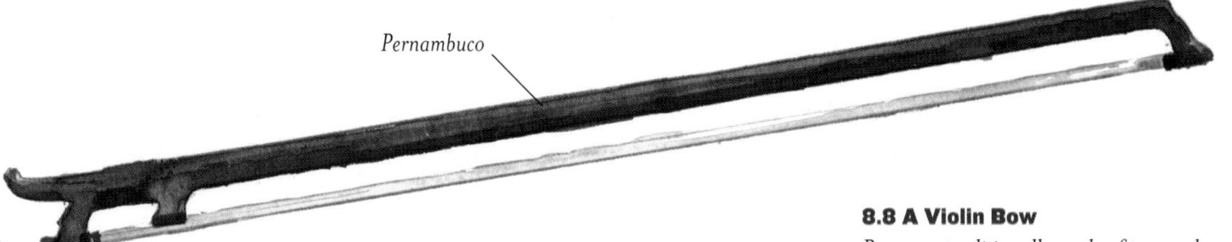

8.8 A Violin Bow
Bows are traditionally made of pernambuco with little variation in the design.

Material Selection – Similarity

The physical principles underlying certain uses of materials can be hard to unravel, and this is one. The bow performs several functions. It acts as a spring, keeping the hair under tension, so its stiffness is important; in seeking a substitute we must match this stiffness, requiring a material with a flexural modulus in the range 9.5–11.6 GPa. The bow mass, too, is important: the lowest bow forces are achieved by using the weight of the bow alone, so here we seek to match pernambuco with a material of similar density: about 900 kg/m^3. There are other constraints: strength is obviously one; bows do split but it seems pernambuco copes with this adequately. And one might anticipate that the mechanical damping (measured by the loss coefficient) might be important: a low loss bow may allow crisper bowing

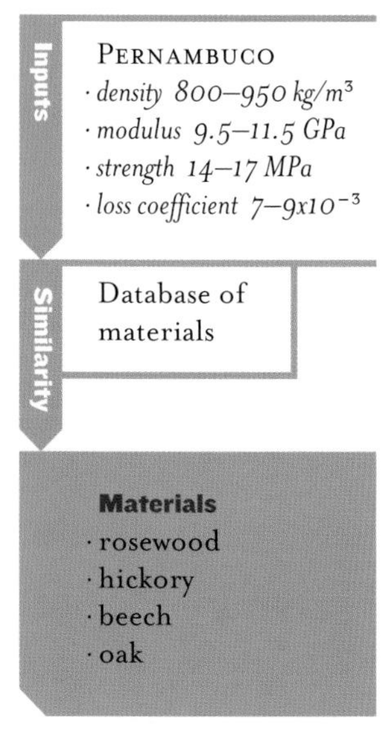

8.9 Selection by Similarity
For the violin, there are very few synthetic materials with a matching profile of attributes.

techniques but, of course, natural frequencies are more easily excited in a poorly damp structure. The answer is to match the damping of pernambuco. The input box of 8.9 summarizes the required match.

Here the strategy is a simple one: simply seek materials with values of these four critical properties that lie within the given ranges; then examine these, seeking those that are not threatened by extinction. The results, easily retrieved via a search of wood handbooks[4] or with current selection software,[5] are listed in the output box of 8.9. Interestingly, all are woods; no other class of materials offers this combination of attributes.

But perhaps we are being too hasty. We don't really know that density, modulus, strength and loss coefficient determine bow performance; all we really know is that the bow performs a mechanical function. It is worth trying a match of a looser kind, one that uses all the mechanical properties in the form of an MDS map (4.5). Seeking a substitute for the bow material by proximity on this map gives – as expected – other woods; but it also suggests that carbon or glass fiber reinforced polymers (such as bulk molding compound) are, in a general way, similar.

So there are alternatives, all widely available and easily sustained. All, it turns out, have been tried at one time or another. But despite this good match, the overwhelming preference remains that for pernambuco. Why?

Some obvious questions remain unanswered. How well can the alternative woods be bent over heat? How well do they retain their curvature? How easily are they worked? And how regular is their growth? How does a bow of BMC perform? The way to find out is to make trial bows from these candidates and test them – and this, as already said, has been done with apparent success. Why, then, are the substitutes resisted?

Materials for musical instruments are chosen not only for mechanical and acoustic behaviors but also for reasons of tradition and aesthetics; a wood that is unfamiliar in color or texture is "wrong," purely for those reasons. So we are brought back to the attributes of perception. Perceptions influence choice, and particularly so in a market as strongly traditional as violin-making. The only way forward is one of persuasion –

4. Handbooks such as the Handbook of Wood and Wood-based Materials of the US Forest Product Laboratory, Forest Service, U.S. Department of Agriculture tabulate wood properties (Hemisphere Publishing Corp, New York, 1989).

5. CES 4 (2002), Granta Design, Cambridge (www.grantadesign.com).

perhaps achieved by inducing great players to use and commend the new bow, encouraging others to perceive it in a new, more desirable, light.

Ice Axes

Mountain climbing equipment serves two primary markets – industrial climbing (the construction industry) and recreational climbing (mountains, rocks, glaciers). The gear is specialized: ropes, harnesses, karabiners, rock screws, backpacks, tents – and ice axes. The focus here is on ice axes for recreational climbing. Ice axes date from the time when climbing first became a recreation – around 1870 – evolving from tools used by the Alpine peasant. Early axes had iron heads and shafts made of hickory, ash or bamboo. Modern axes have heads made of low alloy (Cr-Mo) steel, with aluminum alloy shafts encased in polymer grips. This combination works, but are there alternatives? Could something be learnt from other products that might lead to a better axe?

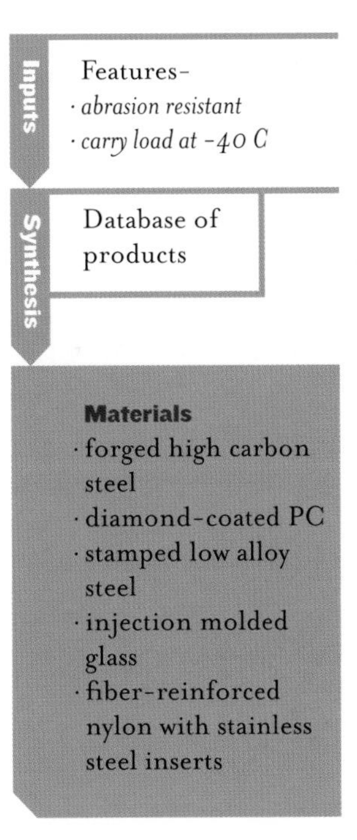

8.10 Selection by Synthesis
For ice axes, synthesis of related products requires access to a database of products and materials that make them.

Material Selection – Synthesis

The requirements for the head are resistance to abrasion and wear, to brittleness when cold, and the ability to take a high polish and a discretely textured manufacturer's logo. Those for the handle are strength, low weight, low thermal conductivity and a good grip. Here we seek ideas drawn from other products that meet similar constraints, even though the application may be quite different; particularly those that are designed for use at low temperatures: ice hockey skates, figure skate blades, fish hooks, crampons, hiking boots, sunglasses. The ice hockey skate has a stainless steel blade that is molded into a nylon/glass fiber composite sole. The figure skate blade and the fish hooks are both made of high carbon steel that has been hardened by deformation itself (forging for the hook, or stamping for the blade) and by subsequent heat treatment. The crampon is made of low alloy steel that has been stamped and bent. The harness that holds the crampon to the hiking boot is made of a polyurethane-coated fabric. The sunglasses are injection molded polycarbonate with a diamond-like coating to protect the lenses from abrasion. Selection by synthesis proceeds by choosing elements from among these that have possibilities in the design of an ice axe, combining them to give a final product specification (8.10).

The ductile-to-brittle transition temperature of carbon steel can lie above -30°C; it is adequate for fish hooks and figure skates that rarely cool below -4°C, but risky for an ice axe that might be used at much lower temperatures. Both stainless steels and low alloy steels meet the toughness requirement; of the two, the low alloy steel is stronger. There is no case here for a change of material for the head, but certain design features such as the barb of the hook or the teeth of the skate might be used to give the axe greater grip in dense snow. The end pick of an ice axe must stay sharp even under severe abrasion; here a ceramic or diamond-like carbon coating like that of the sunglasses could improve performance. The handle offers greater scope for innovation: the use of glass fiber reinforced nylon offers the possibility of lower weight and much lower thermal conductivity – an important consideration when it must be gripped in the hand; and the addition of a soft polyurethane coating could help ensure that it remained free of frozen snow and ice and was comfortable to hold. An example design is shown in 8.11.

This, clearly, could be taken further. There are many variants of each product that was described above – some using other materials, coatings and processes that those listed here, from which more could be learnt. Even a brief survey such as this is sufficient to demonstrate that the method has potential.

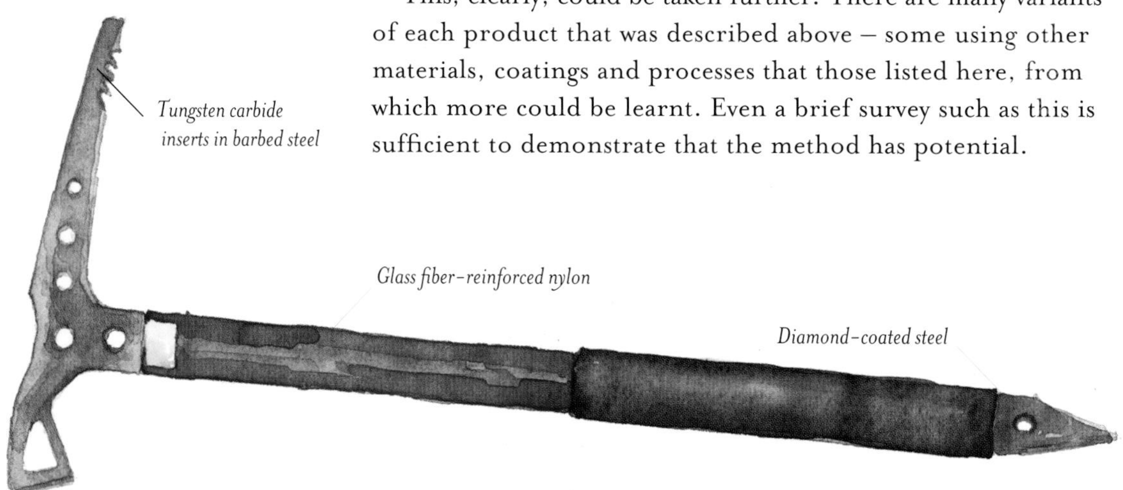

8.11 An Ice Axe
Glass fiber-reinforced nylon forms the handle, the pick has an amorphous diamond coating for increased wear resistance, the pick has slightly barbed steel (perhaps with inserts of tungsten carbide).

Inline Skates

Roller-skating has existed since the 1880s. Inline skates are much more recent, first commercialized in the 1980s by the Rollerblade company. The first commercially successful Rollerblade skate was an ice hockey boot fitted with polyurethane wheels, a rubber heel brake and a fiberglass runner. Since then, the skates have evolved and differentiated, some with rugged wheels for "mountain" skating, others with detachable boots allowing quick conversion from shoes to inline wheels and some with elements that appeal specifically to women or children.

Material Selection – Similarity

And now the challenge: a request to design an inline skate boot that will be used in training for cross-country skiing. It must be stiff but flexible, lightweight and impact resistant. Stiffness is needed so that the boot can be attached to the wheel platform and provide adequate rigidity for ankle support; flexibility is required because, when used for cross-country training, the heel will lift off the wheel platform and the boot will flex; and lightness is required to limit the weight that the athlete must carry. Impact resistance is important to cope with collisions with objects or the ground. The designer starts with the pre-conception (drawn from experience with ice hockey boots) that nylon or polyethylene could be possible material solutions. The set of possible material solutions is expanded by a search for materials that match critical attributes of the two materials. Stiffness and impact resistance depend on the material properties modulus and fracture toughness; if, in addition, lightweight is wanted it is the values of these per unit weight that count. Thus, we seek to match modulus/density and fracture toughness/density (8.12).

To match the performance of nylon and polyethylene, it is important that the designer selects materials that span the range of each attribute that has been specified. The boot of an inline skate could be made of a toe and heel (SAN) with flexible connection between (EVA). The boot must now be designed so that it can be attached to the platform at one (toe only) or two

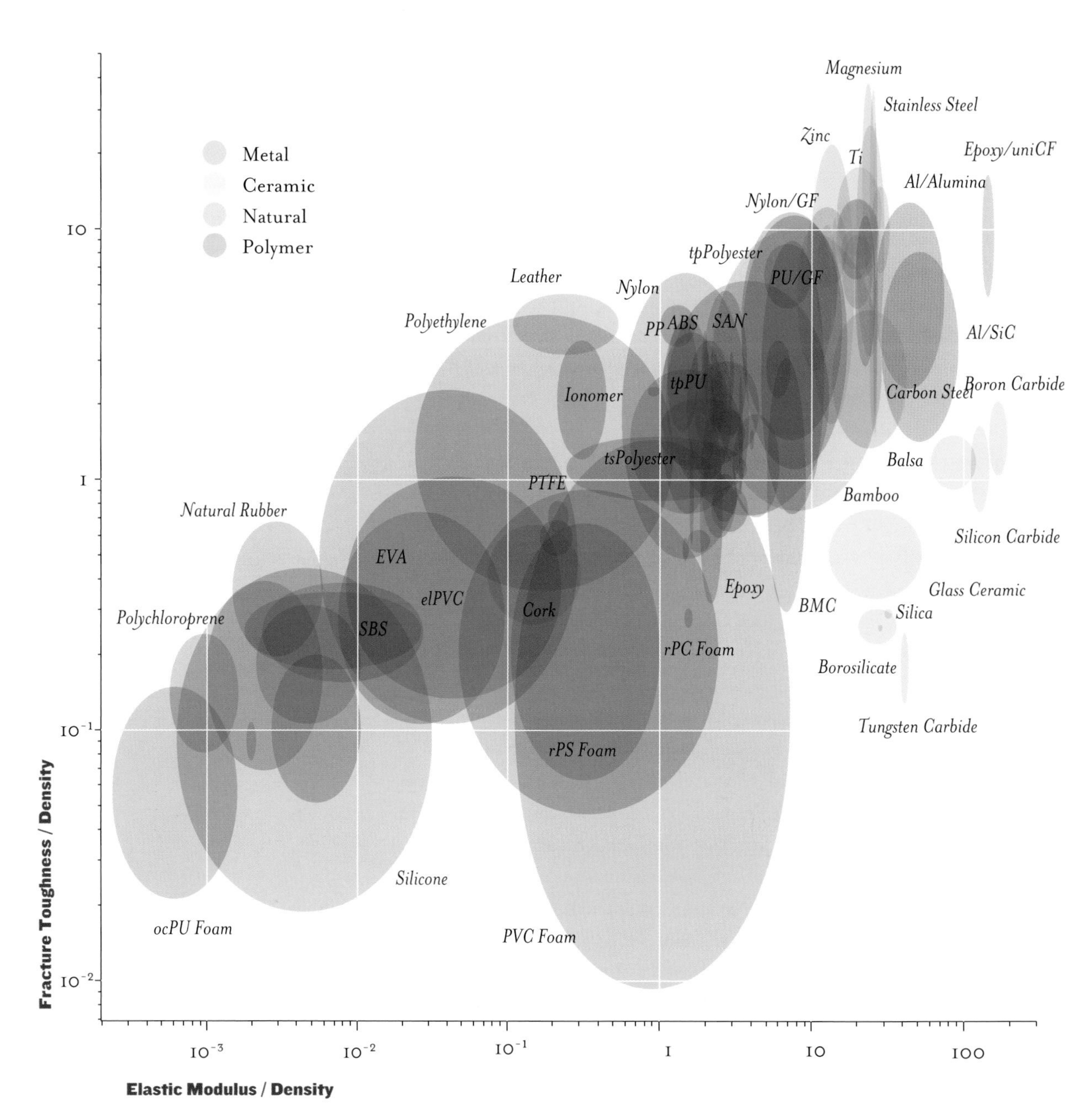

8.12 Impact Resistance and Stiffness at Low Weight

Inline skates balance the constraints of impact resistance and stiffness — both at low weight and with similar behavior to nylon or PE (CES 4,2002).

points (toe and heel) so that cross-training for cross-country skiing is possible, but that is a detail of design that is not considered here. The path of material selection for this case study and an example design for an inline skate are shown in 8.13 and 8.14, respectively.

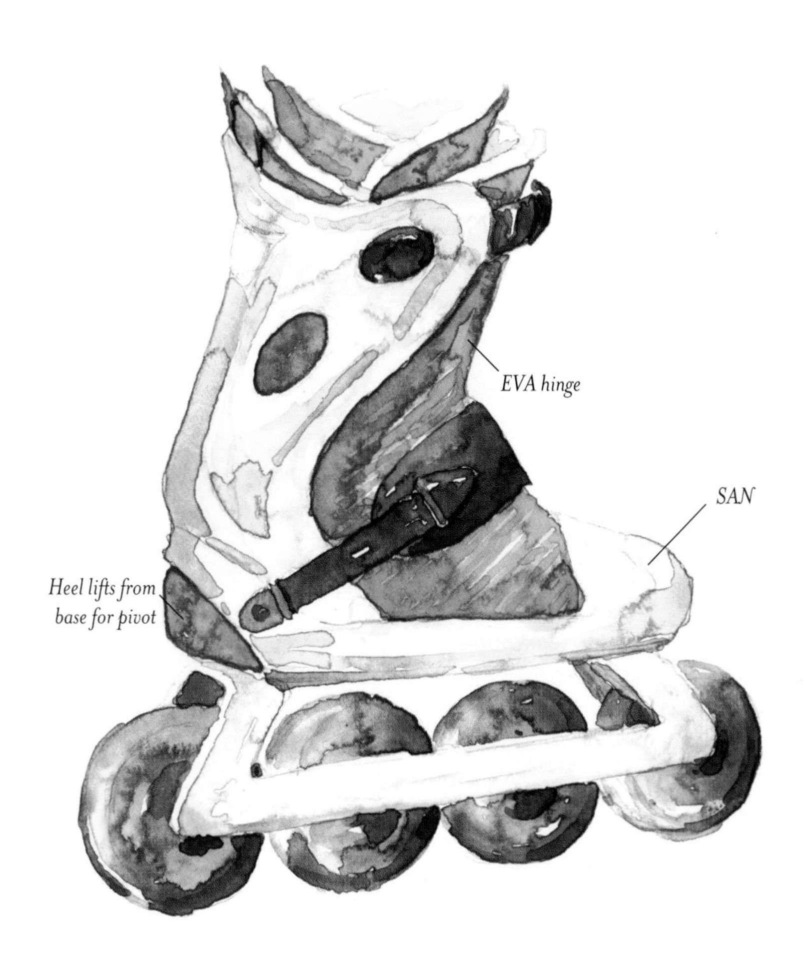

Inputs

NYLON,
POLYETHYLENE
· *modulus*
· *density*
· *fracture toughness*

Similarity

Database of
materials

Materials
· SAN
· EVA
· Ionomers
· PTFE

8.13 Selection by Property Matching
The selected materials have modulus/density and toughness/density that match those of (a) nylon and (b) polyethylene.

8.14 An Inline Skate
A SAN polymer shell is integrated with an accordian EVA hinge that allows flexibility and provides adequate stiffness.

Conclusions

These case studies are brief and the details of each design are not fully resolved; they serve primarily as demonstrations of each of the methods of selection introduced in earlier chapters, and of the way they can be combined to suggest innovative material solutions.

Chapter 9

New Materials – The Potential for Innovation

"The world seems to expect a never-ending supply of new material options."

That was George Beylerian of MaterialConnexion speaking,[1] and he should know – his company makes its money by peddling information about new materials. Whether the whole world is expectant may be debatable, but designers, certainly, are always on the prowl for what is new. It can be found in several different forms:

1. George Beylerian of MaterialConnexion (2001).

- *Materials in research* – light-emitting polymers; nano composites.
- *Materials in early commercialization* – amorphous metals; metal foams; shape-memory alloys.
- *Materials in combination* – wood-filled polypropylene, co-mingled PP and carbon fibers.
- *Materials in unexpected places* – titanium in eyeglass frames; paper and glass as structural materials.

9.1 Materials in Unexpected Places
The unusual use of beech for the hand grips of this high-performance camera suggest hand-crafted quality and an exceptional attention to aesthetics. (Courtesy ALPA of Switzerland)

New materials have their genesis in the laboratories of universities, governments and industry. The technologies they develop are spun-off, appearing first in "demonstration" products, then absorbed into wider markets. As production volume increases the material price falls, and its increased familiarity to designers and consumers broadens the base of its use. The material progressively approaches a kind of maturity, but can revive its "novelty" by combination with other materials, or by new ways of processing to create composites or hybrid materials – sandwich structures, clad systems, graded or coated structures. The use of an old material in a new setting can make it seem new, re-stimulating excitement in its potential.

To a designer, a new material offers both opportunities and risks. Opportunities derive from the new or improved technical or aesthetic behavior that it offers. The risks – and they can be great – lie in the incomplete characterization and the lack of design or manufacturing experience. It is usually the "difficult" time-dependent properties – those relating to long-term integrity (resistance to corrosion, fatigue, creep and wear) – that are least well-documented and cause the most trouble. There is no historical body of design experience on which to draw, with the result that confidence in the new material is low and investment in it is seen as risky. Channels of communication are imperfect, making it difficult for designers to find the information they want. And, early in life, a new material is expensive – the development costs must be paid for – and availability can be restricted.

Yet the interest in new materials and processes remains high because success can have a large payoff. Sports products have improved dramatically through the adoption of carbon fiber reinforced polymers, new elastomers, novel sandwich structures and – most recently – the use of titanium and amorphous metal alloys. Display technology, transformed once by the development of liquid crystal polymers, is about to be transformed again by LEPs – light-emitting polymer films. Advances in medical products have been made possible by new bio-compatible ceramics and polymers; and these, when successful, are very profitable. Hydro-forming and "tailor" blanking of metals, or bladder molding and braiding of polymer-based composites allow shapes that are both material-efficient and aesthetically pleasing. Laser-based processes for cutting and welding give fast, clean joints. An increasing range of rapid prototyping processes allows the designer to realize a concept easily and quickly. And, of course, there is the excitement – the buzz – that a new material or process can generate both in the mind of the designer and that of the consumer.

The Adoption of New Materials

The success of a new material depends on its ability to attract a sequence of early adopters. The simple picture of materials adoption, sketched as the bold line in 9.2, is in reality the envelope of a set of overlapping S-curves describing a succession of applications of increasing volume. The early applications are in markets that value performance highly and can accept a degree of risk; the later, larger volume applications tend to have the opposite characteristics.

A number of market forces are at work here. Acceptance of risk is associated with a balance between the value of performance and the cost of failure. The nuclear industry values performance but perceives the cost of failure to be so great that, it is sometimes said, no new material will ever be used inside a nuclear reactor – the cost of qualifying it is simply too high. At the other extreme, the sports equipment industry values performance so highly that it seizes eagerly on new materials that are perceived as offering the smallest gain, even when they are imperfectly characterized and may, in reality, offer nothing. And within any one industrial sector there are applications that are more or less risk sensitive. Thus the successful introduction of polymer-matrix composites in aerospace was initially confined to non-critical components; only now (in 2002) is an all-composite wing for a large civil aircraft a serious possibility. In civil engineering, the use of advanced materials such as PTFE-coated glass-fiber roof membranes, is limited both by cost and by uncertain long-term durability to buildings in which the dramatic spatial possibilities they offer outweigh the risk of possible failure.[2]

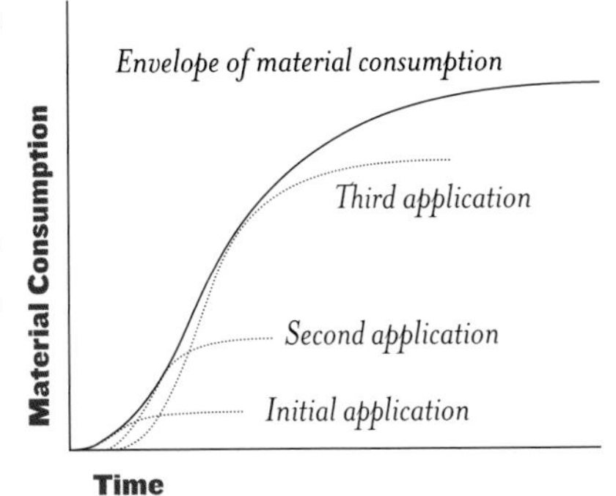

9.2 Material Consumption over Time
The entry of a new material into the market requires a low volume, performance driven opportunity that is not risk-averse.

At this point it is useful to introduce the idea of *material-cost sensitivity* (9.3a,b). The cost of a golf tee is little more than twice that of the material from which it is made; the value-added in manufacturing is low. If material cost doubles, the cost of the golf tee must rise significantly; the product is material-cost sensitive. At the other extreme, the cost of a golf club can be as much as 20 times that of the materials that make it up. Here a doubling of material costs increases the product cost only slightly; the value added is high and the product is material-cost insensitive. New materials are most readily adopted in industries that are material-cost insensitive: high-end appliances and automobiles, sports equipment, aerospace and biomedical equipment (try the calculation[3] for a brand-named toothbrush!).

Among these, the sports equipment industry is one of the most receptive and most visible – a good combination when seeking showcase products for a new material or process.

2. London's Millennium Dome and the Schlumberger building in Cambridge (shown in 6.2k) are examples.

3. A toothbrush weighs roughly 10 grams, can cost up to $5, and typically, is made of polypropylene or PMMA with nylon bristles. These materials cost between $1 and $3 per kilogram.

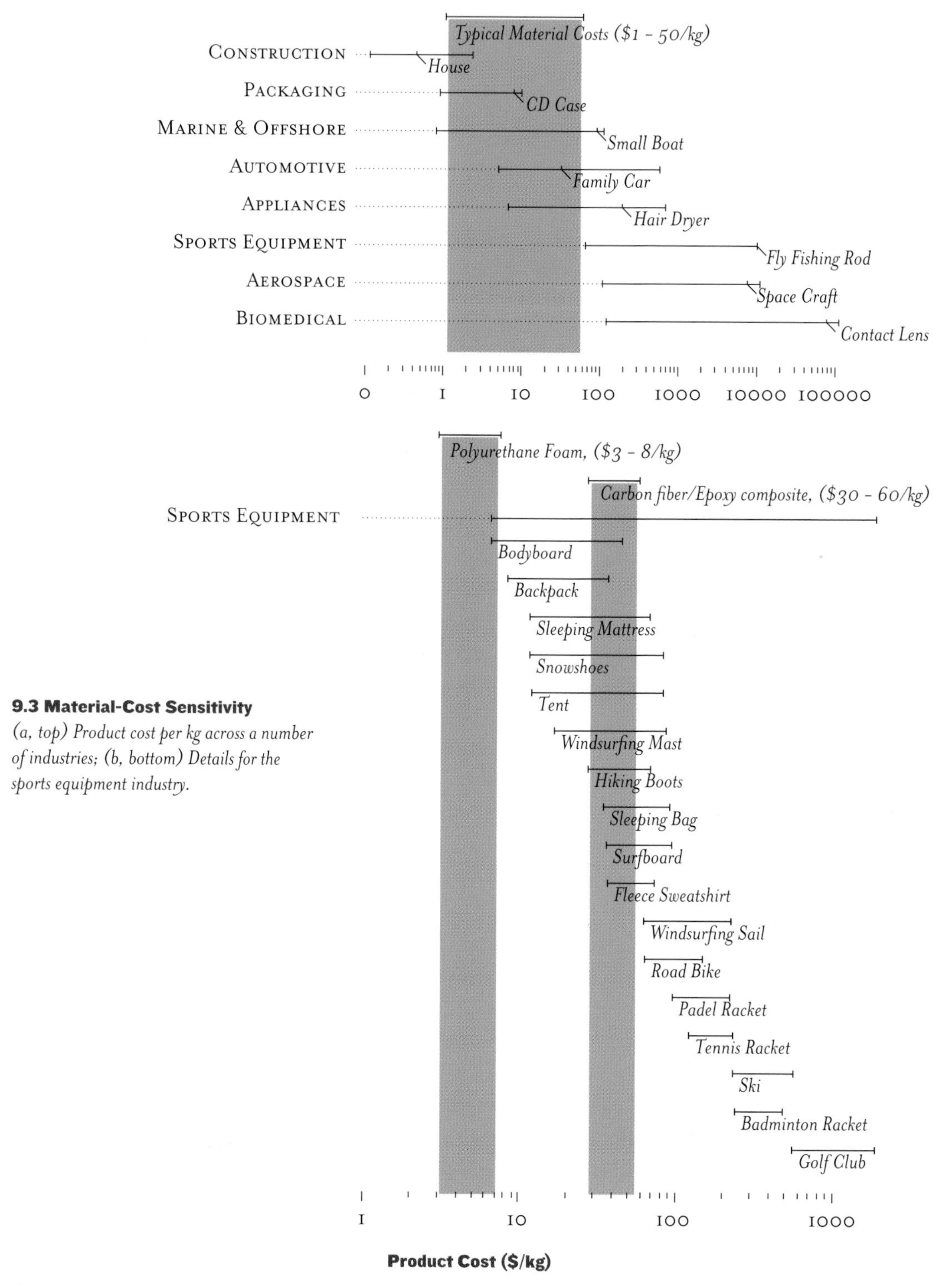

9.3 Material-Cost Sensitivity
(a, top) Product cost per kg across a number of industries; (b, bottom) Details for the sports equipment industry.

Fiberglass, first developed for radar nose cones for WWII aircraft, quickly displaced wood in boat hulls and surfboards and found further applications in high-profile designs like the chairs of Charles and Ray Eames (1950). Carbon fiber/epoxy and metal matrix composites, titanium and high strength aluminum alloys, all developed with defense and space applications in mind, are better known to the average consumer for their use in clubs, racquets and bicycle frames. From there they spread out into other products – watches, lightweight computer housings, furniture, kitchen and bathroom equipment. Materials research, motivated in the past by military and aerospace applications, is now directed more at the consumer than before, giving the designer – and in particular the industrial designer – significant influence. The adoption of translucent polycarbonate for the housing of the iMac and of titanium for that of the iBook – examples of one company's use of old materials in new ways – creates new markets for the materials themselves, and at the same time brings them to the attention of designers of other products.

But the challenges remain. We focus on two: the information void encountered by product designers seeking to use new materials and the difficulty designers have in stimulating suppliers to develop materials with the attributes they want. Central to this is the task of communication, both from developer to designer, and in the opposite direction.

Information about New Materials

Most new materials emerge through the commercialization of research, that is through science-driven development. The developer communicates information about the material through advertising, press releases, profiles and datasheets (9.4). The communication, if successful, stimulates designers to use the material in creative ways. But for this to work two things are necessary. The first: that the information includes that required for product design – and as we know, that means much more than just technical attributes. And second – that

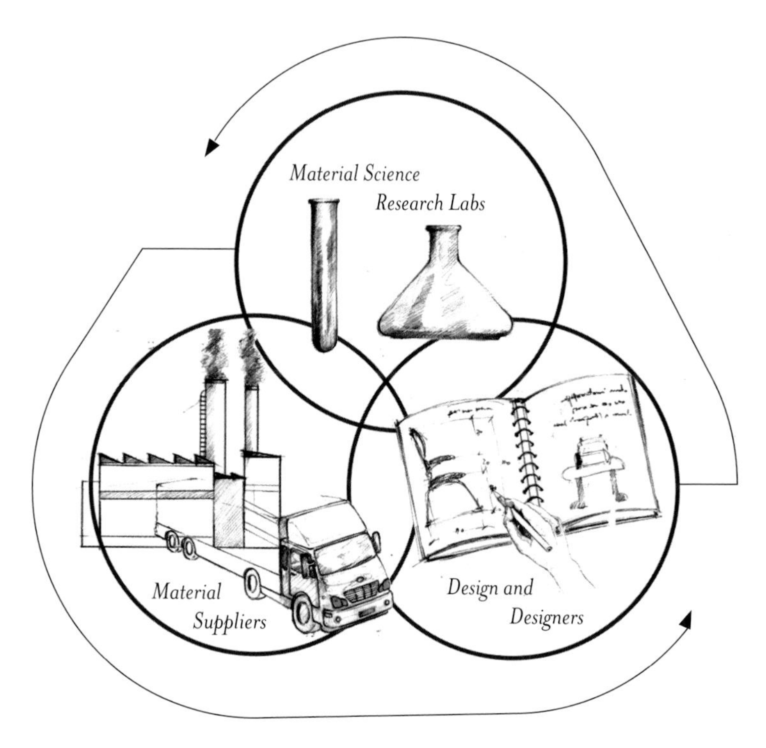

9.4 Positive Feedback
The interaction of materials research, suppliers and designers.

the language in which it is expressed has meaning both for the supplier and the designer, requiring a vocabulary to express design requirements and material behavior that both parties can understand. If information is flowing in the forward direction, it can also flow back: the designer influences the development of materials by suggesting or requesting specific technical, processing or aesthetic behavior.

The need, then, is one of communication. The usual information sources for new materials – press releases, supplier data sheets, producer catalogs – report what is good about them but seldom what is bad, and they generally focus on technical attributes. The attributes that bear on industrial design are far harder to find, with the consequence that materials that are technically familiar are often unknown to the industrial designer. Here the need is for visual and tactile attributes – those that help create the associations and perceptions of a product. The desire to fill this need motivates occasional exhibitions, books and services[4] but some of these are not easily

4. Antonelli's (1995) catalog of the MOMA exhibition "Mutant Materials," Juracek (1996) book of images and textures and Manzini's (1989) book "Material of Invention" are examples. The New York based consultancy MaterialConnexion provides a material information service for industrial designers.

found and none provide structured methods of selection. An example of the information that one of these offers is shown in 9.5. It describes the technical or perceived attributes of a new material – a metal foam – hinting at the possibilities of energy absorption and lightweight design.

Vague rhapsodies about the wonders of a new material can serve to stimulate interest, certainly. But if you want to design with the material you need to know the bad news as well as the good. Metal foams provide a case study. What does a metal foam cost? (A lot.) Does it corrode easily? (Yes.) Is it easy to shape? (You can cast it and machine it but not much else.) Are there any commercially successful consumer applications? (Not yet.) And – since you want to make something out of it – what is its strength? Its thermal conductivity? The datasheet doesn't tell you. Emerging materials can have an awkward adolescence during which their true character only slowly emerges. Mature materials are familiar in all their aspects. Manzini[5] cites the example of wood: it has been touched, smelt, bent, broken, cut, strained, stressed, dried, wetted, burned and maybe even tasted by most humans; we know what wood is and what it does. A new material is not embedded in the designer's experience in the same way. We need a way of assembling and communicating information about it that paints a fuller picture of its emerging character. As the designer Richard Seymour puts it, *"We need a system where the technology can meet the application. Where materials manufacturers can propagate their ideas directly to the very people who can apply the necessary imagination to utilize their properties."*[6]

What can be done? Two possibilities are explored here. The first is to develop profiles of new materials in a more complete form, attempting to capture both technical and aesthetic attributes. The second is the idea of material workshops – direct interaction between those who make and characterize materials and those who design with them.

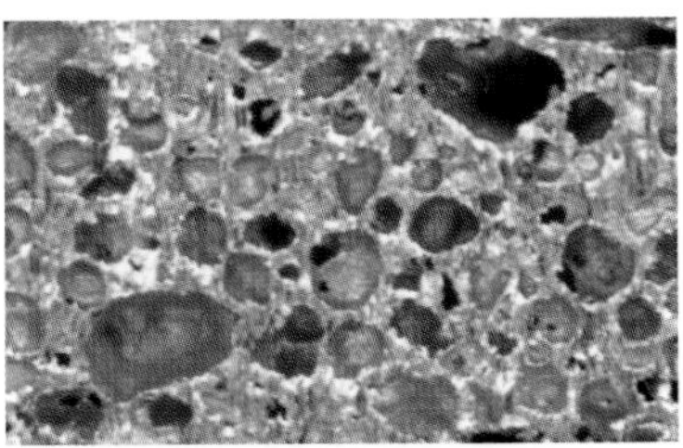

9.5 Introducing a New Material
Aluminum foam description from Materials and Ideas (Material Connexion): The process of aluminum foaming combines powder aluminum with a foaming agent and produces complex shaped foam parts, metal foam sandwich panels and foam filled hollow profiles. The parts have a closed cell microstructure and a high fraction of porosity (40–90%), with a relatively homogeneous and isotropic pore structure. The foam has a high specific stiffness-to-weight ratio, high impact and energy absorption, and high thermal and acoustic insulation. In automotive applications, it is used as a firewall behind the engine, and as the autobody structure, which promotes reduction in weight and thus savings in fuel. The process of foaming metal can therefore lead to new developments in structural furniture construction for the foam's strength and lightweight properties.

5. *Manzini (1989).*

6. *Richard Seymour (1998) writing in "Winning the Design of Sports."*

Profiles for New Materials

The material profiles of the later section of this book have a standard format, reporting technical, perceived and visual behavior, design guidelines and typical uses. For *new* materials some of this information is missing – most notably, guidelines for design and established uses. It is still possible to assemble much of the rest, and even to indicate where applications might lie. Three examples are given at the end of the reference for material profiles; they are *metal foams, amorphous* or *"glassy" metals,* and *shape-memory alloys.*

There is enough information in these profiles to plot them onto the material property and MDS maps described in Chapter 4, allowing them to be compared with the "world" of materials. On the stiffness and density map of 4.4 metal foams lie above rigid polymer foams – a little heavier and a little stiffer – and of course, they can be used at higher temperatures and are much more durable.[7] The same can be done with MDS maps, and here, too, there are interesting insights. Plotting them on the MDS map of 4.5, we find that they lie closer to woods and to ceramic foams than to polymers – a perception that led a researcher to liken the material to "metallic wood." The comparisons are enlightening, suggesting possible applications of metal foams that exploit their stiffness and low weight and their ability to absorb energy, found by pirating applications of polymer foams or of woods in which additional durability or high temperature performance is wanted: cores for sandwich panels, lightweight fireproof partitioning, and certain sorts of packaging.

7. The same thing can be done using the other eight material property charts in the Appendix at the end of this book.

A Material Workshop

The challenge of communication can be addressed in another way – that of a workshop bringing together experts who have developed and characterized the material with the designers who wish to use it. We have tried a number of these, most recently organizing a meeting at SeymourPowell, London (a design consultancy) attended by three materials scientists and four designers. The materials people presented a portrait of the material – aluminum foams again – using samples of metal foams and material property charts to put it in context, addressing the questions:

- *What is it?*
- *What processing possibilities and limits exist?*
- *What is its character?*
- *How does it behave?*
- *What are the competing materials?*
- *Where has it been used before?*
- *Where – from a technical viewpoint – might applications lie?*

This stimulated a brainstorming session in which the designers took the lead, seeking technical information from the materials people when needed. All of the technical applications referred to earlier were explored. The visual aspects of the material, particularly, intrigued the designers – open-cell foams look opaque and solid in reflected light, but they transmit light when back-lit. This, and the interesting natural textures of the skinned foams, suggested applications in light fixtures, furniture, as loud speaker casings (exploiting its damping characteristics) and as architectural paneling.

Meetings like this break down many barriers and open channels for communication of a sort that even the best profiles cannot achieve.

Further Reading

Antonelli, P. (1995) Mutant Materials in Contemporary Design, Museum of Modern Art, New York, USA. ISBN 0-87070-132-0 and 0-8109-6145-8. *(A MOMA publication to accompany their extensive 1995 review of materials in product.)*

Ashby, M.F., Evans, A.G., Fleck, N.A., Gibson, L.J., Hutchinson, J.W. and Wadley, H.N.G. (2000) "Metal Foams: A Design Guide," Butterworth Heinemann, Oxford, UK. ISBN 0-7506-7219-6. *(A study in assembling information for metal foams at an early stage in their commercialization, with the aim of stimulating their take-up and use.)*

Juracek, J.A. (1996) "Surfaces, visual research for artists, architects and designers," Thames and Hudson, London, UK. ISBN 0-500-01758-1. *(An intriguing compilation of images of textured, weathered and deformed surfaces of hundreds of materials.)*

Lesko, J. (1999) "Materials and Manufacturing Guide: Industrial Design," John Wiley and Sons, New York, NY, USA. ISBN 0-471-29769-0. *(Brief descriptions, drawings and photographs of materials and manufacturing processes, with useful matrices of characteristics, written by a consultant with many years of experience of industrial design.)*

Manzani, E. (1989) "The Material of Invention," The Design Council, London, UK. ISBN 0-85072-247-0. *(Intriguing descriptions of the role of material in design and invention.)*

MaterialConnexion (2001) A New York based materials information service *(www.MaterialConnexion.com)*

Seymour, R. (1998) A contribution to the exhibition "Winning the Design of Sports," Glasgow, UK, published by Laurence King, London, UK. ISBN 1-85669-1527. *(A compilation of articles by designers summarizes their methods and ideas.)*

Chapter 10

Conclusions

Inspiration – the ability to stimulate creative thinking – has many sources. One of these is the stimulus inherent in materials. It is one that, since the beginning of time, has driven humans to take materials and make something out of them, using their creativity to choose function and form in ways that best exploit their attributes. The most obvious of these attributes are the engineering properties – density, strength, resilience, thermal conductivity and such; it is these that enable the safe and economical design of products. The enormous economic importance of technical design in any developed society has given material and process development to meet technical needs a high priority; there are established methods to select them, widely taught and extensively documented in texts and software. But a material has other attributes too: color, texture, feel, a sort of "character" deriving from the shapes to which it can be formed, its ability to integrate with other materials, the way it ages with time, the way people feel about it. These, too, can stimulate creativity – the kind of creativity that gives a product its personality, making it satisfying, even delightful.

We have sought, in writing this book, to draw together lines of thinking about the selection of materials to serve both technical and industrial design. Selection for technical design is the subject of the companion volume; here the emphasis is more heavily on the second. What do we learn? First, that a material has many dimensions: a technical dimension, the one seen by the engineer; an eco-dimension, that seen by the environmentalist; an aesthetic dimension, the one encountered by the senses of sight, touch, hearing; and a dimension that derives its features from the way in which the material is perceived, its traditions, the culture of its use, its associations, its personality.

Attributes of aesthetics and perception are less easy to pin down than those that are technical, yet it is essential to capture them in some way if their role – and it is obviously an important one – is to be communicated and discussed. There are words to describe visual, tactile and acoustic attributes; they can even, to some extent, be quantified. Perceptions are more difficult. A few, perhaps, can be identified – gold is, almost universally, associated with wealth, steel with strength, granite with permanence, plastics with the modernity... well, even these are uncertain. Perceptions depend on time, on culture, demographics, taste, and more. But the consumer encounters materials only in the products in which they are used, and these are designed to appeal to a specific culture, demographic and taste; within these boundaries perceptions are much sharper. Words can be attached to them with a significant degree of general agreement; it is through the shaping, joining and surfacing of materials that they are created. So, we conclude: materials have an intrinsic personality, but one that is hard to see until brought into sharp focus by sensitive product design.

The first step in product design is that of identification of concept – the principles on which the product will be based. In the second – visualization – the desired features are developed, using sketching, model-making and computer graphics to sharpen the constraints on configuration, size, functionality and personality, as described in Chapter 6. It is in the third – materialization – that the choice of materials and processes is made, prototypes built and tested, and the final design agreed.

This choice is guided not only by technical requirements but also by requirements of product aesthetics and personality. To achieve it we need selection methods that are flexible and can deal fluently with information of many different types. This reasoning led to the information structure introduced in Chapter 7. A material can be characterized by its name, technical and aesthetic attributes. It can be indexed, so to speak, by the processes able to shape it and by products in which it is used. Products are designed with certain broad intentions in mind. These intentions condition every design decision, including the choice of materials; thus materials can be linked to intentions through products. Products also have aesthetic and perceived attributes that are a deliberate part of the design, materials can be linked to these too. This is not the only way to organize the information product designers need, but it is one that works – it is both practical and inspirational. As a structure for material selection, this allows a range of methods.

Product design has a technical component and this is of vital importance – no one wants a product that does not work;

established methods of analysis deal with this. Analysis is of less help in selection to meet the other aspects of design – these are too heavily dependent on the visual and the perceived. Then methods using judgements of similarity, analogy, or simply directed curiosity ("browsing") are more productive.

The path from design requirement to product specification is rarely linear; almost always it is devious, assembling bits of information by one method, other bits by others, and combining them into a solution that – although based on existing information – can be entirely new. There is structure in it, but there is a degree of chaos too. Without structured methods, material selection gets nowhere. But without a little craziness and chaos, the really novel solutions may be missed. Balancing structure and chaos is key to innovative material selection.

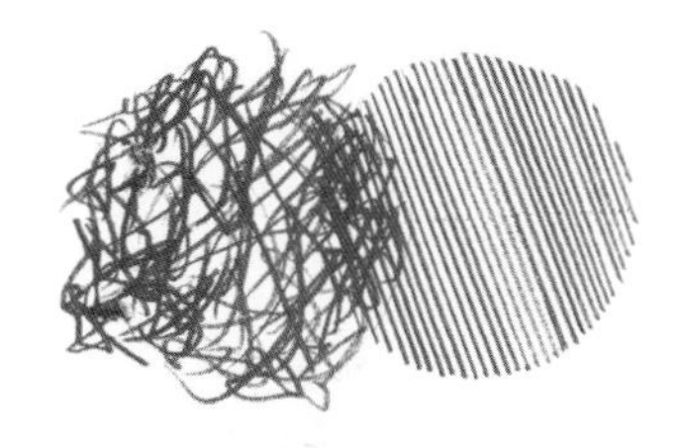

10.1 Structure and Chaos
Balancing structure and chaos is a component of innovative material selection.

A Practical Reference for Inspiration

Engineered products are made from materials, using processes. The numbers of both are very large, but a small subset account for most specifications in product design. This reference assembles profiles for materials and processes; portraits for the designer, so to speak. The profiles, and the information they contain, can be retrieved by using the conventional index at the end of the book or by the hierarchical tree included in the introduction to each set of profiles.

The material profiles start on page 174; those for processes on page 236. Commonly used materials and processes get a full-page profile; less common ones are grouped together and profiled more briefly.

Material Profiles

Materials are the stuff of design, and throughout history have dictated its opportunities and its limits. The ages in which man has lived are named for the materials he used: stone, bronze, iron, plastics and – today – silicon. But today is not the age of just one material; it is the age of an immense range of materials and the combinations these allow. There has never been an era in which the evolution of materials was faster and the sweep of their properties more varied. The menu of materials has expanded so rapidly that designers can be forgiven for not knowing that half of them exist. Yet not-to-know is, for the designer, to risk failure: innovative design, is enabled by the imaginative exploitation of new or improved materials. There is no reason to expect that the pace of material development will slow so things can only get worse (or better!).

Introduction 176
Further Reading 187

Polymers
Polyethylene (PE) 188
Polypropylene (PP) 189
Polystyrene (PS) 190
Acrylonitrile-butadiene-styrene (ABS) 191
Polyamide (PA), Nylon 192
Polymethylmethacrylate (PMMA), Acrylic 193
Polycarbonate (PC) 194
Polyoxymethylene (POM), Acetal 195
Polytetrafluoroethylene (PTFE) 196
Ionomers 197
Celluloses (CA) 198
Polyetheretherketone (PEEK) 199
Polyvinylchloride (PVC) 200
Polyurethane (PU) 201
Silicones 203
Polyesters (PET, PBT) 204
Epoxy 206
Phenolic 207
Elastomers 208
Polymer Foams 212
Polymer Composites 214

Metals
Carbon Steels 216
Stainless Steels 217
Low Alloy Steels 218
Aluminum Alloys 219
Magnesium Alloys 220
Titanium Alloys 221
Nickel Alloys 222
Copper Alloys (Brass, Bronze) 223
Zinc Alloys 225

Ceramics 226
Glass 228
Fibers 229
Natural Materials 231

New Materials
Metal Foams 233
Amorphous Metals 234
Shape-memory Alloys 235

Material Evolution

The figure at the bottom of this page illustrates the pattern and increasing pace of the evolution of materials. The materials of pre-history (>10,000 BC, the Stone Age) were ceramics and glasses, natural polymers and composites. Weapons – always the peak of technology – were made of wood and flint; buildings and bridges of stone and wood, clothing of fur and skin. Naturally occurring gold and silver were available locally but played only a minor role in technology. The discovery of copper and bronze and then iron (the Bronze Age, 3000 BC to 1000 BC and the Iron Age, 1000 BC to 1620 AD) stimulated enormous advances, replacing the older wooden and stone weapons and tools. Cast iron technology (1620s) established the dominance of metals in engineering; and the development of steels (1850 onward), light alloys (1940s) and special alloys since then consolidated their position. By the 1960s, "engineering materials" meant "metals."

There had, of course, been developments in the other classes of material: Portland cement, refractories and fused silica among ceramics; synthetic rubber, Bakelite and polyethylene among polymers; but their share of the total materials market was small. Since 1960 all that has changed. The rate of development of new metallic alloys is now slow; demand for steel and cast iron has in some countries actually fallen. Polymers have displaced metals in an increasing large number of markets, including those central to the metal-producing industry such as the automobile. The composite industry continues to grow strongly and projections of the growth of production of the new high performance ceramics suggests substantial expansion here also.

The Evolution of Engineering Materials with Time

Relative importance in the Stone and Bronze Age is based on assessments of archeologists; that in 1960 on allocated teaching hours in UK and US universities; that in 2020 on prediction of material usage in automobiles by manufacturers.

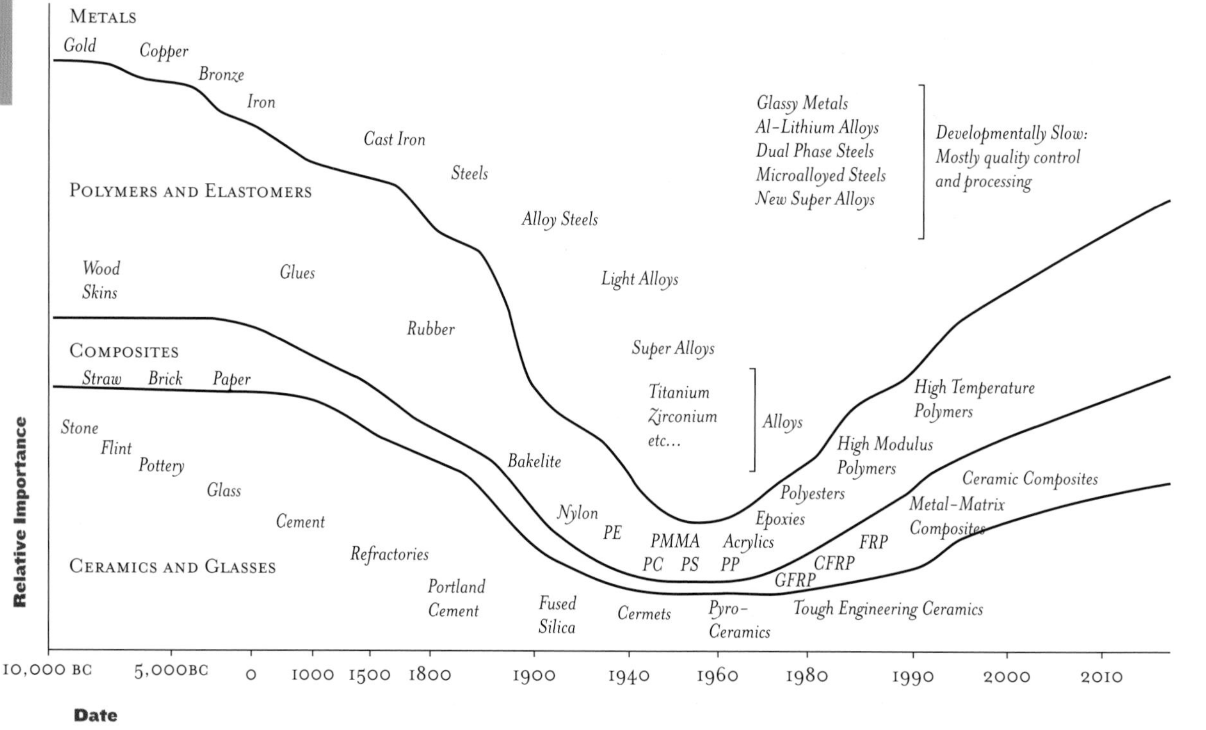

This rapid rate of change offers opportunities that the designer cannot afford to ignore. The following pages contain character sketches of the common polymer, metal, ceramic, composite and natural materials, highlighting aspects of importance to design. Specific material profiles follow.

Polymers

Although nature's polymers – woods, wool, leather – are the oldest of the world's materials, the commodity polymers of today have little that is natural about them; they are the chemist's contribution to the materials world. Almost all are synthesized from oil (although they don't have to be) and are made from the simplest of atoms: carbon, hydrogen, oxygen, chlorine and an occasional whiff of nitrogen or fluorine.

Thermoplastics

Thermoplastics soften when heated and harden again to their original state when cooled. This allows them to be molded to complex shapes. Most accept coloring agents and fillers, and many can be blended to give a wide range of physical, visual and tactile effects. Their sensitivity to sunlight is decreased by adding UV filters, and their flammability is decreased by adding flame retardants. The common thermoplastics are listed in the ajacent table. They include polyolefins (polyethylene, polypropylene), PVCs, polystyrenes, acrylics and certain polyesters (PET and PBT). Some are crystalline, some amorphous, some a mixture of both. The properties of thermoplastics can be controlled by chain length (measured by molecular weight), by degree of crystallinity and by blending and plasticizing. As the molecular weight increases, the resin becomes stiffer, tougher, and more resistant to chemicals, but it is more difficult to mold to thin-wall sections. For thin-walls, choose a low molecular weight resin; for higher performance, choose one with higher molecular weight. Crystalline polymers tend to have better chemical resistance, greater stability at high temperature and better creep resistance than those that are amorphous. For transparency the polymer must be amorphous; partial crystallinity gives translucency. The most transparent polymers are acrylics, PC, PS and PET.

Some polymers crystallize faster than others: polyethylenes crystallize quickly but polyesters do so more slowly – they remain amorphous under normal cooling rates. Crystalline polymers have a more or less sharp melting point, which must be exceeded for molding. Amorphous polymers do not; instead they progressively soften and become more fluid as temperature increases above the glass transition temperature; they must be heated above this temperature for extrusion and injection molding. The processing force required to generate flow decreases slowly as temperature rises above the glass transition temperature. Amorphous polymers have greater impact strength and relatively low mold shrinkage. Semi-crystalline polymers have higher shrinkage because of the volume change on crystallization.

Holes and ribs reduce the effect of shrinkage in a thermoplastic part. Areas near the filling gate tend to shrink less than areas farther away. Shrinkage increases with wall thickness and decreases with higher molding pressures. Fiber filled polymers shrink less in the direction of flow because

Thermoplastic Polymers
Acrylonitrile-butadiene-styrene (ABS)
Cellulose
Ionomers
Polyamide (Nylon, PA)
Polycarbonate (PC)
Polyetheretherkeytone (PEEK)
Polyethylene (PE)
Polymethylmethacrylate (PMMA)
Polyoxymethylene (POM)
Polypropylene (PP)
Polystyrene (PS)
Polytetrafluoroethylene (PTFE)
Polyvinylchloride (tpPVC)
Polyurethanes (tpPU)
Polyesters (PET, PETE, PBT)

the fibers line up in this direction; the shrinkage in the cross-flow direction is 2–3 times more than in the flow direction. High service temperatures can cause shrinking in some semi-crystalline materials. Fillers, or additives, are used to tailor certain properties of the composite such as density, color, flame/smoke retardance, moisture resistance and dimensional stability. Most thermoplastics can be recycled.

Thermosetting Polymers

Epoxy
Phenolic
Polyester
Polyurethane (tsPU)
Polyvinylchloride (tsPVC)

Thermosets

If you are a do-it-yourself type, you have Araldite in your toolbox – two tubes, one a sticky resin, the other an even more sticky hardener. Mix and warm them and they react to give a stiff, strong, durable polymer, stuck to whatever it was put on. Araldite typifies thermosets – resins that polymerize when catalyzed and heated; when reheated they do not melt – they degrade. The common thermosets are listed in the table here. The first commercial thermoset was Bakelite, a trade-name for a phenolic resin. Polyurethane thermosets are produced in the highest volume; polyesters come second; phenolics, epoxies and silicones follow, and – not surprisingly – the cost rises in the same order. Epoxies are two-part system that – when mixed – undergo a mildly exothermic reaction that produces cross-linking. Phenolics are cross-linked by the application of heat or heat and pressure. Vulcanization of rubber, catalyzed by the addition of sulfur, can change the soft rubber of a latex glove to the rigid solid of ebonite, depending on the level of cross-linking. Once shaped, thermosets cannot be reshaped.

Thermosets have greater dimensional stability than thermoplastics; they are used where there is a requirement for high temperature resistance and little or no creep. Most are hard and rigid, but they can be soft and flexible (like natural and synthetic rubber, as described above). Phenolics are most used where close-tolerance applications are necessary, polyesters (often combined with glass fibers) where high strength with low shrinkage is wanted.

Thermosets are shaped by compression molding, resin transfer molding, injection molding, pultrusion and casting. They duplicate the mold finish and are relatively free from flow lines and sink marks, depending on the mold design – high gloss, satin or sand-blasted finishes are possible, and raised lettering can be molded in. Molding can be adapted to low volume production by using low cost molds; but higher production volumes, up to a million or greater, are economical only with expensive molds that allow fast heating, cooling and extraction. Phenolics can only be molded in black or brown; urea, melamine, alkyd and polyester compounds are available in a wider range of colors. The fluidity of some thermosets before molding allows them to take up fine detail, and to penetrate between fibers to create composites. Most high-performance polymer composites have thermosetting matrix materials. Dough and sheet molding compounds (DMC and SMC) use polyesters; filament-wound carbon or glass use epoxies as the matrix to give the highest performance of all. Thermosets cannot be recycled.

Elastomers

Elastomers were originally called "rubbers" because they could rub out pencil marks – but that is the least of their many remarkable and useful properties. Unlike any other class of solid, elastomers remember their shape when they are stretched – some, to five or more times their original length – and return to it when released. This allows conformability – hence their use for seals and gaskets. High damping elastomers recover slowly; those with low damping snap back, returning the energy it took to stretch them – hence their use for springs, catapults, and bouncy things. Conformability gives elastomers high friction on rough surfaces, part of the reason (along with comfort) that they are used for pneumatic tires and footwear. Elastomers are easy to foam, giving them the comfort of cushions, and increasing even further their ability to conform to whatever shape is pressed against them.

Almost all engineering solids have elastic moduli (measuring their stiffness) between 1 and 1000 GPa. Elastomers are much less stiff – between 0.0001 and 1 GPa. This low stiffness, their ability to stretch and to remember their original shape all derive from their structure. The molecules in an elastomer are long chains of linked carbon (or, in silicones, silicon-oxygen chains), with hydrogen, nitrogen, chlorine or fluorine attached to the sides. The carbon atoms that link to form the chain are strongly bonded to each other, but the side-branches of one molecule are only weakly attracted to those of another – indeed, at room temperature these molecule-to-molecule bonds in an elastomer have melted. In this state the elastomer is a very viscous liquid, its molecules tangled like a plate of cooked spaghetti, and it can be molded. It is then cured; the curing creates occasional strong links between molecules, freezing the tangle in its molded shape. Most of the length of any one molecule can still slither over its neighbors, allowing stretch, but when released, the widely spaced attachment points pull the tangle back to its original shape. In the case of natural rubber, curing is achieved by heating with sulphur ("vulcanization"); in synthetic rubbers the curing process is more complex but the effect is the same. This means that elastomers are thermosets – once cured, you can't remold them, or reshape or recycle them, a major problem with car tires. Tires are the single biggest use of elastomers; the second is footwear, followed by industrial rollers, belts, cushions, clothes and sports-equipment. Elastomers are processed by casting, calendering, extrusion, and foaming.

There are probably more than 1000 grades of elastomer. The simpler ones contain only carbon, hydrogen, nitrogen and oxygen (hydrocarbon elastomers). They tend to be vulnerable to oil, chemicals and UV radiation. Filling with carbon-black (soot) helps protect against UV. Replacing some of the hydrogens with chlorine or fluorine (chloro/fluoro-carbon elastomers) gives greater chemical stability. Even greater stability comes with replacing carbon by silicon (silicone elastomers), but they are expensive. Elastomers – as we have said – are thermosets, but tricks can be used to make them behave in some ways like thermoplastics. Blending or co-polymerizing elastomer molecules with a thermoplastic like polypropylene, PP, gives – if done properly – separated clumps of elastomer stuck together by a film of PP (Santoprene). The material behaves like an elastomer, but if heated so that the PP melts, it can be remolded and even recycled.

Elastomers

Acrylic elastomers
Butyl rubbers (NR)
Chlorinated elastomers (Neoprene)
Ethylene-propylene (EPDM)
Ethylene-vinyl-acetate (EVA)
Fluorocarbon elastomers (Viton)
Isoprene
Natural rubber
Nitrile (NBR, BUNA-N)
Polybutadiene elastomers
Polysulphide elastomers
Silicone
Styrene-butadiene (SBS)
Thermoplastic elastomers (TPE, TPO)

Polymers that are Foamed

Phenolic
Polyethylene
Polypropylene
Polystyrene
Polyurethane

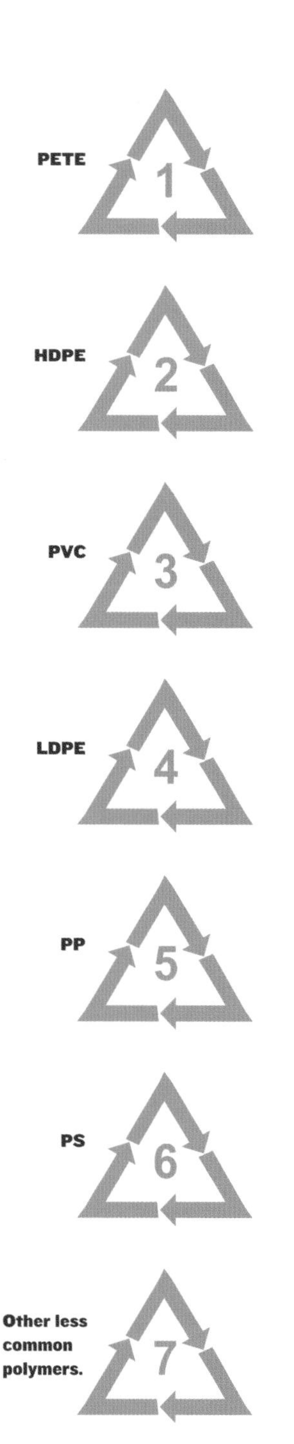

Want to Recycle?
Then these are important. Polymers are hard to identify. These marks identify the six most common polymers.

Design Notes

Attitudes towards polymers – plastics – have changed over the last 70 years. In the 1930s, when Bakelite was already established and cellophane, PVC, polystyrene, Plexiglas and nylons were first introduced, their freedom of form and color inspired young designers. By the 1950s they were plentiful and cheap, and the low cost of both the materials and of their processing led to an era of cheap, poorly designed disposable products that gave plastics a bad name. Since 1970, however, the use of polymers in high quality clothing, footwear, household products and transportation systems has created a market served by innovative designers exploiting the immense range of form, color, surface finish, translucency, transparency, toughness and flexibility of modern polymers. Their combination with fillers and fibers (glass, carbon, Kevlar) has given a range of light materials with stiffness and strength comparable to that of metals, allowing their penetration into the automobile, aerospace and marine sectors.

Polymers are familiar as fibers (nylon and polyester thread, polypropylene rope), as film (clinging film, polyethylene bags), as bulk moldings (plastic garden furniture, computer housings) and as foam (polystyrene packaging, bicycle helmets). Polymer fibers are much stronger and stiffer than their equivalent in bulk form because the drawing process by which they are made orients the polymer chains along the fiber axis. Thus, drawn polypropylene, polyethylene – and above all, aramid – have strengths, relative to their weight, that exceed that of steel. Polymer fibers are rarely used on their own. They are used as reinforcement in polymer resin, woven into fabrics or bundled for ropes.

Ease of molding allows shapes that, in other materials, could only be built up by expensive assembly methods. Their excellent workability allows the molding of complex forms, allowing cheap manufacture of integrated components that previously were made by assembling many parts. Blending allows properties to be "tuned" to meet specific design requirements of stiffness, strength and processability. Smaller additions allow other adjustments of properties: plasticizing additives give leathery behavior; flame retardant additives reduce flammability. Some polymers, like PPO, are used almost exclusively in blends (PPO is difficult to process on its own); others are used in "pure" form.

Many polymers are cheap both to buy and to shape. Most resist water, acids and alkalis well, though organic solvents attack some. All are light, and many are flexible. Their color and freedom of shape allows innovative design. Thermoplastics can be recycled, and most are non-toxic, though their monomers are not. The properties of polymers change rapidly with temperature. Even at room temperatures many creep, slowly changing their shape under load; and when cooled they may become brittle. Polymers generally are sensitive to UV radiation and to strongly oxidizing environments, requiring special protection. They have exceptionally good electrical resistance and dielectric strength. When foamed they provide insulation materials with a thermal conductivity almost as low as that of still air.

The basis of polymers in oil (a non-renewable resource) and the difficulty of disposing of them at the end of their life (they don't easily degrade) has led to a view that polymers are environmentally unfriendly. There is some truth in this, but the present problems are soluble. Using

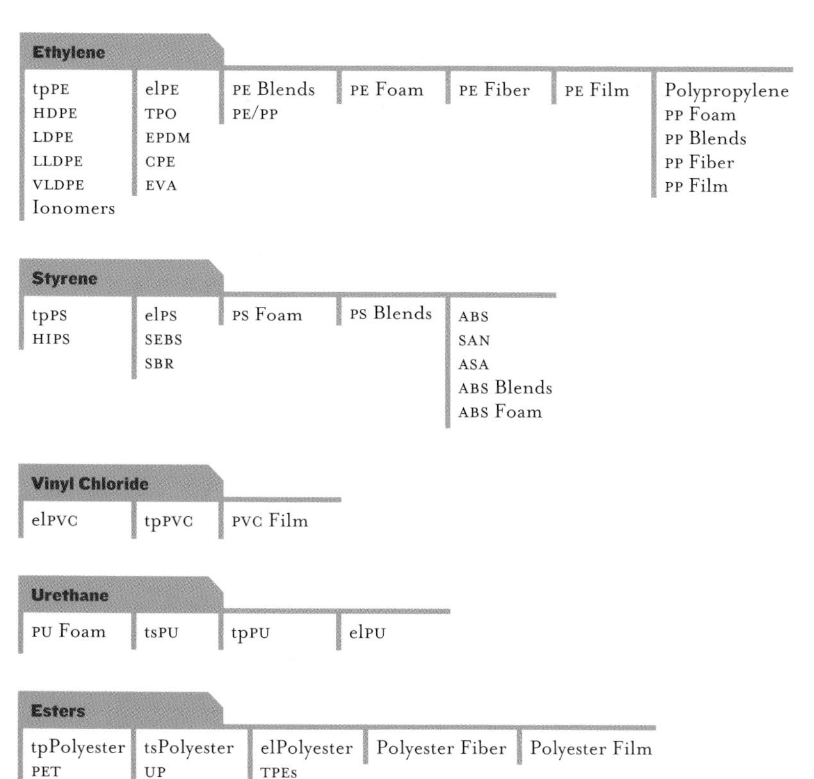

Polymer Classes
More than 95% of all polymer consumption comes from these five classes.

oil to make polymers is a better primary use than just burning it for heat; the heat can still be recovered from the polymer at the end of its life. There are alternatives to oil: polymer feed stocks can be synthesized from agricultural products (notably starch and sugar, via methanol and ethanol). And thermoplastics – provided they are not contaminated – can be (and are) recycled.

Polymers and composites based on them are the most rapidly growing sector of the market for structural and decorative material. Research on biodegradable polymers, and polymers synthesized from agricultural bi-products such as lignin, are leading to new generations of eco-friendly products. Polymers with greater thermal stability, higher stiffness, strength and toughness are all under development. Most exciting is the increasing ability to build functionality into polymers. Electro-active polymers carry an electric dipole moment and thus are influenced by an electric field; they can be used for microphones and loud speakers. Optically-active polymers emit light when excited electrically; they allow large area, flexible, displays. High-temperature polymers, though expensive, are sufficiently stable that they can be used for inlet manifolds and other components on internal combustion engines.

There are many types of polymers and their relationships between each other can be confusing. Some of these relationships are shown above. Polyethylene is the simplest of the polyolefins but even it can be manipulated in many ways. It starts life as ethylene, C_2H_4. Polymerization gives chains of CH_2-CH_2-CH_2-... Depending on the catalyst, the chains can be short or long, linear or branched. Once polymerized, they can be blended, foamed and drawn into fibers or film. And by starting with propylene, C_3H_6, instead of ethylene, a similar diversity of polypropylenes can be synthesized.

Polystyrene has similarly complicated possibilities; it begins as styrene-ethylene with one hydrogen replaced by a benzene ring. It can be modified as a thermoplastic or elastomer, foamed or blended; and, by chemical modifications, become ABS and its derivatives.

Polyesters, PVCs and polyurethanes, too, have many personalities; they overlap the bounds of thermoplastic, elastomer and thermoset more than other resins. PVCs can be a thermoplastic or elastomer; polyurethanes and polyesters can be any of the three; in addition, polyurethane can be foamed and polyester can be made into film or fibers. This diversity can be confusing, but the wide range of properties it allows is the source of great freedom of design

Polymer Composites

Polymer Composites

CFRP
GFRP
KFRP

Composites are one of the great material developments of the 20th century. Those with the highest stiffness and strength are made of continuous fibers (glass, carbon or Kevlar, an aramid) embedded in a thermosetting resin (polyester or epoxy). The fibers carry the mechanical loads, while the matrix material transmits loads to the fibers and provides ductility and toughness as well as protecting the fibers from damage caused by handling or the environment. It is the matrix material that limits the service temperature and processing conditions. Polyester-glass composites (GFRPs) are the cheapest; epoxy-carbon (CFRPs) and Kevlar-epoxy (KFRPs) the most expensive. A recent innovation is the use of thermoplastics as the matrix material, either in the form of a co-weave of cheap polypropylene and glass fibers that is thermoformed, melting the PP, or as expensive high-temperature thermoplastic resins such as PEEK that allow composites with high temperature and impact resistance. If continuous fiber CFRPs and GFRPs are the kings and queens of the composite world, the ordinary workers are polymers reinforced with chopped glass or carbon fibers, or with particulates (fillers) of silica sand, talc or wood flour. They are used in far larger quantities, often in products so ordinary that most people would not guess that they were made of a composite: body and interior panels of cars, household appliances, furniture and fittings. It would, today, be hard to live without them.

The properties of long fiber composites are strongly influenced by the choice of fiber and matrix and the way in which these are combined: fiber-resin ratio, fiber length, fiber orientation, laminate thickness and the presence of fiber/resin coupling agents to improve bonding. Glass offers high strength at low cost; carbon has very high strength, stiffness and low density; Kevlar has high strength and low density, is flame retardant and transparent to radio waves (unlike carbon). Polyesters are the most widely used matrices as they offer reasonable properties at relatively low cost. The superior properties of epoxies and the temperature performance of polyimides can justify their use in certain applications, but they are expensive. The strength of a composite is increased by raising the fiber-resin ratio, and orienting the fibers parallel to the loading direction. The longer the fibers, the more efficient is the reinforcement at carrying the applied loads, but shorter fibers are easier to process and hence cheaper. Increased laminate thickness leads to reduced composite strength and modulus as there is an increased likelihood of entrapped voids. Coupling agents generally increase tensile strength. Environmental conditions affect the performance of composites: fatigue loading, moisture and heat all reduce allowable strength.

Design Notes

Polymer composites can be formed by closed or open mold methods. All the closed mold methods produce fiber orientation parallel to the mold surfaces (for extrusion, it is parallel to the inside surface of the orifice die). Of the open mold methods, all allow multidirectional fiber orientation parallel to the mold or mandrel, except pultrusion, where the fibers are oriented parallel to the laminate surface and the mold plates, and calendering, where they are parallel to the sheet surface. Lay-up methods allow complete control of fiber orientation; they are used for large one-off products that do not require a high fiber-resin ratio. Lamination and calendering form sheets, pultrusion is used to make continuous shapes of constant cross-section and filament winding produces large hollow items such as tubes, drums or other containers.

Joints in long-fiber composite materials are sources of weakness because the fibers do not bridge the joint. Two or more laminates are usually joined using adhesives and, to ensure adequate bonding, an overlap length of 25 mm for single- and double-lap joints or 40–50 mm for strap, step and scarf joints is necessary. Holes in laminates dramatically reduce the failure strength making joining with fasteners difficult. Composite manufacture is labor intensive. It is difficult to predict the final strength and failure mode because defects are easy to create and hard to detect or repair.

Metals

Most of the elements in the periodic table are metals. Metals have "free" electrons – electrons that flow in an electric field – so they conduct electricity well, they reflect light and – viewed with the light behind them – they are completely opaque. The metals used in product design are, almost without exception, alloys. Steels (iron with carbon and a host of other alloying elements to make them harder, tougher, or more corrosion resistant) account for more than 90% of all the metals consumed in the world; aluminum comes next, followed by copper, nickel, zinc, titanium, magnesium and tungsten.

Metals

Aluminum alloys
Copper alloys
Magnesium alloys
Nickel alloys
Steels – carbon
Steels – low alloy
Steels – stainless
Titanium alloys
Zinc alloys

Design Notes

Compared to all other classes of material, metals are stiff, strong and tough, but they are heavy. They have relatively high melting points, allowing some metal alloys to be used at temperatures as high as 2200 C. Only one metal – gold – is chemically stable as a metal; all the others will, given the chance, react with oxygen or sulfur to form compounds that are more stable than the metal itself, making them vulnerable to corrosion. There are numerous ways of preventing or slowing this to an acceptable level, but they require maintenance. Metals are ductile, allowing them to be shaped by rolling, forging, drawing and extrusion; they are easy to machine with precision; and they can be joined in many different ways. This allows a flexibility of design with metals that is only now being challenged by polymers.

The development of metals and alloys continues. Superplastic alloys have the unique property that, in sheet form, they can be vacuum or thermoformed like thermoplastics. Shape-memory alloys have the capacity to remember their initial shape even when deformed very heavily, so that they spring back like a rubber – they are used in thermostats, and as the frames for eye glasses, brassieres and small actuators. Metal matrix composites (such as aluminum with silicon carbide) extend the property range of metals, usually to make them stiffer, lighter, and more tolerant of heat, but their cost limits their applications. Techniques for foaming metals are emerging that have the potential to capture new market. Nonetheless, the status of metals has been eroded during the past few decades by other materials: polymers in small scale structures like household appliances, polymer-based composites in cars, aircraft and boats; and ceramics in certain engine parts and cutting blades. But as Guy Nordenson of the engineering firm Arup, observes, "Materials development and usage are historically cyclical. Work with metals is probably more out of touch than it is behind."

Primary production of metals is energy intensive. Many, among them aluminum, magnesium and titanium, require at least twice as much energy per unit weight (or 5 times more per unit volume) than commodity polymers. But metals can generally be recycled, and the energy required to do so is much less than that required for primary production. Some are toxic, particularly the heavy metals – lead, cadmium, mercury. Some, however, are so inert that they can be implanted in the body: stainless steels, cobalt alloys and certain alloys of titanium, for example.

Material Profiles

Ceramics

Ceramics and Glass

Alumina
Boron carbide
Silicon carbide
Tungsten carbide
Borosilicate glass
Silica glass
Soda lime glass
Glass ceramic

Ceramics are materials both of the past and of the future. They are the most durable of all materials – ceramic pots and ornaments survive from 5000 BC. And it is their durability, particularly at high temperatures, that generates interest in them today. They are exceptionally hard (diamond – a ceramic – is the hardest of them all) and can tolerate higher temperatures than any metal. Ceramics are crystalline (or partly crystalline) inorganic compounds; they include both traditional, pottery-based ceramics and the high-performance technical ceramics. All are hard and brittle, have generally high melting points and low thermal expansion coefficients, and most are good electrical insulators. Traditional ceramics, based on clay, silica and feldspar, are soft and easily molded before they are fired; firing creates a glassy phase that binds the other components together to create brick, stoneware, porcelain and tile. Technical ceramics consist of pure or nearly pure compounds, synthesized by chemical reactions; the commonest are alumina, silicon carbide, silicon nitride, boron carbide, boron nitride, and tungsten carbide.

Ceramics have certain unique properties. Their low atomic packing factors and high melting points give them a low expansion coefficient. Those that are pure and completely crystalline have a high thermal conductivity; impurities and glassy phases greatly reduce it. When perfect they are exceedingly strong, but tiny flaws, hard to avoid, propatate as cracks when the material is loaded in tension or bending, drastically reducing the strength; the compressive strength, however, remains high (8 to 18 times the strength in tension). Impact resistance is low and stresses due to thermal shock are not easily alleviated by plastic deformation so large temperature gradients or thermal shock can cause failure.

Design Notes

The size of commercial ceramic parts can range from small components like spark plug insulators to large nose cones for re-entry vehicles. The high firing temperatures prevent metal inserts being molded-in to ceramics. Shapes should be kept as simple as possible, with liberal tolerances. Shrinkage during drying and firing can be as much as 25%. Edges and corners should have generous radii, large unsupported sections and undercuts should be avoided, symmetrical shapes and uniform wall thicknesses are best, with a draft angle of at least 5 degrees.

Most technical ceramics start as powders that are pressed, extruded, injection molded or cast, using a polymer binder. The resulting "green" part is machined or ground to shape, dried and fired ("sintering"). Metal can be joined to ceramic by adhesives, soldering, brazing or shrink fitting (as long as the metal is on the outside, in tension). Brazing is stronger than adhesives or soldering and more temperature resistant but requires a metallized coating as a base for the brazing alloy. Glaze bonding or diffusion bonding allows joining ceramic to ceramic or metal to ceramic. Glazes can be applied to ceramics for a smooth, glossy surface; patina surfaces are also possible.

Traditional ceramics are familiar as brick, tile, white ware (toilets, baths, sinks) and pottery. Technical ceramics are used as bearings, turbine blades, cams, cutting tools, extrusion dies, nozzles, seals, filters, crucibles and trays, electrical insulators, substrates and heat sinks for electronic circuitry.

Glass

Discovered by the Egyptians and perfected by the Romans, glass is one of the oldest of man-made materials. For most of its long history it was a possession for the rich – as glass beads, ornaments and vessels, and, of course, glass as glaze on pottery. Its use in windows started in the 15th century but it was not wide spread until the 17th. Now, of course, it is so universal and cheap that – as bottles – we throw it away. Glass is a mix of oxides, principally silica, $SiO2$, that does not crystallize when cooled after melting. When pure it is crystal-clear, but it is easy to color by adding metal oxides.

Design Notes

Glass is formed by pressing, blow molding, centrifugal casting, drawing or rolling and must be cooled at a controlled rate to avoid residual

Material Profiles

stresses. It can be strengthened by rapid air cooling of the surface – known as "tempered" glass; tensile stresses build up inside the glass, creating compressive stresses in the surface, increasing the impact strength by a factor of 4. Electric light bulbs are molded on a machine that converts a fast moving ribbon of glass into as many as 10,000 bulbs per hour. Glass is joined by glaze (melt) bonding, clamping or adhesives. A silver coating gives nearly 100% reflection of visible light – this is used for most mirrors. A gold mirror coating can reflect 90% of IR radiation.

Adding metal oxides produces colored glass. Nickel gives a purple hue, cobalt a blue, chromium a green, uranium a green-yellow, iron a green-blue. The addition of iron gives a material that can absorb wavelengths in the infrared range so that heat radiation can be absorbed. Colorless, non-metallic particles (fluorides or phosphates) are added from 5–15% to produce a translucent or an almost opaque white opalescence in glass and glass coatings. Photochromic glass changes color when exposed to UV (it sometimes fades when heated); photosensitive glass turns from clear to opal when exposed to UV or heated. Filter glass protects from intense light and UV radiation – it is used in visors for welding. Dichroic coatings and thin laminations give color modulation.

Further Reading

Amato, I. (1997) "Stuff, the Materials the World is Made of," Harper Collins, New York, USA. ISBN 0-465-08328-5. *(Amato is a science writer for the news media, and he does it exceptionally well. This is in the news media style – readable, highlighting potential, much not yet realized, and heavily dependent on interviews with those who've agreed to do it.)*

Braddock, S.E. and O'Mahony, M. (1998) "Techno Textiles: Revolutionary Fabrics for Fashion and Design," Thames and Hudson. ISBN 0-500-28096-7. *(This book presents information about polymer textiles and fabrics from the fashion industry to a wider audience of design.)*

Cardarelli, F. (1987) "Materials Handbook," Springer, London. ISBN 1-85233-168-2. *(As a handbook of technical material information, this book provides information on metals, semiconductors, superconductors, ceramics, glasses, polymers, elastomers, minerals, rocks and woods.)*

Colling, D.A. and Vasilos, T. (1995) "Industrial Materials," Volume I and II, Prentice Hall, NJ, USA. ISBN 0-02-323560-8 and 0-02-323553-5. *(A set of books that briefly introduce materials – metals, polymers, ceramics and composites.)*

Dieter, G. (1997), editor "ASM Handbook – Volume 20 Materials Selection and Design," ASM International, OH, USA. ISBN 0-87170-386-6. *(One of the latest in a series that covers a wide range of technical data about materials, now available on CD for ease of browsing and searching.)*

Emsley, J. (1998) "Molecules at an Exhibition," Oxford University Press, Oxford, UK. ISBN 0-19-286206-5. *(Popular science writing at its best: intelligible, accurate, simple, and clear. The book is exceptional for its range. The message is that molecules – sometimes meaning materials – influence our health, our lives, the things we make and the things we use.)*

Fiell, C. and Fiell, P. (2000) "Industrial Design A–Z," Taschen GmbH, Koln, Germany, ISBN 3-8228-6310-5. *(A summary of key designers and design firms that includes profiles of particular design styles and materials that have had significant influence: aluminum, Bakelite, carbon fiber, ceramics, chromium, plastics, plywood, Pyrex, rubber and tubular metal.)*

Harper, C.A. (1999) "Handbook of Materials for Product Design," McGraw-Hill, NY, US. ISBN 0-07-135406-9. *(In a similar style to Bralla – listed above – this handbook contains many technical details for steels, aluminum, titanium, polymers, composites, natural and synthetic rubber, elastomers, ceramics and glasses; there is also some information on surface processes (metal coatings in particular), joining polymers and elastomers and recycling materials.)*

Lesko, J. (1999) "Industrial Design: Materials and Manufacturing Guide," John Wiley & Sons, NY, USA. ISBN 0-471-29769-0. *(A summary of materials and processes that are of interest to industrial designers, focused on the relevant technical information for each.)*

Polyethylene (PE)

Recycled polyethylene

Polyethylene

Attributes of Polyethylene

Price, \$/kg	1.10–4.00
Density, Mg/m^3	0.92–1.4

Technical Attributes

El. modulus, GPa	0.03–1.4
Elongation, %	10–1400
Fr. toughness, $MPa \cdot m^{1/2}$	0.40–5.16
Vickers hardness, H_v	5–8
Yld. strength, MPa	8–31
Service temp., C	-40–100
Specific heat, J/kg·K	1559–1916
Th. conduct., W/m·K	0.12–0.50
Th. expansion, 10^{-6}/K	106–450

Eco-Attributes

Energy content, MJ/kg	104–114
Recycle potential	High

Aesthetic Attributes

Low to high pitch, 0–10	3–7
Muffled to ringing, 0–10	1–3
Soft to hard, 0–10	5–7
Warm to cool, 0–10	4–5
Gloss, %	5–136
Transparent to opaque	

Features Relative to Other Polymers

✓ Corrosion resistant
✓ Damping
✓ Elastic
✓ FDA approval
✓ Light
✓ Resilient
✓ Slippery
✗ Strong
✓ Tough
✗ UV resistant

What is it? Polyethylene $(CH_2)n$, first synthesized in 1933, looks like the simplest of molecules, but the number of ways in which the $-CH_2-$ units can be linked is large. It is the first of the polyolefins, the bulk thermoplastic polymers that account for a dominant fraction of all polymer consumption. Polyethylene is inert, and extremely resistant to fresh and salt water, food, and most water-based solutions. Because of this it is widely used in household products and food containers.

Design Notes PE is commercially produced as film, sheet, rod, foam and fiber. Drawn PE fiber has exceptional mechanical stiffness and strength, exploited in geo-textile and structural uses. Polyethylene is cheap, and particularly easy to mold and fabricate. It accepts a wide range of colors, can be transparent, translucent or opaque, has a pleasant, slightly waxy feel, can be textured or metal coated, but is difficult to print on. PE is a good electrical insulator with low dielectric loss, so suitable for containers for microwave cooking.

Typical Uses Oil containers, street bollards, milk bottles, toys, beer crates, food packaging, shrink wrap, squeeze tubes, disposable clothing, plastic bags, Tupperware, chopping boards, paper coatings, cable insulation, artificial joints, and as fibers – low cost ropes and packing tape reinforcement.

Competing Materials Polypropylene, polystyrene, ABS, SAN, EVA, elPVC; aluminum as foil and packaging; steel sheet as cans.

The Environment PE is FDA compliant – indeed it is so non-toxic that it can be embedded in the human body (heart valves, hip-joint cups, artificial artery). PE, PP and PVC are made by processes that are relatively energy-efficient, making them the least energy-intensive of commodity polymers. PE can be produced from renewable resources – from alcohol derived from the fermentation of sugar or starch, for instance. Its utility per kilogram far exceeds that of gasoline or fuel-oil, so that production from oil will not disadvantage it in the near future. Polyethylene is readily recyclable if it has not been coated with other materials, and – if contaminated – it can be incinerated to recover the energy it contains.

Technical Notes Low density polyethylene (LDPE), used for film and packaging, has branched chains which do not pack well, making it less dense than water. Medium (MDPE) and High (HDPE) density polyethylenes have longer, less branched chains, making them stiffer and stronger; they are used for containers and pipes. Linear low-density polyethylene (LLPDE) is less resistant to organic solvents, but even this can be overcome by converting its surface to a fluoro-polymer by exposing it to fluorine gas. Treated in this way (when it is known as "Super PE") it can be used for petrol tanks in cars and copes with oil, cleaning fluid, cosmetics and that most corrosive of substances: cola concentrate. Very low density polyethylene (VLDPE) is similar to EVA and plasticized PVC.

Polypropylene (PP)

What is it? Polypropylene, PP, first produced commercially in 1958, is the younger brother of polyethylene – a very similar molecule with similar price, processing methods and application. Like PE it is produced in very large quantities (more than 30 million tons per year in 2000), growing at nearly 10% per year, and like PE its molecule-lengths and side-branches can be tailored by clever catalysis, giving precise control of impact strength, and of the properties that influence molding and drawing. In its pure form polypropylene is flammable and degrades in sunlight. Fire retardants make it slow to burn and stabilizers give it extreme stability, both to UV radiation and to fresh and salt water and most aqueous solutions.

Design Notes Standard grade PP is inexpensive, light and ductile but it has low strength. It is more rigid than PE and can be used at higher temperatures. The properties of PP are similar to those of HDPE but it is stiffer and melts at a higher temperature (165–170 C). Stiffness and strength can be improved by reinforcing with glass, chalk or talc. When drawn to fiber PP has exceptional strength and resilience; this, together with its resistance to water, makes it attractive for ropes and fabric. It is more easily molded than PE, has good transparency and can accept a wider, more vivid range of colors. Advances in catalysis promise new co-polymers of PP with more attractive combinations of toughness, stability and ease of processing. Mono-filaments fibers and multi-filament yarn or rope have high abrasion resistance and are almost twice as strong as PE fibers.

Typical Uses Ropes, general polymer engineering, automobile air ducting, parcel shelving and air-cleaners, garden furniture, washing machine tanks, wet-cell battery cases, pipes and pipe fittings, beer bottle crates, chair shells, capacitor dielectrics, cable insulation, kitchen kettles, car bumpers, shatter proof glasses, crates, suitcases, artificial turf, thermal underwear.

Competing Materials Polyethylene, polystyrene, ABS, SAN; aluminum as foil and packaging; steel sheet as cans.

The Environment PP is exceptionally inert and easy to recycle, and can be incinerated to recover the energy it contains. PP, like PE and PVC, is made by processes that are relatively energy-efficient, making them the least energy-intensive of commodity polymers. Its utility per kilogram far exceeds that of gasoline or fuel-oil, so that production from oil will not disadvantage it in the near future.

Technical Notes The many different grades of polypropylene fall into three basic groups: homopolymers (polypropylene, with a range of molecular weights and thus properties), co-polymers (made by co-polymerization of propylene with other olefins such as ethylene, butylene or styrene) and composites (polypropylene reinforced with mica, talc, glass powder or fibers) that are stiffer and better able to resist heat than simple polypropylenes.

Mica-filled polypropylene

Attributes of Polypropylene

Price, \$/kg	0.90–1.00
Density, Mg/m³	0.89–0.92

Technical Attributes

El. modulus, GPa	0.90–1.55
Elongation, %	100–600
Fr. toughness, MPa·m$^{1/2}$	3–4.5
Vickers hardness, H_v	6–11
Yld. strength, MPa	20.7–37.2
Service temp., C	-40–120
Specific heat, J/kg·K	1870–1956
Th. conduct., W/m·K	0.11–0.17
Th. expansion, 10^{-6}/K	122–180

Eco-Attributes

Energy content, MJ/kg	76–84
Recycle potential	High

Aesthetic Attributes

Low to high pitch, 0–10	6–7
Muffled to ringing, 0–10	3–4
Soft to hard, 0–10	6–7
Warm to cool, 0–10	4–4
Gloss, %	20–94
Transparent to opaque	

Features Relative to Other Polymers

- ✓ Corrosion resistant
- ✓ Elastic
- ✓ FDA approval
- ✓ Hot temperatures
- ✓ Impact resistant
- ✓ Light
- ✓ Resilient
- ✓ Slippery
- ✓ Tough
- ✗ UV resistant

Polystyrene (PS)

Expanded polystyrene foam

Polystyrene

Attributes of Polystyrene

Price, \$/kg	1.30–1.60
Density, Mg/m^3	1.04–1.05

Technical Attributes

El. modulus, GPa	2.28–3.34
Elongation, %	1.2–3.6
Fr. toughness, $MPa{\cdot}m^{1/2}$	0.7–1.1
Vickers hardness, H_v	9–16
Yld. strength, MPa	28.72–56.2
Service temp., C	-18–100
Specific heat, J/kg·K	1691–1758
Th. conduct., W/m·K	0.12–0.13
Th. expansion, 10^{-6}/K	90–153

Eco-Attributes

Energy content, MJ/kg	101–110
Recycle potential	High

Aesthetic Attributes

Low to high pitch, 0–10	7–7
Muffled to ringing, 0–10	4–4
Soft to hard, 0–10	7–7
Warm to cool, 0–10	4–4
Gloss, %	9–96
Optically clear to opaque	

Features Relative to Other Polymers

- ✓ Flame retardant
- ✗ Impact resistant
- ✓ Light
- ✗ Resilient
- ✗ Tough

What is it? In its simplest form PS is brittle. Its mechanical properties are dramatically improved by blending with polybutadiene, but with a loss of optical transparency. High impact PS (10% polybutadiene) is much stronger even at low temperatures (meaning strength down to -12 C).

Design Notes PS comes in 3 guises: as the simple material ("general purpose PS"); as the high impact variant, blended with polybutadiene; and as polystyrene foam, the most familiar and cheapest of all polymer foams. All are FDA approved for use as food containers and packaging. General purpose PS is easy to mold. Its extreme clarity, ability to be colored, and high refractive index give it a glass-like sparkle, but it is brittle and cracks easily (think of CD cases). It is used when optical attractiveness and low cost are sought, and the mechanical loading is light: cosmetic compacts, transparent but disposable glasses, cassettes of all kinds. Medium and high impact polystyrenes trade their optical for their mechanical properties. Medium impact PS, translucent, appears in electrical switch gears and circuit breakers, coat hangers and combs. High impact PS – a blend of PPO and PS, is opaque, but is tough and copes better with low temperatures than most plastics; it is found in interiors of refrigerators and freezers, and in food trays such as those for margarine and yogurt. PS can be foamed to a very low density (roughly 1/3 of all polystyrene in foamed). These foams have low thermal conduction and are cheap, and so are used for house insulation, jackets for water boilers, insulation for disposable cups. They crush at loads that do not cause injury to delicate objects (such as TV sets or to the human body), making them good for packaging.

Typical Uses Disposable cups; light fittings; toys; pens; models; in expanded form – packaging, thermal insulation and ceiling tiles; TV cabinets; wall tiles; disposable dishes; furniture; molded parts and containers; CD covers, disposable glass, razors, hot drink cups.

Competing Materials High density polyethylene, Polypropylene and polymer-coated paper.

The Environment The flammability of PS foam, and the use of CFCs as blowing agents in the foaming process was, at one time, a cause for concern. New flame retardants allow PS foams to meet current fire safety standards, and CFC blowing agents have been replaced by pentane, CO_2 or HFCs which do not have a damaging effect on the ozone layer. PS can be recycled. The monomer, styrene, is irritating to the eyes and throat, but none survives in the polymer.

Technical Notes Polystyrene, PS, is – like PE and PP – a member of the polyolefin family of moldable thermoplastics. In place of one of the H-atoms of the polyethylene it has a C_6H_5-benzene ring. This makes for a lumpy molecule which does not crystallize, and the resulting material is transparent with a high refractive index. The benzene ring absorbs UV light, exploited in the PS screening of fluorescent lights, but this also causes the polymer to discolor in sunlight.

Acrylonitrile-butadiene-styrene (ABS)

What is it? ABS is tough, resilient, and easily molded. It is usually opaque, although some grades can now be transparent, and it can be given vivid colors. ABS-PVC alloys are tougher than standard ABS and, in self-extinguishing grades, are used for the casings of power tools.

Design Notes ABS has the highest impact resistance of all polymers. It takes color well. Integral metallics are possible (as in GE Plastics' Magix). ABS is UV resistant for outdoor application if stabilizers are added. It is hygroscopic (may need to be oven dried before thermoforming) and can be damaged by petroleum-based machining oils. ASA (acrylic-styrene-acrylonitrile) has very high gloss; its natural color is off-white but others are available. It has good chemical and temperature resistance and high impact resistance at low temperatures. UL-approved grades are available. SAN (styrene-acrylonirtile) has the good processing attributes of polystyrene but greater strength, stiffness, and chemical and heat resistance. By adding glass fiber the rigidity can be increased dramatically. It is transparent (over 90% in the visible range but less for UV light) and has good color. Depending on the amount of acrylonitrile that is added this can vary from water white to pale yellow, but without a protective coating, sunlight causes yellowing and loss of strength, slowed by UV stabilizers. All three can be extruded, compression molded or formed to sheet that is then shaped by vacuum thermoforming. They can be joined by ultrasonic or hot-plate welding, or bonded with polyester, epoxy, isocyanate or nitrile-phenolic adhesives.

Typical Uses ABS: cases for computers and TVs, telephones, food mixers, vacuum cleaners, baths, shower trays, pipes, luggage shells, RV parts, shower stalls, cassette holders, automotive parts, safety hard hats, legos, computer mice, razors, handles, shavers, chairs. SAN: telephone cases, food processing bowls, medical syringes, mixing bowls, beakers, coffee filters, cassettes, industrial battery cases, toothbrushes, cosmetic packs, dinnerware, food containers. ASA: appliance panels and knobs, toys, medical instruments, rear view mirrors, garden tables and chairs, hose fittings, garden tools, letter boxes, boat shells, windsurfing boards.

Competing Materials High density polyethylene, polypropylene, butyl rubber, nylon.

The Environment The acrylonitrile monomer is nasty stuff, almost as poisonous as cyanide. Once polymerized with styrene it becomes harmless. Some grades of ABS are FDA compliant and can be recycled.

Technical Notes ABS is a terpolymer – one made by co-polymerizing 3 monomers: acrylonitrile, butadiene and styrene. The acrylonitrile gives thermal and chemical resistance, rubber-like butadiene gives ductility and strength, the styrene gives a glossy surface, ease of machining and a lower cost. In ASA, the butadiene component (which gives poor UV resistance) is replaced by an acrylic ester. Without the addition of butyl, ABS becomes SAN – a similar material with lower impact resistance or toughness.

Formed ABS sheet

Attributes of ABS	
Price, \$/kg	1.50–2.80
Density, Mg/m^3	1.01–1.21
Technical Attributes	
El. modulus, GPa	1.1–2.9
Elongation, %	1.5–100
Fr. toughness, $MPa{\cdot}m^{1/2}$	1.2–4.2
Vickers hardness, H_V	6–15
Yld. strength, MPa	18.5–51
Service temp., C	-18–90
Specific heat, J/kg·K	1386–1919
Th. conduct., W/m·K	0.18–0.33
Th. expansion, 10^{-6}/K	84.6–234
Eco-Attributes	
Energy content, MJ/kg	95–104
Recycle potential	High
Aesthetic Attributes	
Low to high pitch, 0–10	6–7
Muffled to ringing, 0–10	3–4
Soft to hard, 0–10	6–7
Warm to cool, 0–10	4–5
Gloss, %	10–96
Transparent to opaque	

Features Relative to Other Polymers

- ✓ FDA approval
- ✓ Flame retardant
- ✓ Impact resistant
- ✓ Resilient
- ✓ Tough

ABS

Polyamide (PA), Nylon

Flocked nylon sheet

Nylon

Attributes of Nylon

Price, \$/kg	2.90–11.50
Density, Mg/m^3	1–1.42

Technical Attributes

El. modulus, GPa	0.67–4.51
Elongation, %	4–1210
Fr. toughness, $MPa \cdot m^{1/2}$	0.58–8.03
Vickers hardness, H_V	6–28
Yld. strength, MPa	20.7–101.6
Service temp., C	-80–120
Specific heat, J/kg·K	1421–2323
Th. conduct., W/m·K	0.18–0.35
Th. expansion, 10^{-6}/K	50.4–216

Eco-Attributes

Energy content, MJ/kg	110–120
Recycle potential	High

Aesthetic Attributes

Low to high pitch, 0–10	6–7
Muffled to ringing, 0–10	3–4
Soft to hard, 0–10	6–7
Warm to cool, 0–10	4–5
Gloss, %	65–150
Optically clear to opaque	

Features Relative to Other Polymers

✓ Elastic
✓ Fatigue resistant
✓ Flame retardant
✓ Impact resistant
✓ Resilient
✓ Slippery
✓ Stiff
✓ Tough
✓ Wear resistant

What is it? Back in 1945, the war in Europe just ended, the two most prized luxuries were cigarettes and nylons. Nylon (PA) can be drawn to fibers as fine as silk, and was widely used as a substitute for it. Today, newer fibers have eroded its dominance in garment design, but nylon-fiber ropes, and nylon as reinforcement for rubber (in car tires) and other polymers (PTFE, for roofs) remains important. It is used in product design for tough casings, frames and handles, and – reinforced with glass – as bearings gears and other load-bearing parts. There are many grades (Nylon 6, Nylon 66, Nylon 11....) each with slightly different properties.

Design Notes. Nylons are tough, strong and have a low coefficient of friction. They have useful properties over a wide range of temperature (-80–120 C). They are easy to injection mold, machine and finish, can be thermally or ultrasonically bonded, or joined with epoxy, phenol-formaldehyde or polyester adhesives. Certain grades of nylon can be electroplated allowing metallization, and most accept print well. A blend of PPO/Nylon is used in fenders, exterior body parts. Nylon fibers are strong, tough, elastic and glossy, easily spun into yarns or blended with other materials. Nylons absorb up to 4% water; to prevent dimensional changes, they must be conditioned before molding, allowing them to establish equilibrium with normal atmospheric humidity. Nylons have poor resistance to strong acids, oxidizing agents and solvents, particularly in transparent grades.

Typical Uses Light duty gears, bushings, sprockets and bearings; electrical equipment housings, lenses, containers, tanks, tubing, furniture casters, plumbing connections, bicycle wheel covers, ketchup bottles, chairs, toothbrush bristles, handles, bearings, food packaging. Nylons are used as hot-melt adhesives for book bindings; as fibers – ropes, fishing line, carpeting, car upholstery and stockings; as aramid fibers – cables, ropes, protective clothing, air filtration bags and electrical insulation.

Competing Materials Polypropylene, polyester and ABS.

The Environment Nylons have no known toxic effects, although they are not entirely inert biologically. Nylons are oil-derivatives, but this will not disadvantage them in the near future. With refinements in polyolefin catalysis, nylons face stiff competition from less expensive polymers.

Technical Notes The density, stiffness, strength, ductility and toughness of nylons all lie near the average for unreinforced polymers. Their thermal conductivities and thermal expansion are a little lower than average. Reinforcement with mineral, glass powder or glass fiber increases the modulus, strength and density. Semi-crystalline nylon is distinguished by a numeric code for the material class indicating the number of carbon atoms between two nitrogen atoms in the molecular chain. The amorphous material is transparent, the semi-crystalline material is opal white.

Polymethylmethacrylate (PMMA), Acrylic

What is it? When you think of PMMA, think transparency. Acrylic, or PMMA, is the thermoplastic that most closely resembles glass in transparency and resistance to weathering. The material has a long history: discovered in 1872, commercialized in 1933, its first major application was as cockpit canopies for fighter aircraft during the second World War.

Design Notes Acrylic, or PMMA, is hard and stiff as polymers go, easy to polish but sensitive to stress concentrations. It shares with glass a certain fragility, something that can be overcome by blending with acrylic rubber to give a high-impact alloy (HIPMMA). PVC can be blended with PMMA to give tough, durable sheets. Acrylic is available as a sheet, rod or tube and can be shaped by casting or extrusion. Cell casting uses plates of glass and gasketing for a mold; it allows clear and colored panels up to 4 inches thick to be cast. Extrusion pushes melted polymer pellets through a die to give a wide variety of shapes, up to 0.25 inches thick for sheet. Clear and colored PMMA sheet lends itself to thermoforming, allowing inexpensive processing. A hybrid sheet manufacturing process, continuous casting, combines the physical benefits of cell casting and the cost efficiency of extrusion. Extruded and continuous cast sheet have better thickness tolerance than cell-cast sheet. PMMA can be joined with epoxy, alpha-cyanoacrylate, polyester or nitrile-phenolc adhesives. It scratches much more easily than glass, but this can be partially overcome with coatings.

Typical Uses Lenses of all types; cockpit canopies and aircraft windows; signs; domestic baths; packaging; containers; electrical components; drafting equipment; tool handles; safety spectacles; lighting, automotive tail lights, chairs, contact lenses, windows, advertising signs, static dissipation products; compact disks.

Competing Materials Polycarbonate, polystyrene, PVC, PET.

The Environment Acrylics are non-toxic and recyclable.

Technical Notes Polymers are truly transparent only if they are completely amorphous – that is, non-crystalline. The lumpy shape of the PMMA molecule ensures an amorphous structure, and its stability gives good weathering resistance. PMMA is attacked by esters, ketones, acids and hydrocarbons, and has poor resistance to strong acids or bases, solvents and acetone.

Edge-glow acrylic sheet

Attributes of Acrylic

Price, $/kg	1.70–2.40
Density, Mg/m^3	1.16–1.22

Technical Attributes

El. modulus, GPa	2.24–3.8
Elongation, %	2 – 10
Fr. toughness, $MPa \cdot m^{1/2}$	0.7–1.6
Vickers hardness, H_v	16–21
Yld. strength, MPa	53.8–72.4
Service temp., C	-50–100
Specific heat, J/kg·K	1485–1606
Th. conduct., W/m·K	0.08–0.25
Th. expansion, 10^{-6}/K	72–162

Eco-Attributes

Energy content, MJ/kg	97–105
Recycle potential	High

Aesthetic Attributes

Low to high pitch, 0–10	7–7
Muffled to ringing, 0–10	4–4
Soft to hard, 0–10	7–7
Warm to cool, 0–10	4–5

Optically clear to transparent

Features Relative to Other Polymers

- ✗ Corrosion resistant
- ✓ FDA approval
- ✗ Impact resistant
- ✓ Stiff
- ✓ Strong
- ✗ Tough
- ✓ UV resistant
- ✗ Wear resistant

Acrylic

Polycarbonate (PC)

Polycarbonate lattice

Attributes of Polycarbonate

Price, \$/kg	3.80–4.30
Density, Mg/m^3	1.14–1.21

Technical Attributes

El. modulus, GPa	2.21–2.44
Elongation, %	70–150
Fr. toughness, $MPa{\cdot}m^{1/2}$	2.1–4.602
Vickers hardness, H_V	17–22
Yld. strength, MPa	59.1–69
Service temp., C	-40–120
Specific heat, J/kg·K	1535–1634
Th. conduct., W/m·K	0.19–0.22
Th. expansion, $10^{-6}/K$	120.1–136.8

Eco-Attributes

Energy content, MJ/kg	120–130
Recycle potential	High

Aesthetic Attributes

Low to high pitch, 0–10	7–7
Muffled to ringing, 0–10	4–4
Soft to hard, 0–10	7–7
Warm to cool, 0–10	4–5

Optically clear to transparent

Features Relative to Other Polymers

- ✓ FDA approval
- ✓ Flame retardant
- ✓ Hot temperatures
- ✓ Impact resistant
- ✓ Stiff
- ✓ Strong
- ✓ Tough
- ✓ Wear resistant

What is it? PC is one of the "engineering" thermoplastics, meaning that they have better mechanical properties than the cheaper "commodity" polymers. The family includes the plastics polyamide (PA), polyoxymethylene (POM) and polytetrafluorethylene (PTFE). The benzene ring and the -OCOO- carbonate group combine in pure PC to give it its unique characteristics of optical transparency and good toughness and rigidity, even at relatively high temperatures. These properties make PC a good choice for applications such as compact disks, safety hard hats and housings for power tools. To enhance the properties of PC even further, it is possible to co-polymerize the molecule with other monomers (improves the flame retardancy, refractive index and resistance to softening), or to reinforce the PC with glass fibers (giving better mechanical properties at high temperatures).

Design Notes The optical transparency and high impact resistance of PC make it suitable for bullet-resistant or shatter-resistant glass applications. It is readily colored. PC is usually processed by extrusion or thermoforming (techniques that impose constraints on design), although injection molding is possible. When designing for extrusion with PC, the wall thickness should be as uniform as possible to prevent warping, and projections and sharp corners avoided – features like hollows and lone unsupported die sections greatly increase the mold cost. The stiffness of the final part can be improved by the incorporation of corrugations or embossed ribs. PC can be reinforced using glass fibers to reduce shrinkage problems on cooling and to improve the mechanical performance at high temperatures. It can be joined using adhesives, fasteners or welding.

Typical Uses Compact disks, housings for hair dryers, toasters, printers and power tool housings, refrigerator linings, mechanical gears, instrument panels, motorcycle helmets, automotive bumpers and body parts, riot shields.

Competing Materials Acetal, acrylic and polyester.

The Environment The processing of engineering thermoplastics requires a higher energy input than that of commodity plastics, but otherwise there are no particular environmental penalties. PC can be recycled if unreinforced.

Technical Notes The combination of the benzene ring and carbonate structures in the PC molecular structure give the polymer its unique characteristics of high strength and outstanding toughness. PC can be blended with ABS or polyurethane. ABS/PC gets its flame retardance and UV resistance from polycarbonate at a lower cost than that of ABS. PU/PC gets its rigidity from polycarbonate and flexibility and ease of coating from polyurethane.

Polyoxymethylene (POM), Acetal

What is it? POM was first marketed by DuPont in 1959 as Delrin. It is similar to nylon but is stiffer, and has better fatigue and water resistance – nylons, however, have better impact and abrasion resistance. It is rarely used without modifications: most often filled with glass fiber, flame retardant additives or blended with PTFE or PU. The last, POM/PU blend, has good toughness. POM is used where requirements for good moldability, fatigue resistance and stiffness justify its high price relative to mass polymers, like polyethylene, which are polymerized from cheaper raw materials using lower energy input.

Design Notes POM is easy to mold by blow molding, injection molding or sheet molding, but shrinkage on cooling limits the minimum recommended wall thickness for injection molding to 0.1 mm. As manufactured, POM is gray but it can be colored. It can be extruded to produce shapes of constant cross-section such as fibers and pipes. The high crystallinity leads to increased shrinkage upon cooling. It must be processed in the temperature range 190–230 C and may require drying before forming because it is hygroscopic. Joining can be done using ultrasonic welding, but POM's low coefficient of friction requires welding methods that use high energy and long ultrasonic exposure. Adhesive bonding is an alternative. POM is also an electrical insulator. Without co-polymerization or the addition of blocking groups, POM degrades easily.

Typical Uses Automobile carburettors and door handles, videocassette parts, gears and bearings, tool handles, plumbing parts, clothing zips.

Competing Materials Nylon, polyester, PTFE.

The Environment Acetal, like most thermoplastics, is an oil derivative, but this poses no immediate threat to its use.

Technical Notes The repeating unit of POM is -(CH2O)n , and the resulting molecule is linear and highly crystalline. Consequently, POM is easily moldable, has good fatigue resistance and stiffness, and is water resistant. In its pure form, POM degrades easily by de-polymerization from the ends of the polymer chain by a process called "unzipping." The addition of "blocking groups" at the ends of the polymer chains or co-polymerization with cyclic ethers such as ethylene oxide prevents unzipping and hence degradation.

Machined acetal blocks

Attributes of Acetal

Price, \$/kg	2.70–4.00
Density, Mg/m^3	1.39–1.43

Technical Attributes

El. modulus, GPa	2.35–6.27
Elongation, %	10–75
Fr. toughness, $MPa{\cdot}m^{1/2}$	1.71–4.2
Vickers hardness, H_V	14–24
Yld. strength, MPa	48.6–72.4
Service temp., C	-30–110
Specific heat, J/kg·K	1364–1433
Th. conduct., W/m·K	0.22–0.35
Th. expansion, 10^{-6}/K	75–200

Eco-Attributes

Energy content, MJ/kg	115–121
Recycle potential	High

Aesthetic Attributes

Low to high pitch, 0–10	7–7
Muffled to ringing, 0–10	4–5
Soft to hard, 0–10	7–7
Warm to cool, 0–10	4–5
Opaque	

Features Relative to Other Polymers

- ✓ Fatigue resistant
- ✗ Flame retardant
- ✓ Heavy
- ✓ Slippery
- ✓ Stiff
- ✓ Strong
- ✓ Wear resistant

Acetal

Polytetrafluoroethylene (PTFE)

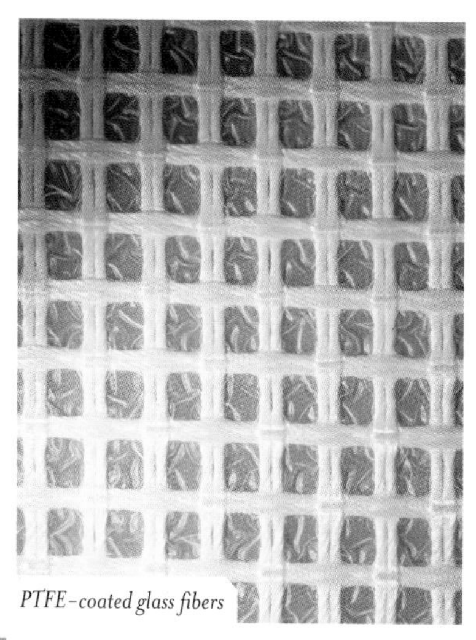
PTFE-coated glass fibers

PTFE

Attributes of PTFE

Price, \$/kg	13.90–15.90
Density, Mg/m^3	2.14–2.2

Technical Attributes

El. modulus, GPa	0.4–0.55
Elongation, %	200–400
Fr. toughness, $MPa{\cdot}m^{1/2}$	5–7
Vickers hardness, H_V	59–65
Yld. strength, MPa	19.7–21.7
Service temp., C	-270–250
Specific heat, J/kg·K	1014–1054
Th. conduct., W/m·K	0.24–0.26
Th. expansion, 10^{-6}/K	126–216

Eco-Attributes

Energy content, MJ/kg	180–195
Recycle potential	High

Aesthetic Attributes

Low to high pitch, 0–10	5–5
Muffled to ringing, 0–10	2–3
Soft to hard, 0–10	6–6
Warm to cool, 0–10	5–5
Gloss, %	85–95
Transparent to opaque	

Features Relative to Other Polymers

- ✓ Cold temperatures
- ✓ Corrosion resistant
- ✓ Damping
- ✓ Flame retardant
- ✓ Heavy
- ✓ Hot temperatures
- ✓ Slippery
- ✗ Strong

What is it? PTFE (Teflon)is a member of the fluoroplastic family, which includes chlorotrifluoroethylene, CTFE or CFE, polyvinyl fluoride, PVF, and polyvinylidene fluoride PVF2. PTFE has exceptionally low friction, is water repellant, and extremely stable. It was first commercialized in the late 1940s as Teflon. Non-stick cooking utensils (Tefal = teflon coated aluminum) exploit its chemical inertness, its thermal stability and its non-wettability – the reason nothing sticks to it. It is expensive as polymers go, but it is used in high-value applications (non-stick pans; GoreTex rain gear; artificial arteries).

Design Notes PTFE is 2.7 times denser than polyethylene and 12 times more expensive. But it is much more resistant to chemical attack; it can safely be used from –270 to +250 C. It has remarkably low friction; and it has an exceptional ability to resist wetting. All fluoroplastics are white, and to some degree, translucent. They give long-term resistance to attacks of all sorts, including ultraviolet radiation. PTFE itself has a characteristically soft, waxy feel, partly because of the low coefficient of friction. It is an excellent electrical insulator, with low dielectric loss. It can be "foamed" to give a light, micro-porous film that rejects liquid water but allows water vapor to pass – the principle of GoreTex. The mechanical properties of PTFE are not remarkable, but it can be made more abrasive resistant by filling with inert ceramic, and it can be reinforced with glass, nylon or Kevlar fibers to give a leather-like skin of exceptional toughness, strength and weather-resistance (exploited in tensile roofs). Bonding PTFE is difficult; thermal or ultrasonic methods are good; epoxy, nitrile-phenolic and silicone adhesives can be used.

The use of GoreTex derivatives in fabrics is expanding. The pore size in these fabrics can be controlled to reject not merely water, but bacteria, with potential for protective clothing for surgeons, and against certain kinds of biological weapons.

Typical Uses Wire and cable covers; high-quality insulating tape; corrosion resistant lining for pipes and valves; protective coatings; seals and gaskets; low friction bearings and skis; translucent roofing and weather protection for other polymers (e.g. ABS); non-stick cooking products; water repellent fabrics.

Competing Materials Polyethylene for low friction, PVC for external use for roofs and cladding.

The Environment PTFE is non-flammable and FDA approved. Like all thermoplastics, simple PTFE can be recycled. But in making it into non-stick surfaces, or in transforming it into GoreTex, additives are made which prevent further recycling.

Technical Notes Fluorine is the most reactive of gases, yet combined with carbon to form polytetrafluoroethylene, PTFE, it becomes the most stable of molecules, resistant to practically everything except excessive heat. It has an exceptionally low coefficient of friction against steel, making it attractive for bearings and – in dispersed form – as a lubricant.

Ionomers

What is it? Ionomers, introduced by DuPont in 1964, are flexible thermoplastics but they have ionic cross-links, from which they derive their name. Their thermoplastic character allows them to be processed by blow molding, injection molding and thermoforming, and to be applied as coatings. But cooled below 40 C they acquire the characteristic of thermosets: high strength, good adhesion and chemical stability.

Design Notes Ionomers are very tough, they have high tensile strength and excellent impact, tear, grease and abrasion resistance. Optical clarity is also quite high. They are most often produced as film. Ionomers have outstanding hot tack (10 times that of LDPE). Their resistance to weather is similar to that of PE, and like PE, they can be stabilized with the addition of carbon black. Permeability is also similar to that of PE, except for carbon dioxide where the permeability is lower. Low temperature flexibility is excellent but they should not be used at temperatures above 70 C. Because of the ionic nature of the molecules, ionomers have good adhesion to metal foil, nylon and other packaging films. Foil and extrusion coating are common. Ionomers have higher moisture vapor permeability (due to the low crystallinity) than polyethylene, are easily sealed by heat and retain their resilience over a wide temperature range.

Typical Uses Food packaging, athletic soles with metal inserts, ski boots, ice skate shells, wrestling mats, thermal pipe insulation, license plate holders, golf ball covers, automotive bumpers, snack food packaging, blister packs, bottles.

Competing Materials LDPE, polyisoprene.

The Environment Ionomers have properties that resemble thermosets yet they can be recycled – an attractive combination.

Technical Notes Ionomers are co-polymers of ethylene and methacrylic acid. Some grades contain sodium and have superior optical properties and grease resistance; some contain zinc and have better adhesion. The ionic cross-links are stable at room temperature, but break down upon heating above about 40 C. The advantages of cross-linking are seen in the room temperature toughness and stiffness. At high temperatures the advantages of linear thermoplastics appear – ease of processing and recyclability.

Coated and dimpled ionomer

Attributes of Ionomers

Price, $/kg	1.40–1.60
Density, Mg/m^3	0.93–0.96

Technical Attributes

El. modulus, GPa	0.2–0.42
Elongation, %	300–700
Fr. toughness, $MPa{\cdot}m^{1/2}$	1.14–3.43
Vickers hardness, H_v	2–5
Yld. strength, MPa	8.27–15.9
Service temp., C	-30–70
Specific heat, J/kg·K	1814–1887
Th. conduct., W/m·K	0.24–0.28
Th. expansion, 10^{-6}/K	180–306

Eco-Attributes

Energy content, MJ/kg	115–120
Recycle potential	High

Aesthetic Attributes

Low to high pitch, 0–10	5–6
Muffled to ringing, 0–10	2–3
Soft to hard, 0–10	6–6
Warm to cool, 0–10	4–5
Gloss, %	20–143
Transparent	

Features Relative to Other Polymers

- ✓ Cold temperatures
- ✓ Elastic
- ✓ Impact resistant
- ✓ Light
- ✓ Resilient
- ✓ Strong
- ✓ Tear resistant
- ✓ Tough
- ✓ Wear resistant

Celluloses (CA)

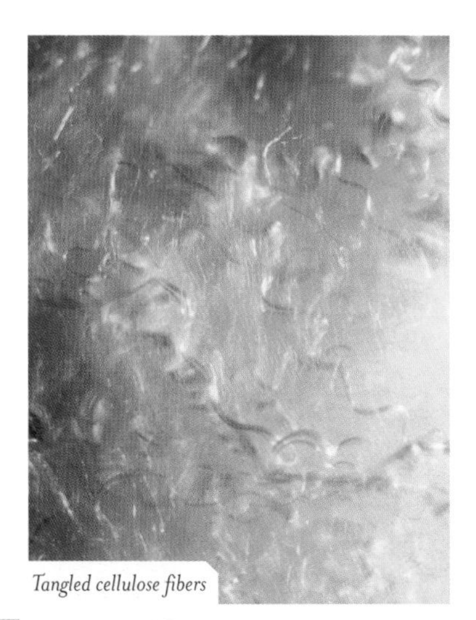
Tangled cellulose fibers

Attributes of Celluloses

Price, $/kg	3.40–3.90
Density, Mg/m³	0.98–1.4

Technical Attributes

El. modulus, GPa	0.75–4.1
Elongation, %	5–100
Fr. toughness, MPa·m$^{1/2}$	0.85–3.20
Vickers hardness, H_v	18–15
Yld. strength, MPa	24.8–52.4
Service temp., C	-10–60
Specific heat, J/kg·K	1386–1665
Th. conduct., W/m·K	0.13–0.34
Th. expansion, 10^{-6}/K	118.8–360

Eco-Attributes

Energy content, MJ/kg	120–126
Recycle potential	High

Aesthetic Attributes

Low to high pitch, 0–10	6–7
Muffled to ringing, 0–10	3–4
Soft to hard, 0–10	6–7
Warm to cool, 0–10	4–5
Transparent to translucent	

Features Relative to Other Polymers

✕ Hot temperatures
✕ Flame retardant
✕ Tough

Celluloses

What is it? Cellulose is starch; it is one of the main structural polymers in plants (specifically that of wood or cotton). These natural fibers are treated with acids to produce a resin, a process called "esterification." Cellulose acetate (CA), cellulose acetate butyrate (CAB) and cellulose acetate propionate (CAP) are three common cellulose materials. CA combines toughness, transparency and a natural surface texture. Some cellulose resins are biodegradable allowing their use for envelopes with transparent windows that can be disposed of as if they were paper alone.

Design Notes Cellulose materials have good optical transparency (up to 90%) and mechanical properties; they are often used in products where surface finish is important. They can be translucent, transparent or opaque and come in a range of colors including pearlescent, opaque or metallic. Cellulose materials are anti-static and so dust does not stick to the surface. They have glossy surfaces and a self-polishing effect that allows surface scratches to disappear with use. They can be formulated for outdoor use. Cellulose acetates have a high vapor permeability and limited heat resistance; their electrical insulation and dimensional stability are poor; water absorption is high compared with PVC. Compared with CA, CAB and CAP are slightly softer, of lower densities and heat distortion temperatures and they flow more easily. CAB and CAP have a wider operating temperature range (0 to 60 C). In bulk, cellulose acetate often has a light yellow-brown color but it can also attain water, white transparency. Cellulose is easily made into fibers, fabrics and films, known under the trade name of Rayon. It has excellent flow properties and so is well-suited to injection molding but the resin must be kept dry. Stiffer flow grades are available for extrusion to produce film and sheet. These polymers are also some of the best for cutting, turning and milling. CAB is well-suited for thermoforming.

Typical Uses Film packaging, sunglasses, safety glasses, eye glass frames, laminated brochures, tool handles, hammer heads, electric screwdrivers, photographic film, typewriter keys, blister packaging, combs, hair decoration.

Competing Materials Acrylics, polyethylene, polycarbonate.

The Environment Cellulose – a natural, renewable material – forms one ingredient in cellulose-based polymers. The processing, however, involves chemicals that create a problem of disposal. Most cellulose-based polymers burn easily, requiring protection from naked flame. Some are biodegradable.

Technical Notes Grades of cellulose materials are varied by increasing the level of acid-treatment or by adding plasticizers. Increasing the plasticizer increases flow and toughness but reduces the creep resistance. But this lack of creep resistance can be used to an advantage in molding where parts can be cooled quickly because the stresses will relax quite quickly.

Polyetheretherketone (PEEK)

What is it? PEEK is a high-performance thermoplastic, meaning that – among thermoplastics – is has exceptionally high stiffness, strength and resistance to heat. This comes at a price: PEEK is 50 times more expensive than PP, and 10 to 20 times more than nylon. This limits its use to applications in which technical performance is paramount.

Design Notes PEEK can be used up to temperatures of 300 C for a short time and 250 C for a long time. It offers high hardness and therefore abrasion resistance; it has excellent fatigue properties and good creep resistance. It has a low coefficient of friction, a low flammability, and low smoke emission during combustion. Chemical resistance is very good (and retained to the same high temperatures) and there is very low water absorption. Unreinforced PEEK offers the highest elongation and toughness of all PEEK grades. Glass-reinforcement significantly reduces the expansion rate and increases the flexural modulus. PEEK can be used as a matrix in continuous carbon fiber composites. Carbon-reinforced PEEK has high compressive strength and stiffness and low expansion coefficient, and its thermal conductivity can be 3 times better than pure PEEK.

Processing PEEK is not difficult, despite its high heat resistance, provided the temperature is held at 375 C. It can be injection molded, extruded (into rod, profile, film or wire insulation) and compression molded. It is available as extruded film and sheet in thicknesses from approximately 0.025 to 1 mm.

Typical Uses Electrical connectors, hot water meters, F1 engine components, valve and bearing components, wire and cable coatings, film and filament for specialized applications, pump wear rings, electrical housing, bushings, bearings.

Competing Materials PTFE for applications where strength is not critical, epoxy for electrical housing, bronze for bushings, bearings.

The Environment PEEK can be recycled if uncontaminated.

Technical Notes PEEK is a semi-crystalline thermoplastic. It has a high glass transition temperature (Tg = 150 C) and can be used well above this temperature, but its stiffness falls and its expansion coefficient rises above Tg.

Natural PEEK sheet

Attributes of PEEK

Price, $/kg	66.40–70.60
Density, Mg/m^3	1.3–1.32

Technical Attributes

El. modulus, GPa	3.76–3.95
Elongation, %	30–150
Fr. toughness, $MPa \cdot m^{1/2}$	2.73–4.30
Vickers hardness, H_v	25–28
Yld. strength, MPa	87–95
Service temp., C	-30–250
Specific heat, J/kg·K	1443–1501
Th. conduct., W/m·K	0.24–0.26
Th. expansion, 10^{-6}/K	72–194.4

Eco-Attributes

Energy content, MJ/kg	305–326
Recycle potential	High

Aesthetic Attributes

Low to high pitch, 0–10	7–7
Muffled to ringing, 0–10	4–4
Soft to hard, 0–10	7–7
Warm to cool, 0–10	5–5
Opaque	

Features Relative to Other Polymers

- ✓ Fatigue resistant
- ✓ Flame retardant
- ✓ Hot temperatures
- ✓ Slippery
- ✓ Strong
- ✓ UV resistant
- ✓ Wear resistant

Polyvinylchloride (PVC)

PVC tubes

PVC

Attributes of tpPVC

Price, $/kg	1.00–1.20
Density, Mg/m³	1.3–1.58

Technical Attributes

El. modulus, GPa	2.14–4.14
Elongation, %	11.93–80
Fr. toughness, MPam$^{1/2}$	1.46–5.12
Vickers hardness, H_v	10–15
Yld. strength, MPa	35.4–52.1
Service temp., C	-20–70
Specific heat, J/kg·K	1355–1445
Th. conduct., W/m·K	0.15–0.29
Th. expansion, 10^{-6}/K	1.8–180

Eco-Attributes

Energy content, MJ/kg	77–83
Recycle potential	High

Aesthetic Attributes

Low to high pitch, 0–10	6–7
Muffled to ringing, 0–10	4–4
Soft to hard, 0–10	7–7
Warm to cool, 0–10	4–5
Transparent to opaque	

Features Relative to Other Polymers

- ✓ Corrosion resistant
- ✓ Damping
- ✓ FDA approval
- ✓ Flame retardant
- ✓ Heavy
- ✗ Impact resistant
- ✓ Resilient
- ✓ Stiff
- ✗ Strong
- ✗ Tough
- ✓ UV resistant

What is it? PVC – Vinyl – is one of the cheapest, most versatile and – with polyethylene – the most widely used of polymers and epitomizes their multi-faceted character. In its pure form – as a thermoplastic, tpPVC – it is rigid, and not very tough; its low price makes it a cost-effective engineering plastic where extremes of service are not encountered. Incorporating plasticizers creates flexible PVC, elPVC, a material with leather-like or rubber-like properties, and used a substitute for both. By contrast, reinforcement with glass fibers gives a material that is sufficiently stiff, strong and tough to be used for roofs, flooring and building panels. Both rigid and flexible PVC can be foamed to give lightweight structural panels, and upholstery for cars and domestic use. Blending with other polymers extends the range of properties further: vinyl gramophone records were made of a vinyl chloride/acetate co-polymer; blow molded bottles and film are a vinyl chloride/acrylic co-polymer.

Design Notes In its pure form, PVC is heavy, stiff and brittle. Plasticizers can transform it from a rigid material to one that is almost as elastic and soft as rubber. Plasticized PVC is used as a cheap substitute for leather, which it can be made to resemble in color and texture. It is less transparent than PMMA or PC, but it also costs much less, so it is widely used for transparent, disposable containers. PVC is available as film, sheet or tube. It can be joined with polyester, epoxy or polyurethane adhesives. It has excellent resistance to acids and bases and good barrier properties to atmospheric gases, but poor resistance to some solvents.

Typical Uses tpPVC: pipes, fittings, profiles, road signs, cosmetic packaging, canoes, garden hoses, vinyl flooring, windows and cladding, vinyl records, dolls, medical tubes. elPVC: artificial leather, wire insulation, film, sheet, fabric, car upholstery.

Competing Materials Polyethylene and polypropylene, PTFE for high-performance roofing.

The Environment The vinyl chloride monomer is thoroughly nasty stuff, leading to pressure to discontinue production. But properly controlled, the processing is safe, and the polymer PVC has no known harmful effects. Disposal, however, can be a problem: thermal degradation releases chlorine, HCl and other toxic compounds, requiring special high-temperature incineration for safety.

Technical Notes PVC can be a thermoplastic or a thermoset. There are many types of PVC: expanded rigid PVC, type I, type II, CPVC, acrylic/PVC blend, clear PVC.

Polyurethane (PU)

Polyurethane-coated polyester fabric

What is it? Think of polyurethanes and you think of soft, stretchy, materials and fabrics (Lycra or Spandex), but they can also be leathery or rigid. Like PVC, polyurethanes have thermoplastic, elastomeric and thermosetting grades. They are easily foamed; some 40% of all PU is made into foam by mixing it with a blowing agent. The foams can be open- or closed-cell, microcellular or filter grades. PU is a versatile material.

Design Notes PU foams are cheap, easy to shape, and have good structural performance and resistance to hydrocarbons. Most foamable PUs are thermosets, so they are shaped by casting rather than heat-molding, giving a high surface finish and the potential for intricate shapes. In solid form PUs can be produced as sheet or bulk shapes. For load-bearing applications as power-transmission belts and conveyer belts tpPUs are reinforced with nylon or aramid fibers, giving flexibility with high strength. tpPUs can have a wide range of hardness, softening point and water absorption. They are processed in the same way as nylon, but are considerably more expensive. tpPU fibers are hard, wiry and have a low softening point compared to nylon; they have been used as bristles on brushes. elPU fibers are much more common – they are used in clothing and flexible products under the trade-name of Spandex or Lycra. elPU foams are used for mattresses, seating of furniture and packaging; more rigid foams appear as crash protection in cars, and, in low density form (95% gas) as insulation in refrigerators and freezers. These flexible resins are good in laminate systems where damping is required. elPU is amorphous, tpPU is crystalline; elPU is com-monly cast or drawn, tpPU is commonly injection molded or extruded. Polyurethane elastomers have exceptional strength (up to 48 MPa) and abrasion resistance, low compression set and good fuel resistance. They have useful properties from -55 C to 90 C. PU foam is usually processed by reaction injection molding: the resin and hardener are mixed and injected into a mold where they react and set. PU can be bonded with polyurethane, nitrile, neoprene, epoxies and cyan-acrylates adhesives. It has good resistance to hydrocarbons, degrades in many solvents and is slow burning in fire.

Typical Uses Cushioning and seating, packaging, running shoe soles, tires, wheels, fuel hoses, gears, bearings, wheels, car bumpers, adhesives, fabric-coatings for inflatables, transmission belts, diaphragms, coatings that are resistant to dry-cleaning, furniture, thermal insulation in refrigerators and freezers; as elastomers: truck wheels, shoe heals, bumpers, conveyor belts and metal forming dyes.

Competing Materials Nylon, acetal, natural rubber, polystyrene foams.

The Environment PU is synthesized from diisocyanate and a polyester or polyether. The diisocyanate is toxic, requiring precautions during production. PU itself is inert and non-toxic. The flammability of PU foam, and the use of CFCs as blowing agents in the foaming process were, at one time, a cause for concern. New flame retardants now mean that PU foams meet current fire safety standards, and CFCs have been replaced by CO_2 and HFCs which do not have a damaging effect on the ozone layer.

Thermoplastic PUs can be recycled (thermosetting PUs cannot), and when all useful life is over, incinerated to recover heat. Legislation for return of packaging and disposal problems may disadvantage PU.

Technical Notes Almost all polyurethanes are co-polymers of linked polyester, alcohols and isocyanate groups. Depending on the mix, polyurethanes can be soft and elastic (Lycra, Spandex) or nearly rigid (track-shoe soles, floor tiles), making PU one of the most versatile of polymers.

Attributes of PU

	tpPU	elPU	tsPU
Price, $/kg	5.44–6.01	2.86–4.01	2.15–2.43
Density, Mg/m^3	1.12–1.24	1.02–1.25	1.04–1.06

Technical Attributes

	tpPU	elPU	tsPU
El. modulus, GPa	1.31–2.07	0.002–0.03	4.1–4.3
Elongation, %	60–550	3.8–7.2	3–6
Fr. toughness, MPam$^{1/2}$	1.84–4.97	0.2–0.4	1.38–1.65
Vickers hardness, H_v	16–22		16–18
Yld. strength, MPa	40–53.8	55.2–60.6	
Service temp., C	-30–80	-55–90	-30–100
Specific heat, J/kg·K	1554–1616	1650–1700	1685–1752
Th. conduct., W/m·K	0.24–0.24	0.28–0.3	0.30–0.32
Th. expansion, 10^{-6}/K	90–144	150–165	90–91

Eco-Attributes

	tpPU	elPU	tsPU
Energy content, MJ/kg	110–118	130–142	105–112
Recycle potential	High	Low	Low

Aesthetic Attributes

	tpPU	elPU	tsPU
Low to high pitch, 0–10	6–7	2–4	7–7
Muffled to ringing, 0–10	3–4	1–1	4–4
Soft to hard, 0–10	7–7	3–4	7–7
Warm to cool, 0–10	4–5	5–5	5–5
Gloss, %	100–100		
Transparent to opaque			

Features Relative to Other Polymers

	tpPU	elPU	tsPU
Cold temperatures	✓		✓
Corrosion resistant	✗		✗
Damping	✓		
Elastic	✓		
Fatigue resistant	✓		
Flame retardant	✓	✓	
Hot temperatures	✓	✓	
Impact resistant	✗	✓	
Resilient	✓		
Strong	✓	✓	
Wear resistant	✓	✓	

Polyurethane

Silicones

What is it? Silicones are high-performance, high cost materials. They have poor strength, but can be used over an exceptional range of temperature (-100–300 C), have great chemical stability, and an unusual combination of properties (Silly Putty is a silicone elastomer – it bounces when dropped but flows if simply left on the desk).

Design Notes Silicones are the most chemically stable of all elastomers, with good electrical properties, but relatively low strength (up to 5 MPa). Silicone resins are expensive and they are difficult to process. They have properties like those of natural rubber, but have a completely different chemical structure. Glass fibers and other fillers are commonly used as reinforcement. The resulting parts are relatively low in strength but have high heat resistance. For glass fiber composites, the mechanical properties are better with a phenolic or melamine resin, but the electrical properties are better with silicone. Electrical and high temperature applications dominate their use. They are chemically inert, do not absorb water and can be used in surgical or food processing equipment and seals. Silicones can be produced as fluids, adhesives, coatings, elastomers, molding resins and release agents. But each suffers from a short shelf life (3–6 months). Silicone fluids were the earliest commercial silicones, used as lubricants over a wide range of temperature (-75–450 C). Silicone adhesives can be made as liquids or pastes.

RTV silicone was first developed for its rapid mold filling – a few seconds at high temperatures. Silicone elastomers can be air-curing, cold-curing (by the addition of a catalyst) or heat-curing; they may be pure or loaded with carbon black to give conductivity. Silicone molding resins are compounded with inert fillers to allow the production of flexible parts with high heat resistance.

Typical Uses Wire and cable insulation, mold release agents, lens cleaning tissue coatings, seals, gaskets, adhesives, o-rings, insulation, encapsulation and potting of electronic circuitry, surgical and food processing equipment, baby bottle tips, breast implants, swimming caps and seals for swimming goggles.

Competing Materials Polyurethanes, EVA.

The Environment Silicones are energy intensive – although they are not oil-derivatives. They cannot be recycled.

Technical Notes Silicone and fluoro-silicone elastomers have long chains of linked O-Si-O-Si- groups (replacing the -C-C-C-C- chains in carbon-based elastomers), with methyl (CH_3) or fluorine (F) side chains. Some are relatively stiff solids, others are elastomers.

Rolled silicone fabric

Attributes of Silicone

Price, \$/kg	7.20–17.20
Density, Mg/m^3	1.1–2.3

Technical Attributes

El. modulus, GPa	0.001–0.05
Elongation, %	0.8–5.3
Fr. toughness, MPa·m$^{1/2}$	0.03–0.7
Yld. strength, MPa	2.4–5.5
Service temp., C	-100–300
Specific heat, J/kg·K	1050–1300
Th. conduct., W/m·K	0.2–2.55
Th. expansion, 10^{-6}/K	250–300

Eco-Attributes

Energy content, MJ/kg	175–190
Recycle potential	Low

Aesthetic Attributes

Low to high pitch, 0–10	1–4
Muffled to ringing, 0–10	0–1
Soft to hard, 0–10	3–4
Warm to cool, 0–10	4–6
Transparent to opaque	

Features Relative to Other Polymers

- ✓ Cold temperatures
- ✓ Corrosion resistant
- ✓ Damping
- ✓ FDA approval
- ✓ Flame retardant
- ✓ Heavy
- ✓ Hot temperatures
- ✗ Impact
- ✓ Resilient
- ✗ Strong
- ✗ Wear resistant

Polyesters (PET, PBT)

Recycled fabric from PET bottles

Polyesters

What is it? The name polyester derives from a combination of "Polymerization" and "esterification." Saturated polyesters are thermoplastic – examples are PET, PBT, PETG; they have good mechanical properties to temperatures as high as 160 C. PET is crystal clear, impervious to water and CO_2, but a little oxygen does get through. It is tough, strong, easy to shape, join and sterilize – allowing reuse. When its first life comes to an end, it can be recycled to give fibers and fleece materials for clothing and carpets. Unsaturated polyesters are thermosets; they are used as the matrix material in glass fiber/polyester composites. Polyester elastomers are resilient and stretch up to 45% in length; they have good fatigue resistance and retain flexibility at low temperatures.

Design Notes There are four grades of thermoplastic polyesters: unmodified, flame retardant, glass-fiber reinforced and mineral-filled. Unmodified grades have high elongation; flame retardant grades are self-extinguishing; glass-fiber reinforced grades (like Rynite) are some of the toughest polymers but there are problems with dimensional stability; and mineral-filled grades are used to counter warping and shrinkage although some strength is lost. The PET used in carbonated drink containers is able to withstand pressure from within, it is recyclable and lighter than glass. The permeability to oxygen is overcome by sandwiching a layer of polyethylvinylidene-alcohol between two layers of PET giving a multi-layer material that can still be blow molded. Polyester can be optically transparent, clear, translucent, white or opaque; the resin is easily colored.

Thermosetting polyesters are the cheapest resins for making glass or carbon fiber composites, but they have lower strength than epoxies. They can be formulated to cure at or above room temperature. Modifications can improve the chemical resistance, UV resistance and heat resistance without too much change in the ease of processing. Polyester elastomers have relatively high moduli and are stronger than polyurethanes. They have good melt flow properties, low shrinkage, good resistance to oils and fuels. Polyester can be made conductive by adding 30% carbon fiber. As a tape, Mylar is used for magnetic sound recording. Unfilled polyester thermosetting resins are normally used as surface coatings but they tend to be brittle. A warning: thermosetting polyester corrodes copper.

Typical Uses Decorative film, metallized balloons, photography tape, videotape, carbonated drink containers, oven-proof cookware, windsurfing sails, credit cards, carpets, clothing, brushes, boats, fishing rods, automobile body panels.

Competing Materials Glass, acrylic, polystyrene, polycarbonate.

The Environment PET bottles take less energy to make than glass bottles of the same volume, and they are much lighter – saving fuel in delivery. Thick-walled bottles can be reused; thin-walled bottles can be recycled – and are, particularly in the US.

Technical Notes Polyesters are made by a condensation reaction of an alcohol like ethyl alcohol (the one in beer) and an organic acid like acetic acid (the one in vinegar). The two react, releasing water, and forming an ester. PET, PBT and PCT are not cross-linked and thus are thermoplastic. The polyesters that are used as the matrix polymer in bulk and sheet molding compounds are thermosets.

Attributes of Polyester		
	tsPolyester	tpPolyester
Price, $/kg	1.44–2.51	1.25–2.5
Density, Mg/m^3	01–1.40	1.19–1.81
Technical Attributes		
El. modulus, GPa	0.30–4.41	1.6–4.4
Elongation, %	2–310	1.3–5
Fr. toughness, $MPam^{1/2}$	1.01–1.70	1.05–9.16
Vickers hardness, H_v	10–21	11–40
Yld. strength, MPa	8–40	30–40
Service temp., C	-30–130	-20–160
Specific heat, J/kg·K	1506–1695	1160–1587
Th. conduct., W/m·K	0.15–0.30	0.28–0.58
Th. expansion, 10^{-6}/K	99–180	115–170
Eco-Attributes		
Energy content, MJ/kg	90–96	89–95
Recycle potential	Low	High
Aesthetic Attributes		
Low to high pitch, 0–10	5–7	6–8
Muffled to ringing, 0–10	2–4	4–6
Soft to hard, 0–10	6–7	7–8
Warm to cool, 0–10	4–5	4–5
Optically clear to opaque		
Features Relative to Other Polymers		
Hot temperatures	✓	✓
Flame retardant	✓	
Fatigue resistant	✓	
Heavy	✓	
Impact resistant	✓	
Resilient	✗	✓
Stiff	✓	
Strong	✗	✓
Tough	✗	✓
UV resistant	✓	
Wear resistant	✓	

Epoxy

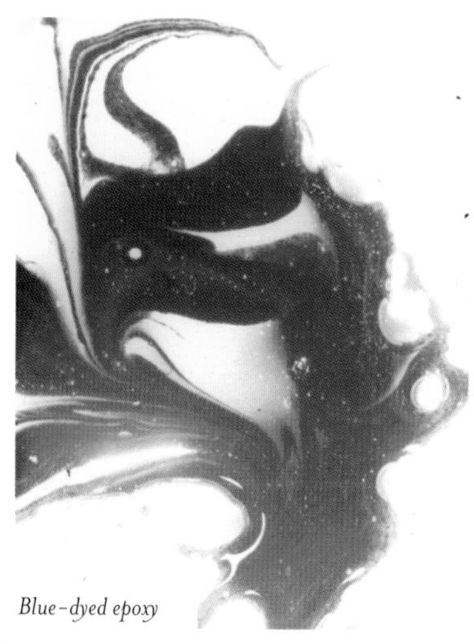
Blue-dyed epoxy

Epoxy

Attributes of Epoxy

Price, $/kg	2.70–3.70
Density, Mg/m³	1.11–1.4

Technical Attributes

El. modulus, GPa	2.35–3.08
Elongation, %	2–10
Fr. toughness, MPa·m$^{1/2}$	0.4–2.22
Vickers hardness, H_V	11–22
Yld. strength, MPa	36–71.68
Service temp., C	-20–175
Specific heat, J/kg·K	1494–2000
Th. conduct., W/m·K	0.18–0.5
Th. expansion, 10^{-6}/K	58–117

Eco-Attributes

Energy content, MJ/kg	120–128
Recycle potential	Low

Aesthetic Attributes

Low to high pitch, 0–10	7–7
Muffled to ringing, 0–10	4–4
Soft to hard, 0–10	7–7
Warm to cool, 0–10	4–5
Gloss, %	62
Transparent to opaque	

Features Relative to Other Polymers

✓ Hot temperatures
✓ Fatigue resistant
✓ Flame retardant
✓ Resilient
✓ Strong
✗ Tough

What is it? Epoxies are thermosetting polymers with excellent mechanical, electrical and adhesive properties and good resistance to heat and chemical attack. They are used for adhesives (Araldite), surface coatings and, when filled with other materials such as glass or carbon fibers, as matrix resins in composite materials. Typically, as adhesives, epoxies are used for high-strength bonding of dissimilar materials; as coatings, they are used to encapsulate electrical coils and electronic components; and, when filled, for tooling fixtures for low-volume molding of thermoplastics.

Design Notes Epoxy molding compounds are supplied in liquid or granular form. They can be shaped by transfer molding at low pressures (350–700 kPa). When designing with epoxy, as with any thermosetting material, allowance must be made for shrinkage; perfectly flat molded surfaces are not achievable, and the minimum wall thickness for average-sized epoxy molded parts is 2.0 mm.

Unmodified epoxies have a high viscosity; they are shaped by transfer molding. Diluted epoxy resins have a lower viscosity and cure slowly, but can be cast or used to impregnate a mat or weave of fibers. The addition of fillers gives epoxies improved machinability, hardness, impact resistance and thermal conductivity; thermal expansion and mold shrinkage are both reduced. Plasticizers and flexibilizers increase flexibility and toughness. Epoxy is also commonly used as a pattern or mold material. Epoxy resin laminates are formed using a wide range of processes, from batch techniques such as hand lay up and bulk molding compound (BMC) molding, producing, for example, mechanical components like gears and distributor caps, to continuous processes such as filament winding, pultrusion and continuous laminating, making rods or girder stock.

Typical Uses Pure epoxy molding compounds: the encapsulation of electrical coils and electronics components; epoxy resins in laminates: pultruded rods, girder stock, special tooling fixtures, mechanical components such as gears; adhesives, often for high-strength bonding of dissimilar materials; patterns and molds for shaping thermoplastics.

Competing Materials Phenolic, polyester, silicone.

The Environment Both resin and hardener are irritants; their vapors are potentially toxic. Ventilation and skin protection are important, but both are achievable. Thermosets cannot be recycled, though it may be possible to use them as fillers. Cutting and machining of glass and carbon fiber composites requires special forced-air ventilation to remove the fine glass or carbon dust that is damaging if inhaled.

Technical Notes Most epoxies are formed by the combination of bisphenol-A and epichlorohydrin in the presence of a catalyst. Catalysts include amines and acid anhydrides. The curing temperature, ranging from room- to high-temperature, is determined by the type of catalyst, which also affects the properties of the final product.

Phenolic

What is it? Bakelite, commercialized in 1909, triggered a revolution in product design. It was stiff, fairly strong, could, to a muted degree, be colored, and – above all – was easy to mold. Products that, earlier, were hand-crafted from woods, metals or exotics such as ivory, could now be molded quickly and cheaply. At one time the production of phenolics exceeded that of PE, PS and PVC combined. Now, although the ratio has changed, phenolics still have a unique role. They are stiff, chemically stable, have good electrical properties, are fire-resistant and easy to mold – and they are cheap.

Design Notes Phenolic resins are hard, tolerate heat and resist most chemicals except the strong alkalis. Phenolic laminates with paper have excellent electrical and mechanical properties and are cheap; filled with cotton the mechanical strength is increased and a machined surface is finer; filled with glass the mechanical strength increases again and there is improved chemical resistance. Fillers play three roles: extenders (such as wood flour and mica) are inexpensive and reduce cost; functional fillers add stiffness, impact resistance and limit shrinkage; reinforcements (such as glass, graphite and polymer fibers) increase strength, but cost increases too. Unfilled phenolics are susceptible to shrinkage when exposed to heat over time, glass-filled resins are less susceptible. They have good creep resistance, and they self-extinguish in a fire.

Phenolics can be cast (household light and switch fittings) and are available as rod and sheet. Impregnated into paper (Nomex) and cloth (Tufnol), they have exceptional durability, chemical resistance and bearing properties. Phenolics accept paint, electroplating, and melamine overlays.

Typical Uses Switchboards, insulating washers, intricate punched parts (phenolic with paper laminate), gears, pinions, bearings, bushings (phenolic with cotton laminate), gaskets and seals (phenolic with glass), used to bond friction materials for automotive brake linings, beads, knife handles, paperweights, billiard balls, domestic plugs and switches, telephones, fuse box covers, distributor heads, saucepan handles and knobs, golf ball heads for typewriters, toilet seats. As a foam, phenolic resin is used in paneling for building work; its fine resistance is a particular attraction.

Competing Materials Epoxy, polyester, silicone.

The Environment Phenolics, like all thermosets, cannot be recycled.

Technical Notes Phenolic resins are formed by a condensation, generating water in the process, involving a reaction between phenol and formaldehyde to form the A-stage resin. Fillers, colorants, lubricants and chemicals to cause cross-linking are added to form the B-stage resin. This resin is then fused under heat and pressure converting to the final product – a C-stage resin – or completely cross-linked polymer.

Phenolic laminates

Attributes of Phenolic	
Price, $/kg	0.90–1.20
Density, Mg/m³	1.24–1.32

Technical Attributes	
El. modulus, GPa	2.76–4.83
Elongation, %	1.5–2
Fr. toughness, MPa·m$^{1/2}$	0.79–1.21
Vickers hardness, H_v	8–15
Yld. strength, MPa	27.6–49.68
Service temp., C	-20–160
Specific heat, J/kg·K	1467–1526
Th. conduct., W/m·K	0.141–0.152
Th. expansion, 10^{-6}/K	120.1–124.9

Eco-Attributes	
Energy content, MJ/kg	94–100
Recycle potential	Low

Aesthetic Attributes	
Low to high pitch, 0–10	7–7
Muffled to ringing, 0–10	4–4
Soft to hard, 0–10	7–7
Warm to cool, 0–10	4–4
Opaque	

Features

- ✓ Corrosion resistant
- ✓ Flame retardant
- ✓ Hot temperatures
- ✓ Stiff
- ✓ Strong
- ✗ Tough
- ✓ UV resistant
- ✓ Wear resistant

Phenolic

Elastomers

Latex on linen fabric

Natural Rubber

Natural rubber was known to the natives of Peru many centuries ago, and is now one of Malaysia's main exports. It made the fortune of Giles Macintosh who, in 1825, devised the rubber-coated waterproof coat that still bears his name. Latex, the sap of the rubber tree, is cross-linked (vulcanized) by heating with sulfur; the amount of the cross-linking determines the properties. It is the most widely used of all elastomers – more than 50% of all produced.

- *Design Notes* Natural rubber is an excellent, cheap, general purpose elastomer with large stretch capacity and useful properties from -50 C to 115 C, but with poor oil, oxidation, ozone and UV resistance. It has low hysteresis – and is thus very bouncy.
- *Typical Uses* Tires, springs, seals, shock-mounts, and toys.

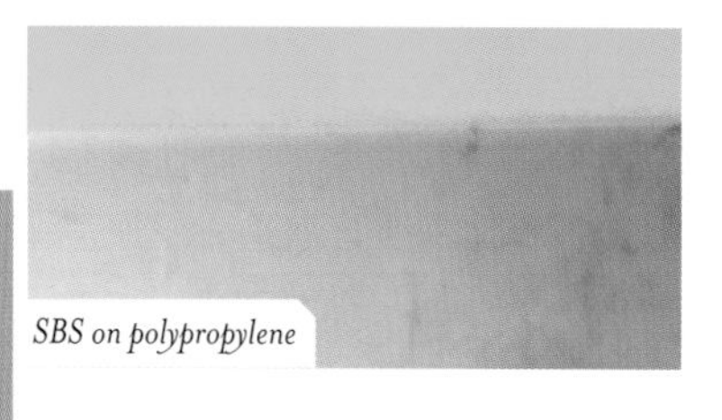
SBS on polypropylene

Styrene-Butadiene Elastomers (SBS, SEBS, SBR, BUNA-S)

Styrene-butadiene elastomers are co-polymers of butadiene and styrene. They exceed all other synthetic rubbers in use-volume.

- *Design Notes* Styrene-butadiene elastomers have low cost, with properties like natural rubber, though lower in strength, thus often requiring reinforcement.
- *Typical Uses* Predominantly tires, hoses and seals.

Cork/butyl composite

Butyl Rubbers (NR)

Butyl rubbers are synthetics that resemble natural rubber in properties.

- *Design Notes* Butyl rubbers have good resistance to abrasion, tearing and flexing, with exceptionally low gas permeability and useful properties up to 150 C. They have low dielectric constant and loss, making them attractive for electrical applications.
- *Typical Uses* Inner tubes, belts, hoses, cable insulation, encapsulation.

Isoprene

Isoprene is synthetic natural rubber, and is processed in the same way as butyl rubber.

- *Design Notes* Isoprene has low hysteresis and high tear resistance, making it bouncy and tough.
- *Typical Uses* The same as butyl rubber.

Acrylate Elastomers

Acrylate elastomers are made from butyl or ethyl acetate.

- *Design Notes* Acrylate elastomers have good oil resistance but low strength and tear resistance; they have useful properties up to 200 C.
- *Typical Uses* Automobile transmission seals, gaskets, o-rings.

Nitrile Elastomers (NBR, BUNA-N)

Nitrile elastomers are co-polymers of acrylonitrile and butadiene.

- *Design Notes* Nitrile eastomers have excellent resistance to oil and fuels, and retain good flexibility both at low (-30 C) and elevated temperatures.
- *Typical Uses* Oil-well parts, automobile and aircraft fuel hoses, flexible couplings, printing blankets, rollers, pump diaphragms.

Elastomers

Polybutadiene Elastomers

Polybutadiene elastomers have exceptional low-temperature performance.

- *Design Notes* Polybutadiene has exceptional resilience and abrasion resistance, retained to -70 C, but with poor chemical resistance.
- *Typical Uses* Generally blended with other polymers to improve low-temperature performance; cores for solid golf balls.

Polysulphide Elastomers

Polysulphide elastomers have the highest resistance to oil and gasoline.

- *Design Notes* Polsulphides have excellent chemical and ageing resistance, low gas permeability, though with poor strength.
- *Typical Uses* Oil and fuel hoses, gaskets, washers, diaphragms, sealing and caulking, and adhesives.

Ethylene-Propylene Elastomers (EP, EPM, EPDM)

Ethylene-Propylene elastomers are co-polymers of the two monomers that make up their name.

- *Design Notes* Ethylene-propylene elastomers have good resilience, minimal compression-set and good resistance to chemicals, weathering and UV. Useful properties up to 177 C.
- *Typical Uses* Electrical insulation, footwear, belts, hoses.

Ethylene-Vinyl-Acetate Elastomers (EVA)

Ethylene-Vinyl-Acetate elastomers are built around polyethylene. They are soft, flexible and tough, and retain these properties down to -60 C. Fillers improve both hardness and stiffness, but with some degradation of other properties. EVAs blend well with PE because of their chemical similarity.

- *Design Notes* EVA is available in pastel or deep hues, it has good clarity and gloss. It has good barrier properties, little or no odor, has UV resistance and FDA-approval for direct food contact. The toughness and flexibility is retained even at low temperatures and it has good stress-crack resistance and good chemical resistance. EVA can be processed by most normal thermoplastic processes: co-extrusion for films, blow molding, rotational molding, injection molding and transfer molding.
- *Typical Uses* Medical tubes, milk packaging, beer dispensing equipment, bags, shrink film, deep freeze bags, co-extruded and laminated film, closures, ice trays, gaskets, gloves, cable insulation, inflatable parts, running shoes.

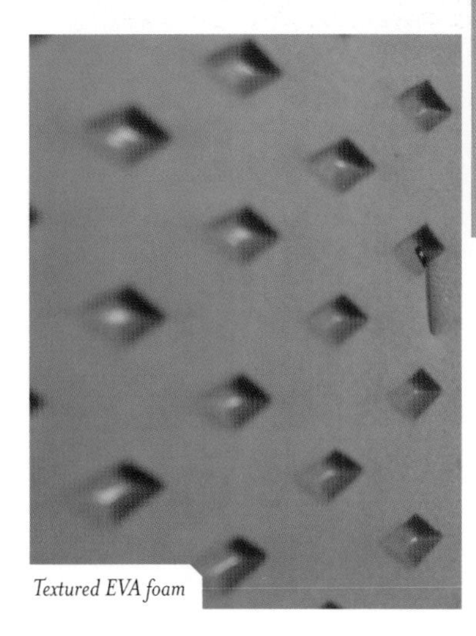

Textured EVA foam

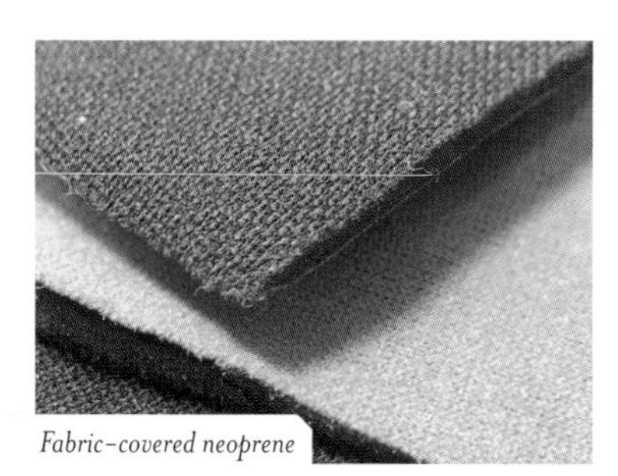
Fabric-covered neoprene

Chlorinated Elastomers (Hypalon, Neoprene)
Chlorinated hydrocarbon and chlorinated polyethylene are the leading non-tire elastomers.

- *Design Notes* Chlorinated hydrocarbons are characterized by exceptional chemical resistance, ability to be colored, and useful properties up to 175 C. Some have low gas permeability and low hysteresis, mini-mizing heating when cyclically loaded, and resist burning.
- *Typical Uses* Tank linings, conveyor and timing belts, shoe soles, O-rings, seals and gaskets, cable jackets, vibration control mounts, diaphragms, tracked vehicle pads, under-hood hoses and tubing, footwear, wetsuits.

Fluoro-Carbon Elastomers (Viton, Fluorel, Kalres and Perfluoro)
Fluoro-Carbon elastomers are based on polyolefins with some of the hydrogen atoms replaced by fluorine. They are heavier than simple hydrocarbons, but much more resistant to heat and chemicals.

- *Design Notes* Fluorocarbon elastomers have exceptional resistance to chemicals, oxidation, solvents and heat – but they are also expensive. They have useful properties up to 250 C, but become brittle below -20 C and have only modest strengths.
- *Typical Uses* Brake seals, diaphragms, hoses and O-rings.

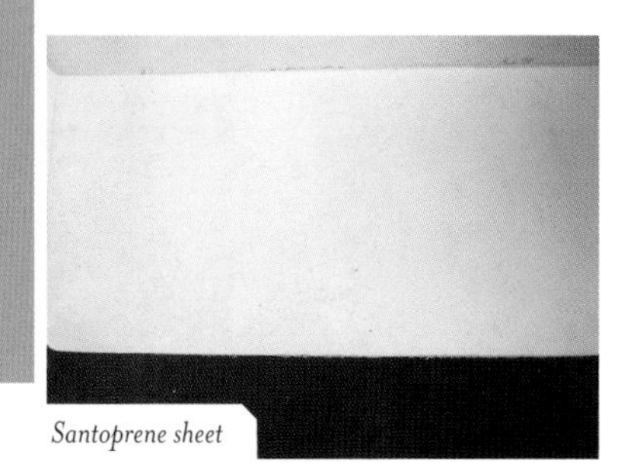
Santoprene sheet

Thermoplastic Elastomers (TPE)
Thermoplastic Elastomers are exceptional in that they can be molded and extruded in standard polymer processing equipment, their scrap can be remelted, and products made from them can be recycled. They include styrene co-polymers (Kraton), polyurethanes (Hytrol, Ritefeox, Ecdel) and polypropylene blended with elastomers (Santoprene, Geolast).

- *Design Notes* TPEs allow rapid processing by standard thermoplastic methods, and recyclability.
- *Typical Uses* Bumpers, sports shoes, hoses, diaphragms, rollers, seals for automotive and architectural use.

Attributes of Selected Elastomers

	EVA	Polychloroprene	SBS
Price, $/kg	1.15–1.29	1.4–4.29	2.50–2.70
Density, Mg/m^3	0.93–0.96	1.23–1.25	0.94–0.95

Technical Attributes

El. modulus, GPa	0.007–0.09	0.001–0.02	.002–0.01
Elongation, %	730–900	100–800	450–500
Fr. toughness, MPa·m$^{1/2}$	0.1–0.7	0.1–0.3	0.1–0.3
Yld. strength, MPa	9.5–19	3.4–24	12–21
Service temp., C	-20–60	-50–110	-50–120
Specific heat, J/kg·K	1900–2200	2000–2200	2000–2200
Th. conduct., W/m·K	0.3–0.4	0.1–0.12	0.14–1.49
Th. expansion, 10^{-6}/K	160–190	605–625	660–675

Eco-Attributes

Energy content, MJ/kg	95–101	115–124	105–111
Recycle potential	Low	Low	Low

Aesthetic Attributes

Low to high pitch, 0–10	3–5	1–2	2–3
Muffled to ringing, 0–10	1–2	0–1	0–1
Soft to hard, 0–10	4–4	3–3	3–4
Warm to cool, 0–10	5–5	4–4	4–6
Gloss, %	17–123		
Transparent to opaque			

Features Relative to Other Polymers

Cold temperatures	✓		
Damping	✓	✓	✓
Fatigue resistant	✗	✗	✗
Flame retardant	✓	✓	
Hot temperatures	✓	✓	
Light	✓	✓	
Resilient	✓	✓	✓
Stiff	✓		
Strong	✗	✗	✗
Tear resistant	✓	✓	
Tough	✓	✗	✗
Wear resistant	✓		

Polymer Foams

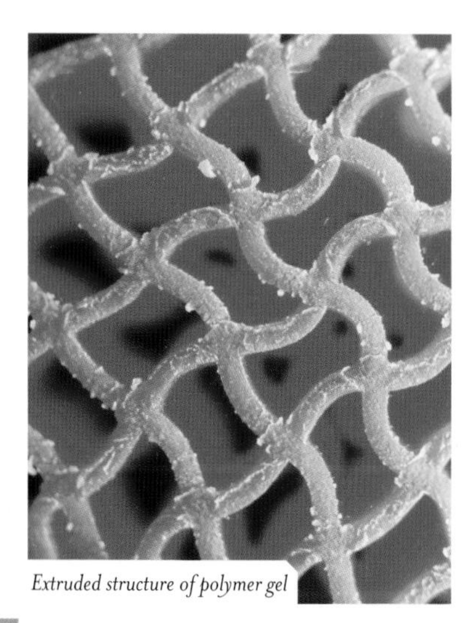
Extruded structure of polymer gel

What are they? Solid materials are much too solid for comfort and safety. If you want to relax in cushioned luxury and avoid self-inflicted injury when you bump into things, then surround yourself with foams. Foams are made by the controlled expansion and solidification of a liquid or melt through a blowing agent; physical, chemical or mechanical blowing agents are possible. The resulting cellular material has a lower density, stiffness and strength than the parent material, by an amount that depends on its relative density – the volume-fraction of solid in the foam. Foams can be soft and flexible, but they can also be hard and stiff. Open-celled foams absorb fluids and can be used as filters; closed cell foams give buoyancy. Self-skinning foams, called "structural" or "syntactic," have a dense surface skin made by foaming in a cold mold.

Design Notes Polymer foams have characteristics that suit them for cushioning, energy management, acoustic control or thermal insulation. For cushioning, the requirements are comfort and long life; polyurethane foams have been commonly used, but concerns about flammability and durability limit their use in furniture. Energy management and packaging requires the ability to absorb energy at a constant, controlled crushing stress; here polyurethane, polypropylene and polystyrene foams are used. Acoustic control requires the ability to absorb sound and damp vibration; polyurethane, polystyrene and polyethylene foams are all used. Thermal insulation requires long life; polyurethane foams were common but are now replaced by phenolics and polystyrenes. When fire-protection is needed phenolic foams are used.

Microcellular elastomeric foams with a small cell size and integral skin are used for shock and vibration control. Examples are expanded EVA and PVC, widely used in shoes. Foamed extrusions have less design restrictions than solid profiles because shrinkage is negligible. The swirled pattern on the surfaces of self-skinning foamed parts, caused by the foaming action, is eliminated by the use of high pressure or by painting.

Structural foam components have high stiffness-to-weight ratios, good sound absorption, high impact strength but poor tensile strength. Stiffening ribs, bosses and handles can be included in parts without causing molding problems. Reinforcement with glass fibers is possible and improves the structural performance but the ease of molding is diminished.

Polymers can be foamed in place, injection molded, extruded, thermoformed, laminated or cast. Low and high-pressure injection molding uses pellets of thermoplastic material to which a blowing agent is added. Reaction injection molding is used with thermosetting polymers, usually polyurethane. Casting of foams, particularly urethanes, uses cheap silicone rubber or epoxy molds, making low-volume production practical. Polyurethanes, epoxies, phenolics and polyesters can all be cast and reaction-injection molded. Foaming reduces material usage and increases bending stiffness without increasing weight.

Polyethylene foam is light, has negligible water absorption and can be used as sheet or film; it is used in thermal and wire insulation and high-quality packaging. Polystyrene foam, much used in packaging, is processed by expanded foam molding, extrusion, injection molding, compression molding, blow molding or thermoforming. Polyurethane, the most widely used elastomer foams, is self-skinning; but it does not age well, cannot be recycled and is no friend of the environment. Developments in polyurethane foam technology, like those of ICI Waterlily, overcome these weaknesses. The most common structural foams are based on ABS, polystyrene, PVC, polyurethane, polyethylene and phenolic.

Typical Uses Rigid foams: Computer housings, pallets, battery cases, chair frames, TV cabinets, oars, dishwasher tubs, automotive fenders, luggage, cheap boats and recreational vehicle bodies, floatation, filtration, thermal insulation. Flexible foams: cushions, packaging, padding, seating.

Attributes of Polymer Foams		
	Flexible foams	Rigid foams
Price, $/kg	1–9	2–50
Density, Mg/m^3	0.03–0.1	0.07–0.5
Technical Attributes		
El. modulus, GPa	0.0001–0.01	0.05–0.5
Elongation, %	10–100	2–10
Fr. toughness, $MPam^{1/2}$	0.002–0.1	0.01–0.1
Vickers hardness, H_V	–	0.1–1.2
Yld. strength, MPa	0.02–0.07	0.4–12
Service temp., C	70–110	80–150
Specific heat, J/kg·K	1750–2260	1120–1910
Th. conduct., W/m·K	0.04–0.08	0.03–0.6
Th. expansion, 10^{-6}/K	115–220	20–70
Eco-Attributes		
Energy content, MJ/kg	150–190	150–180
Recycle potential	low	low
Aesthetic Attributes		
Low to high pitch, 0–10	1–2	3–4
Muffled to ringing, 0–10	1–2	3–4
Soft to hard, 0–10	1–2	3–4
Warm to cool, 0–10	8–9	8–9
Translucent to opaque		
Features Relative to Other Polymers		
Cold temperatures	✗	✓
Corrosion resistant	✗	✗
Ductile	✗	✗
Fatigue resistant	✗	✗
Hot temperatures	✗	✓
Impact resistant	✗	✗
Resilient	✓	✓
Stiff	✗	✗
Strong	✗	✗
Tough	✗	✗
Wear resistant	✗	✗

Polymer Composites

Polyurethane-coated glass fibers

Polymers/Glass Fiber (GFRP), Fiberglass

GFRP, the first of the modern fiber composites, was developed after WWII as a lightweight material with high strength and stiffness. It is made of continuous or chopped glass fibers in an polymer matrix – usually polyester or epoxy. Epoxy is more expensive than polyester, but gives better properties. Glass fibers are much cheaper than carbon or Kevlar, and therefore are widely used despite having lower stiffness than carbon and lower strength than Kevlar (glass accounts for 98% of the fiber market).

- *Design Notes* The stiffness of GFRP is limited by the relatively low modulus of the glass fibers. The matrix material limits the operating environment to less than 120 C. Moisture can degrade properties. Short, discontinuous fibers can give a tensile strength of almost 50% that of the corresponding continuous fiber at a considerably reduced processing cost. GFRPs can be filled and colored to order. To reduce the possibility of delamination, cracking and splitting, holes, abrupt section changes and sharp corners should be avoided. There are several grades of glass fiber, differing in composition and properties. E-glass is the standard reinforcement. C-glass has better corrosion resistance than E; R and S have better mechanical properties than E but are more expensive. AR-glass resists alkalis, allowing it to be used to reinforce cement. Thermosetting resins such as polyesters and epoxies are usually supplied as viscous syrups that wet and penetrate the fibers, setting when mixed with a catalyst. Thermoplastic resin such as polypropylene or polyamide are solids that soften on heating; the difficulty is to get them to mix well with the glass.
- *Typical Uses* Boat hulls, train seats, pressure vessels, automobile components, sports equipment, bathroom and kitchen fixtures.

Polymer Composites

Nylon coated carbon fiber

Polymers/Carbon Fiber (CFRP)

CFRPs are the materials of sport and aerospace. They are made of carbon fibers embedded in a polymer matrix, usually epoxy or polyester. Carbon fibers have very high strength and stiffness, and they are light – much lighter than glass. This makes CFRPs the composite-choice for aerospace, structural and other applications that require high performance. Epoxy is more expensive than the competing polyester, but gives better properties and better adhesion to the fibers, giving the composite excellent fatigue resistance. CFRP is increasingly used to replace aluminum in aircraft structures in order to reduce weight – the US Air Force has been flying aircraft with structural CFRP components since 1972.

- *Design Notes* Carbon has a higher modulus than glass and is much more expensive, limiting CFRP, to applications in which high stiffness is important. It is usually black because of the color of the fibers, but the matrix can be tinted. To reduce the possibility of delamination, cracking and splitting, holes, abrupt section changes and sharp corners should be avoided. CFRPs have good creep and fatigue resistance, low thermal expansion, low friction and wear, vibration-damping characteristics, and environmental stability. They can be electrically conducting, making them opaque to radio waves. They are used mainly as continuous fiber reinforcement for epoxy resins, and, more recently, in thermoplastic resins such as PP or PEEK. The fibers come in four grades: high modulus (HM), high strength (HS), ultra high modulus (UHM) and ultra high strength (UHS) – with cost increasing in that order. Thermosetting resins such as polyesters and epoxies are usually supplied as viscous syrups that wet and penetrate the fibers, setting when mixed with a catalyst. As with CFRP, the matrix limits the maximum use temperature of the composite to 120 C, and it can absorb water – they are not good in hot, wet environments.

· *Typical Uses* Aircraft structures, power boats, racing cars, tennis rackets, fishing rods, golf club shafts, performance bicycles.

Attributes of Quasi-isotropic Polymer Composites

	GFRP	CFRP	KFRP
Price, $/kg	9–20	50–61	120–140
Density, Mg/m³	1.75–1.95	1.55–1.6	1.37–1.40

Technical Attributes

	GFRP	CFRP	KFRP
El. modulus, GPa	21–35	50–60	23–30
Elongation, %	0.8–1	0.3–0.35	0.3–0.4
Fr. toughness, MPam$^{1/2}$	6–25	6–15	6–40
Vickers hardness, H_V	12–22	12–23	10–21
Yld. strength, MPa	200–500	500–1050	130–150
Service temp., C	140–220	140–220	120–200
Specific heat, J/kg·K	1000–1200	900–1000	510–680
Th. conduct., W/m·K	0.4–0.5	1.3–2.6	0.20–0.26
Th. expansion, 10^{-6}/K	8.6–11	2–4	9.4–15

Eco-Attributes

	GFRP	CFRP	KFRP
Energy content, MJ/kg	250–300	600–800	400–500
Recycle potential	Low	Low	Low

Aesthetic Attributes

	GFRP	CFRP	KFRP
Low to high pitch, 0–10	5–6	7–8	5–6
Muffled to ringing, 0–10	4–5	5–6	4–5
Soft to hard, 0–10	6–7	7–8	6–7
Warm to cool, 0–10	5–6	5–6	5–6
Opaque			

Features Relative to Other Polymers

	GFRP	CFRP	KFRP
Corrosion resistant	✓	✓	✓
Ductile	✗	✗	✗
Fatigue resistant	✓	✓	✓
Hot temperatures	✓	✓	✓
Impact resistant	✓	✓	✓
Resilient	✓	✓	✓
Stiff	✓	✓	✓
Strong	✓	✓	✓
Tough	✓	✓	✓
Wear resistant	✓	✓	✓

Carbon Steels

Carbon Steels

Perforated mild steel

Attributes of Carbon Steels

Price, $/kg	0.40–0.60
Density, Mg/m³	7.8–7.9

Technical Attributes

El. modulus, GPa	200–216
Elongation, %	4–47
Fr. toughness, MPa·m$^{1/2}$	12–92
Vickers hardness, H_v	120–650
Yld. strength, MPa	250–1755
Service temp., C	-70–360
Specific heat, J/kg·K	440–520
Th. conduct., W/m·K	45–55
Th. expansion, 10^{-6}/K	10–14

Eco-Attributes

Energy content, MJ/kg	57–72
Recycle potential	High

Aesthetic Attributes

Low to high pitch, 0–10	9
Muffled to ringing, 0–10	6–7
Soft to hard, 0–10	9
Warm to cool, 0–10	9
Reflectivity	59
Opaque	

Features Relative to Other Metals

- ✗ Corrosion resistant
- ✓ Ductile
- ✓ Fatigue resistant
- ✓ Stiff
- ✓ Strong
- ✓ Tough
- ✓ Wear resistant

What are they? Think of steel and you think of railroads, oilrigs, tankers, and skyscrapers. And what you are thinking of is not just steel, it is carbon steel. That is the metal that made them possible – nothing else is at the same time so strong, so tough, so easily formed – and so cheap.

Design Notes Carbon steels are alloys of iron with carbon. Low carbon or "mild" steels have the least carbon – less than 0.25%. They are relatively soft, easily rolled to plate, I-sections or rod (for reinforcing concrete) and are the cheapest of all structural metals – it is these that are used on a huge scale for reinforcement, for steel-framed buildings, ship plate and the like. Medium carbon steel (0.25–0.5% carbon) hardens when quenched – a quality that gives great control over properties. "Hardenability" measures the degree to which it can be hardened in thick sections; plain carbon steels have poor hardenability – additional alloying elements are used to increase it (see Low alloy steels). Medium carbon steels are used on an enormous scale for railroad tracks; there are many other lower-volume applications. High carbon steels (0.5–1.6% carbon) achieve even greater hardness, sufficient for them to be used as cutting tools, chisels and cables, and "piano wire" – the metal strings of pianos and violins. More carbon than that and you have cast iron (1.6–4% carbon), easy to cast, but with properties that are less good than steel.

Typical Uses Low carbon steels are used so widely that no list would be complete. Reinforcement of concrete, steel sections for construction, sheet for roofing, car body panels, cans and pressed sheet products give an idea of the scope. Medium carbon steels are the materials of general construction and engineering, axles and gears, bearings, cranks and shafts. High carbon steels are used for cutting tools, high performance bearings, cranks and shafts, springs, knives, ice axes and ice skates.

Competing Materials In many applications carbon steels have no credible competition. Where higher hardenability and performance are sought, low alloy steels displace carbon steels. Where extreme corrosion resistance is necessary, stainless steels take over – but they are expensive.

The Environment The production energy of steel is comparatively low – per unit weight, about a half that of polymers; per unit volume, though, twice as much. Carbon steels are easy to recycle, and the energy to do so is small.

Technical Notes The two standard classifications for steels, the AISI and the SAE standards, have now been merged. In the SAE-AISI system, each steel has a four-digit code. The first two digits indicate the major alloying elements. The second two give the amount of carbon, in hundredths of a percent. Thus the plain carbon steels have designations starting 10xx, 11xx, 12xx or 14xxx, depending on how much manganese, sulfur and phosphorus they contain. The common low-carbon steels have the designations 1015, 1020, 1022, 1117, 1118; the common medium carbon steels are 1030, 1040, 1050, 1060, 1137, 1141, 1144 and 1340; the common high carbon steels are 1080 and 1095.

Stainless Steels

What are they? Stainless steels are alloys of iron with chromium, nickel, and – often – four of five other elements. The alloying transmutes plain carbon steel that rusts and is prone to brittleness below room temperature into a material that does neither. Indeed, most stainless steels resist corrosion in most normal environments, and they remain ductile to the lowest of temperatures.

Design Notes Stainless steel must be used efficiently to justify its higher costs, exploiting its high strength and corrosion resistance. Economic design uses thin, rolled gauge, simple sections, concealed welds to eliminate refinishing, and grades that are suitable to manufacturing (such as free machining grades when machining is necessary). Surface finish can be controlled by rolling, polishing or blasting. Stainless steels are selected, first, for their corrosion resistance, second, for their strength and third, for their ease of fabrication. Most stainless steels are difficult to bend, draw and cut, requiring slow cutting speeds and special tool geometry. They are available in sheet, strip, plate, bar, wire, tubing and pipe, and can be readily soldered and braised. Welding stainless steel is possible but the filler metal must be selected to ensure an equivalent composition to maintain corrosion resistance. The 300 series are the most weldable; the 400 series are less weldable.

Typical Uses Railway cars, trucks, trailers, food-processing equipment, sinks, stoves, cooking utensils, cutlery, flatware, architectural metalwork, laundry equipment, chemical-processing equipment, jet-engine parts, surgical tools, furnace and boiler components, oil-burner parts, petroleum-processing equipment, dairy equipment, heat-treating equipment, automotive trim.

Competing Materials Carbon steel, plated or painted; nickel based alloys; titanium and its alloys.

The Environment Stainless steels are FDA approved – indeed, they are so inert that they can be implanted in the body. All can be recycled.

Technical Notes Stainless steels are classified into four categories: the 200 and 300 series austenitic (Fe-Cr-Ni-Mn) alloys, the 400 series ferritic (Fe-Cr) alloys, the martensitic (Fe-Cr-C) alloys that also form part of the 400 series, and precipitation hardening or PH (Fe-Cr-Ni-Cu-Nb) alloys with designations starting with S. Austenitic stainless steels have excellent resistance to pitting and corrosion, and are non-magnetic (a way of identifying them). Typical is the grade 304 stainless steel: 74% iron, 18% chromium and 8% nickel. Here the chromium protects by creating a protective Cr_2O_3 film on all exposed surfaces. Nickel stabilizes face-centred cubic austenite, giving ductility and strength both at high and low temperatures.

Woven stainless steel wire

Attributes of Stainless Steels

Price, \$/kg	1.20–8.50
Density, Mg/m^3	7.4–8.1

Technical Attributes

El. modulus, GPa	189–210
Elongation, %	0.5–70
Fr. toughness, MPa·m$^{1/2}$	12–280
Vickers hardness, H_v	130–600
Yld. strength, MPa	170–2090
Service temp., C	-270–850
Specific heat, J/kg·K	400–530
Th. conduct., W/m·K	11–29
Th. expansion, 10^{-6}/K	9–20

Eco-Attributes

Energy content, MJ/kg	83–115
Recycle potential	High

Aesthetic Attributes

Low to high pitch, 0–10	8–9
Muffled to ringing, 0–10	6–7
Soft to hard, 0–10	9
Warm to cool, 0–10	7–8
Reflectivity	60–80
Opaque	

Features Relative to Other Metals

- ✓ Cold temperatures
- ✓ Corrosion resistant
- ✓ Ductile
- ✓ Fatigue resistant
- ✓ Hot temperatures
- ✓ Impact resistant
- ✓ Stiff
- ✓ Strong
- ✓ Tough
- ✓ Wear resistant

Stainless Steels

Low Alloy Steels

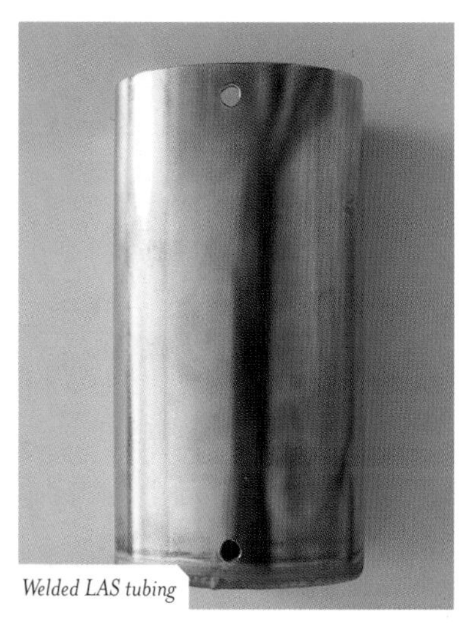
Welded LAS tubing

Attributes of Low Alloy Steels

Price, $/kg	0.40–0.90
Density, Mg/m^3	7.8–7.9

Technical Attributes

El. modulus, GPa	201–217
Elongation, %	3–38
Fr. toughness, MPa·m$^{1/2}$	14–210
Vickers hardness, H_V	140–700
Yld. strength, MPa	245–2255
Service temp., C	-70–660
Specific heat, J/kg·K	410–530
Th. conduct., W/m·K	34–55
Th. expansion, 10^{-6}/K	10.5–13.5

Eco-Attributes

Energy content, MJ/kg	60–83
Recycle potential	High

Aesthetic Attributes

Low to high pitch, 0–10	9–9
Muffled to ringing, 0–10	6–8
Soft to hard, 0–10	9
Warm to cool, 0–10	8–9
Reflectivity	40–60
Opaque	

Features Relative to Other Metals

✗ Corrosion resistant
✓ Ductile
✓ Fatigue resistant
✓ Impact resistant
✓ Resilient
✓ Stiff
✓ Strong
✓ Tough
✓ Wear resistant

What is it? Pure iron is soft stuff. Add carbon and heat-treat it right, and you can get an a material that is almost as hard and brittle as glass, or as ductile and tough as boiler plate. "Heat treat" means heating the steel to about 800 C to dissolve the carbon, then quenching (rapid cooling, often by dropping into cold water) and tempering – reheating it to a lower temperature and holding it there. Quenching turns the steel into hard, brittle "martensite;" tempering slowly restores the toughness and brings the hardness down. Control of tempering time and temperature gives control of properties. It's wonderful what 1% of carbon can do. But (the inevitable "but") the cooling rate in that initial quench has to be fast – more than 200 C/second for plain carbon steels. There is no difficulty in transforming the surface of a component to martensite, but the interior cools more slowly because heat has to be conducted out. If the component is more than a few millimeters thick, there is a problem – the inside doesn't cool fast enough. The problem is overcome by alloying. Add a little manganese, Mn, nickel, Ni, molybdenum, Mo or chromium, Cr, and the critical cooling rate comes down, allowing thick sections to be hardened and then tempered. Steels alloyed for this purpose are called low alloy steels, and the property they have is called "hardenability."

Design Notes Low alloy steels are heat treatable – most other carbon steels are not – and so are used for applications where hardness or strength is an important feature, particularly in large sections. They have greater abrasion resistance, higher toughness and better strength at high temperatures than plain carbon steels. Alloy steels with carbon content of 0.30 to 0.37% are used for moderate strength and great toughness; 0.40–0.42% for higher strength and good toughness; 0.45–0.50% for high hardness and strength with moderate toughness; 0.50–0.62% for hardness (springs and tools); 1% for high hardness and abrasion resistance (ball bearings or rollers).

Typical Uses Springs, tools, ball bearings, rollers; crankshafts, gears, connecting rods.

Competing Materials Carbon steels, stainless steels.

The Environment Steels are not particularly energy intensive to make, and are easily and widely recycled.

Technical Notes The SAE-AISI system for low alloy steels works the same way as that for plain carbon steels. Each steel has a four-digit code; the first two digits indicate the major alloying elements, the second two give the amount of carbon, in hundredths of a percent. Typical are the nickel-chrome-molybdenum steels with the designation 43xx, but the alloying elements can include any of the following: more than 2% silicon, more than 0.4% copper, more than 0.1% molybdenum, more than 0.5% nickel, more than 0.5% chromium.

Aluminum Alloys

What are they? Aluminum was once so rare and precious that the Emperor Napoleon III of France had a set of cutlery made from it that cost him more than silver. But that was 1860; today, nearly 200 years later, aluminum spoons are things you throw away – a testament to our ability to be both technically creative and wasteful. Aluminum, the first of the "light alloys" (with magnesium and titanium), is the third most abundant metal in the earth's crust (after iron and silicon) but extracting it costs much energy. It has grown to be the second most important metal in the economy (steel comes first), and the mainstay of the aerospace industry.

Design Notes Aluminum alloys are light, can be strong, and are easily worked. Pure aluminum has outstanding electrical and thermal conductivity (copper is the only competition here) and is relatively cheap – though still more than twice the price of steel. It is a reactive metal – in powder form it can explode – but in bulk an oxide film (Al_2O_3) forms on its surface, protecting it from corrosion in water and acids (but not strong alkalis). Aluminum alloys are not good for sliding surfaces – they scuff – and the fatigue strength of the high-strength alloys is poor. Nearly pure aluminum (1000 series alloys) is used for small appliances and siding; high strength alloys are used in aerospace (2000 and 7000 series), and extrudable, medium strength alloys are used in the automotive and general engineering sectors (6000 series).

Typical Uses Aerospace engineering; automotive engineering; die cast chassis for household and electronic products; siding for buildings; foil for containers and packaging; beverage cans; electrical and thermal conductors.

Competing Materials Magnesium, titanium, CFRP for lightweight structures; steel for cans; copper for conductors.

The Environment Aluminum ore is abundant. It takes a lot of energy to extract aluminum, but it is easily recycled at low energy cost.

Technical Notes Until 1970, designations of wrought aluminum alloys were a mess; in many countries, they were simply numbered in the order of their development. The International Alloy Designation System (IADS), now widely accepted, gives each wrought alloy a 4-digit number. The first digit indicates the major alloying element or elements. Thus the series 1xxx describe unalloyed aluminum; the 2xxx series contain copper as the major alloying element, and so forth. To these serial numbers are added a suffix F, O, H, or T indicating the state of hardening or heat treatment. In the AAUS designation for cast alloys, the first digit indicates the alloy group. In the 1xx.x group, the second two digits indicate the purity; in the other groups it indicates the alloying elements. In the 2xx.x to 9xx.x groups, the second two digits are simply serial numbers. The digit to the right of the decimal point indicates the product form: 0 means "castings" and 1 means "ingot."

Brushed aluminum sheet

Attributes of Aluminum Alloys

Price, \$/kg	1.30–5.70
Density, Mg/m³	2.50–2.95

Technical Attributes

El. modulus, GPa	68–88.5
Elongation, %	1–44
Fr. toughness, MPa·m$^{1/2}$	18–40
Vickers hardness, H_v	20–150
Yld. strength, MPa	30–510
Service temp., C	-270–180
Specific heat, J/kg·K	857–990
Th. conduct., W/m·K	76–235
Th. expansion, 10^{-6}/K	16–24

Eco-Attributes

Energy content, MJ/kg	235–335
Recycle potential	High

Aesthetic Attributes

Low to high pitch, 0–10	8–9
Muffled to ringing, 0–10	5–8
Soft to hard, 0–10	8–9
Warm to cool, 0–10	9–10
Reflectivity	80–92
Opaque	

Features Relative to Other Metals

- ✓ Corrosion resistant
- ✓ Ductile
- ✓ Light
- ✓ Stiff
- ✓ Strong

Magnesium Alloys

Natural magnesium sheet

Attributes of Magnesium Alloys

Price, $/kg	2.60–11.40
Density, Mg/m³	1.73–1.95

Technical Attributes

El. modulus, GPa	40–47
Elongation, %	1.5–20
Fr. toughness, $MPa{\cdot}m^{1/2}$	12–70
Vickers hardness, H_V	35–135
Yld. strength, MPa	65–435
Service temp., C	-40–300
Specific heat, J/kg·K	950–1060
Th. conduct., W/m·K	50–156
Th. expansion, 10^{-6}/K	24.6–30

Eco-Attributes

Energy content, MJ/kg	300–500
Recycle potential	High

Aesthetic Attributes

Low to high pitch, 0–10	8–9
Muffled to ringing, 0–10	3–6
Soft to hard, 0–10	8
Warm to cool, 0–10	8–9
Reflectivity	68
Opaque	

Features Relative to Other Metals

✓ Damping
✓ Ductile
✓ Light
✓ Strong
✗ Wear resistant

What are they? Magnesium is the second of the light-metal trio (with partners aluminum and titanium) and light it is: a computer case made from magnesium is barely two thirds as heavy as one made from aluminum. It and its partners are the mainstays of aerospace engineering.

Design Notes The push for compact, light, electronics (laptops, mobile phones) and lightweight vehicles (wheels, in-cabin metal parts) has prompted designers to look harder at magnesium alloys than ever before, and has stimulated production and driven prices down. What do they offer? Magnesium has a low density, good mechanical damping, much better thermal conductivity than steel, less good electrical conductivity than copper and aluminum, but still good. It survives well in the protected environment of a house or office, but it corrodes badly in salt water and acids; even sweat is enough to tarnish it. Magnesium is flammable, but this is only a problem when its in the form of powder or very thin sheet. It costs more than aluminum but nothing like as much as titanium. It is easy to machine, but because of its low stiffness, parts must be firmly clamped while doing so. Magnesium alloys are designed for specific forming purposes. Some (like AZ31B) are good for extrusions. Others (AZ63, AZ92 and AM100) have been formulated for investment casting; the AZ91 range are used for die-casting. Most magnesium alloys can be welded using TIG or MIG methods; and soldering and adhesive bonding are both feasible. Spot and seam welds are possible but only in low stress applications; riveting is better, provided aluminum rivets are used to avoid galvanic corrosion.

Typical Uses Aerospace; automotive; sports goods; nuclear fuel cans; vibration damping and shielding of machine tools; engine case castings; automotive wheels; ladders; housings for electronic equipment, office equipment, and lawnmowers.

Competing Materials Titanium, aluminum, CFRP.

The Environment Magnesium is the fifth most abundant metal in the Earth's crust, and the third in its oceans – and it can be extracted economically from both (the Dead Sea, thick with dissolved salts – is the best source of all). But its extraction is very energy intensive, consuming three times more per unit weight than commodity polymers and nearly twice as much as aluminum. It can be recycled, and doing this uses barely one fifth as much energy.

Technical Notes The classification system of the American Society for Testing Materials (ASTM) is the most widely used. In this system, the first two letters indicate the principal alloying elements. The letter corresponding to the element present in the greatest quantity is used first; if they are equal, they are listed alphabetically. The letters are followed by numbers which represent the nominal compositions of the principal alloying elements in weight % rounded to the nearest whole number; thus AZ91 means the alloy 90% Mg, 9% Al and 1% Zn.

Titanium Alloys

What are they? Titanium is the seventh most abundant metal in the Earth's crust, but extracting the metal from the oxide in which it occurs naturally is unusually difficult. This makes titanium, third member of the light alloy trio, by far the most expensive of the three (more than ten times the price of aluminum). Despite this, the use of titanium is growing, propelled by its remarkable properties. It has a high melting point (1660 C), it is light, and – although reactive – its resists corrosion in most chemicals, protected by a thin film of oxide on its surface. Titanium alloys are exceptionally strong for their weight, and can be used at temperatures up to 500 C – compression blades of aircraft turbines are made of them. They have unusually poor thermal and electrical conductivity, and low expansion coefficients.

Design Notes Titanium alloys are expensive, requiring vacuum processing to prevent take up of oxygen, which makes them brittle. But they are unusually strong, light and corrosion resistant, so much so that pure titanium can be implanted in the body to repair broken bones. More usually it is alloyed with aluminum and vanadium (Ti with 8% Al 6%V, or simply Ti–6–4) to give a material that can be forged and worked yet has good resistance to creep. Titanium alloys have limited ductility – sheet cannot easily be bent to radii less than 1.5 times its thickness. They can – with difficulty – be welded, but are easy to diffusion bond. The drive to miniaturize consumer electronics gives titanium a growing importance in product design. The casings of mobile phones and portable computers are now so thin that polymers cannot take the strain – they are not stiff or strong enough. The strength and low density of titanium makes it – despite its cost – an attractive replacement.

Typical Uses Aircraft turbine blades; general aerospace applications; chemical engineering; heat exchangers; bioengineering; medical; missile fuel tanks; heat exchangers, compressors, valve bodies, surgical implants, marine hardware, paper-pulp equipment, casings for mobile phones and portable computers.

Competing Materials Magnesium, aluminum, CFRP

The Environment Extracting titanium from its ores is very energy intensive. It can be recycled provided it is not contaminated with oxygen.

Technical Notes There are four groups of titanium alloys: α alloys, near-α alloys, α-β alloys, and β alloys. The α alloys are hcp, the β alloys are bcc. α alloys are preferred for high temperature applications because of their creep resistance and for cryogenic applications because of their good toughness at low temperatures. A designation system with some logic to it simply lists the quantities of the principal alloying additions; thus "Ti–8–1–1" contains 8% aluminum, 1% molybdenum and 1% vanadium; and "Ti–6–4" means 6% aluminum and 4% vanadium.

Titanium sheet

Attributes of Titanium Alloys

Price, $/kg	21.00–28.00
Density, Mg/m^3	4.36–4.84

Technical Attributes

El. modulus, GPa	90–137
Elongation, %	1–40
Fr. toughness, MPa·m$^{1/2}$	14–120
Vickers hardness, H_v	60–380
Yld. strength, MPa	172–1245
Service temp., C	-40–500
Specific heat, J/kg·K	510–650
Th. conduct., W/m·K	3.8–20.7
Th. expansion, 10^{-6}/K	7.9–11

Eco-Attributes

Energy content, MJ/kg	750–1250
Recycle potential	High

Aesthetic Attributes

Low to high pitch, 0–10	8–9
Muffled to ringing, 0–10	5–8
Soft to hard, 0–10	8–9
Warm to cool, 0–10	7–8
Reflectivity, %	44–53
Opaque	

Features Relative to Other Metals

✓ Corrosion resistant
✓ Resilient
✓ Strong
✓ Tough

Titanium Alloys

Nickel Alloys

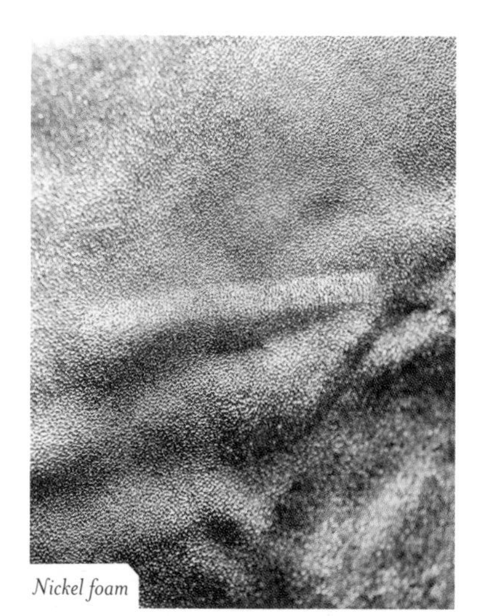
Nickel foam

Attributes of Nickel Alloys

Price, $/kg	4.30–28.60
Density, Mg/m³	7.65–9.3

Technical Attributes

El. modulus, GPa	125–245
Elongation, %	0.3–70
Fr. toughness, MPa·m$^{1/2}$	65–150
Vickers hardness, H_v	75–600
Yld. strength, MPa	70–2100
Service temp., C	-200–1200
Specific heat, J/kg·K	365–565
Th. conduct., W/m·K	8–91
Th. expansion, 10^{-6}/K	0.5–16.5

Eco-Attributes

Energy content, MJ/kg	40–690
Recycle potential	High

Aesthetic Attributes

Low to high pitch, 0–10	8–9
Muffled to ringing, 0–10	3–6
Soft to hard, 0–10	8
Warm to cool, 0–10	8–9
Reflectivity	50–65
Opaque	

Features Relative to Other Metals

✓ Corrosion resistant
✓ Ductile
✓ Fatigue resistant
✓ Heavy
✓ Hot temperatures
✓ Resilient
✓ Stiff
✓ Strong
✓ Tough
✓ Wear resistant

What is it? The US 5¢ piece, as its name suggests, is made of pure nickel. Look at one – much can be learned from it: that nickel is a ductile, silvery metal, easily formed by stamping (or rolling or forging), hard when worked, and with good resistance to corrosion. Not bad for 5¢. What it does not reveal is that nickel has two more remarkable sets of properties. One is the effect it has when alloyed with steel, stabilizing the face-centered cubic structure that gives the ductility and corrosion resistance to stainless steel (its largest single use). The other is its strength at high temperatures, a strength that can be boosted by alloying to make the materials on which all aircraft jet engines depend. These alloys have properties so extreme that they are known as "super alloys."

Design Notes Pure nickel has good electrical conductivity, thermal conductivity, and strength and corrosion resistance; nickel and its alloys are used in marine applications for heat exchanges in other structures. Nickel-iron alloys have high magnetic permeability (good for electronic shielding and magnetic coils) and low thermal expansion (good for glass-to-metal joints). Invar, an alloy based on nickel, has essentially zero thermal expan-sion coefficient near room temperature; a magnetic contraction counter-acts the ordinary thermal expansion, canceling it out. Nickel-chrome-iron alloys have high electrical resistance and are used as heating elements in toasters and industrial furnaces. Bi-metallic sheet of nickel bonded to copper is used as actuators for thermostats and safety devices. Nickel alloys based on the combination nickel-titanium-aluminum have exceptionally high-temperature strength, toughness and resistance to attack by gases. These super alloys, carrying trade names such as Nimonic, Inconel and Hastelloy are used for turbine blades, disks and combustion chambers, chemical engineering equipment and high temperature design.

Typical Uses The principal uses of nickel are an alloying element in stainless steels and superalloys; blades, disks, and combustion chambers in turbines and jet engines, rocket engines, bi-metallic strips, thermocouples, springs, food equipment, heating wires, electroplating for corrosion protection, coinage, and nickel-cadmium batteries.

Competing Materials Iron based super alloys (lower in performance); cobalt based super alloys (more expensive); stainless steels.

The Environment About 10% of the population is sensitive to nickel, causing them to react even to the nickel in stainless steel watch straps. Compounds of nickel can be more toxic; nickel carbonyl, used in the extraction of nickel, is deadly.

Technical Notes 8% of nickel, added to steel, stabilizes austenite, making the steel ductile down to cryogenic temperatures. Nickel in super alloys combines with other alloying elements, particularly titanium and aluminum, to form precipitates of intermetallic Ni_3Al and Ni_3Ti.

Copper Alloys (Brass, Bronze)

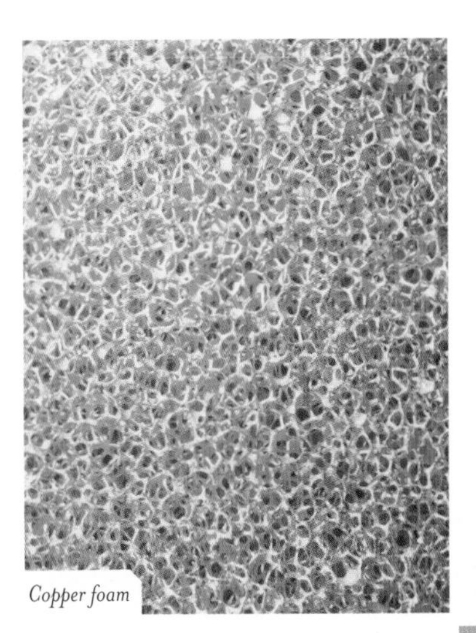

Copper foam

What are they? In Victorian times you washed your clothes in a "copper" – a vat or tank of beaten copper sheet, heated over a fire; the device exploited both the high ductility and the thermal conductivity of the material. Copper has a distinguished place in the history of civilization: it enabled the technology of the Bronze age (3000 BC to 1000 BC). It is used in many forms: as pure copper, as copper-zinc alloys (brasses), as copper-tin alloys (bronzes), and as copper-nickel and copper-beryllium. The designation of "copper" is used when the percentage of copper is more than 99.3%.

Design Notes Copper and its alloys are easy to cast, to roll to sheet, to draw to wire, and to shape in other ways. They resist corrosion in the atmosphere, acquiring an attractive green patina (copper carbonate) in clean air, and a black one (copper sulfide) in one that is not – copper roofs in cities are usually black. The patina of bronze is a rich warm brown, much loved by sculptors. Pure copper has excellent electrical and thermal conductivity, is easy to fabricate and join, has good corrosion resistance and reasonable strength. Where high conductivity is necessary, oxygen-free high-conductivity (OFHC) copper is used. In its annealed form it is soft and ductile; with work hardening the material becomes harder but less ductile. In brasses and bronzes the alloying increases the strength. The commonest are the "cartridge brasses" (used in musical instruments, hardware and – of course – ammunition), "yellow brasses" (used in springs and screws) and "Muntz metals" (used in architectural applications and condenser tubes). Cartridge brass, with 30% zinc, has the highest ductility and is used to draw brass wires. Bronzes are exceptionally liquid when molten, allowing them to be cast to intricate shapes.

Typical Uses Electrical conductors, heat exchangers, coinage, pans, kettles and boilers, plates for etching and engraving, roofing and architecture, cast sculptures, printing wires, heat pipes, filaments, and musical instruments.

Competing Materials Aluminum for conductors and thermal conduction, stainless steel for cooking and protection.

The Environment Copper and its alloys are particularly easy to recycle – in many countries the recycle fraction approaches 90%.

Technical Notes There is now a UNS designation system for copper and its alloys: the letter C (for "copper") followed by a 5 digit number. Only the first digit means anything: C1**** designates almost pure copper, the C2, C3 and C4 series are brasses with increasing zinc content, the C5s are bronzes based on copper and tin, the C6s are other bronzes containing aluminum instead of tin, and the C7s are copper-nickel alloys.

Attributes of Copper, Brass, Bronze

	Copper	Brass	Bronze
Price, $/kg	1.72–1.93	1.43–2.00	3.72–5.44
Density, Mg/m^3	8.93–8.94	7.8–8.8	8.5–9
Technical Attributes			
El. modulus, GPa	121–133	90–120	70–120
Elongation, %	4–50	5–55	2–60
Fr. toughness, MPam$^{1/2}$	40–100	30–86	24–90
Vickers hardness, H_v	44–180	50–300	60–240
Yld. strength, MPa	45–330	70–500	65–700
Service temp., C	-270–180	-270–220	-270–200
Specific heat, J/kg·K	372–388	372–383	382–385
Th. conduct., W/m·K	147–370	110–220	50–90
Th. expansion, 10^{-6}/K	16.8–17.9	16.5–20.7	16.5–19
Eco-Attributes			
Energy content, MJ/kg	100–130	100–120	110–120
Recycle potential	High	High	High
Aesthetic Attributes			
Low to high pitch, 0–10	8–8	8–8	8–8
Muffled to ringing, 0–10	6–9	6–9	5–10
Soft to hard, 0–10	8–9	8–9	8–9
Warm to cool, 0–10	9–10	9–10	9–9
Reflectivity	51–91	50–90	50–80
Opaque			
Features Relative to Other Metals			
Corrosion resistant	✓	✓	✓
Ductile	✓	✓	
Heavy	✓	✓	✓
Strong	✓	✓	
Tough	✓	✓	✓
Wear resistant			✓

Zinc Alloys

What are they? The slang in French for a bar or pub is "le zinc;" bar counters in France used to be clad in zinc – many still are – to protect them from the ravages of wine and beer. Bar surfaces have complex shapes – a flat top, curved profiles, rounded or profiled edges. These two sentences say much about zinc: it is hygienic; it survives exposure to acids (wine), to alkalis (cleaning fluids), to misuse (drunk customers) and it is easy to shape. These remain among the reasons it is still used today. Another is the "castability" of zinc alloys – their low melting point and fluidity gives them a leading place in die casting. The molds are relatively cheap, and details are accurately reproduced.

Design Notes Most zinc is used in galvanizing steel to improve corrosion resistance. Zinc die-casting alloys are strong enough for most consumer products; and the metal itself is cheap. They are the metallic answer to injection molded polymers. Zinc alloys offer higher strength than other die-casting alloys except those of copper. Die cast parts can be held to close tolerances in thin sections and are easily machined. Wrought zinc is available as strip, sheet, foil, plate, rod, wire and blanks for forging or extrusion. Bends in rolled zinc sheet should be at right angles to the grain or rolling direction and should have a radius no less than the sheet thickness.

Typical Uses Roofing, gutters, flashlight reflectors, fruit jar caps, radio shielding, gaskets, photoengraving plates, handles, gears, automotive components, kitchen counter-tops, protective plating.

Competing Materials Product design: die-cast aluminum alloys, thermoplastics that can be injection molded; corrosion protection; nickel, chrome plating, polymer powder coating, or stainless steel.

The Environment Zinc vapor is toxic – if you inhale it you get the "spelter-shakes" – but adequate protection is now universal. In all other ways zinc is a star: it is non toxic, has low energy content, and – in bulk – can be recycled (not, of course, as plating).

Technical Notes Most zinc alloys are die cast; for this, the prime alloys are AG40A and AC41A. Wrought zinc is made by hot rolling cast sheets, by extrusion or by drawing. Zinc foil is made by electroplating zinc on an aluminum drum and then stripping it off. Superplastic zinc alloys can be formed by methods normally used for polymers – vacuum forming, compression molding – as well as traditional metal processes like deep drawing and impact extrusion. Extrusion and forging is done with zinc-manganese alloys. Wrought zinc alloys are easily soldered and spot welded. Zinc is frequently left uncoated. It can be polished, textured, plated or painted.

Die cast zinc

Attributes of Zinc Alloys	
Price, $/kg	0.90–2.90
Density, Mg/m^3	5.5–7.2
Technical Attributes	
El. modulus, GPa	60–110
Elongation, %	1–90
Fr. toughness, $MPa{\cdot}m^{1/2}$	10–130
Vickers hardness, H_v	30–160
Yld. strength, MPa	50–450
Service temp., C	-45–120
Specific heat, J/kg·K	380–535
Th. conduct., W/m·K	95–135
Th. expansion, $10^{-6}/K$	14–40
Eco-Attributes	
Energy content, MJ/kg	50–145
Recycle potential	High
Aesthetic Attributes	
Low to high pitch, 0–10	8–8
Muffled to ringing, 0–10	4–7
Soft to hard, 0–10	8–9
Warm to cool, 0–10	9–9
Reflectivity	74–85
Opaque	

Features Relative to Other Metals

- ✓ Damping
- ✓ Ductile
- ✓ Fatigue resistant
- ✗ Impact resistant
- ✓ Tough

Zinc Alloys

Ceramics

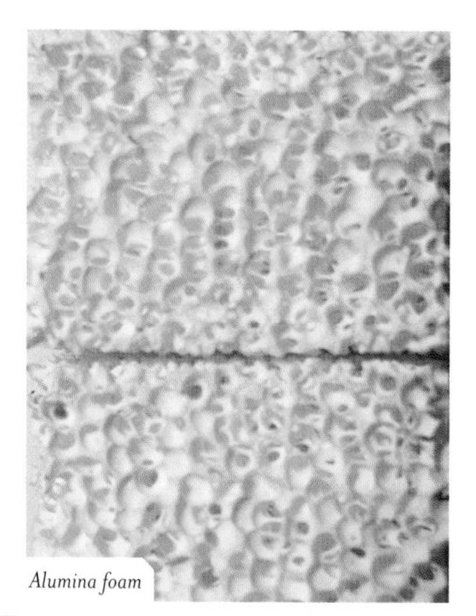
Alumina foam

Attributes of Alumina (97.5%)

Price, \$/kg	4–12
Density, Mg/m^3	3.7–3.8

Technical Attributes

El. modulus, GPa	333–350
Elongation, %	0
Fr. toughness, MPa·m$^{1/2}$	3.6–3.8
Vickers hardness, H_v	1400–1600
Tensile strength	175–200
Service temp., C	-270–1700
Specific heat, J/kg·K	635–700
Th. conduct., W/m·K	25–30
Th. expansion, 10^{-6}/K	6.7–7

Eco-Attributes

Energy content, MJ/kg	150–200
Recycle potential	Low

Aesthetic Attributes

Low to high pitch, 0–10	8–9
Muffled to ringing, 0–10	8–9
Soft to hard, 0–10	8–9
Warm to cool, 0–10	7–8
Translucent to opaque	

Features Relative to Other Ceramics

✓ Corrosion resistant
✗ Ductile
✓ Hot temperatures
✗ Impact resistant
✗ Resilient
✓ Stiff
✗ Tough
✓ Wear resistant

Ceramics

Alumina

Alumina (Al_2O_3) is to technical ceramics what mild steel is to metals – cheap, easy to process, the work horse of the industry. It is the material of spark plugs and electrical insulators. In single crystal form it is sapphire, used for watch faces and cockpit windows of high speed aircraft. More usually it is made by pressing and sintering powder, giving grades ranging from 80 to 99.9% alumina; the rest is porosity, glassy impurities or deliberately added components. Pure aluminas are white; impurities make them pink or green. The maximum operating temperature increases with increasing alumina content. Alumina has a low cost and a useful and broad set of properties: electrical insulation, high mechanical strength, good abrasion and temperature resistance up to 1650 C, excellent chemical stability and moderately high thermal conductivity, but it has limited thermal shock and impact resistance. Chromium oxide is added to improve abrasion resistance; sodium silicate, to improve processability but with some loss of electrical resistance. Competing materials are magnesia, silica and borosilicate glass.

· *Typical Uses* Insulators, heating elements, micro-electronic substrates, radomes, bone replacement, tooth replacement, tank armour, spark plug insulators, dies for wire drawing, nozzles for welding and sandblasting.

Boron Carbide

Boron carbide (B_4C) is nearly as hard as diamond and vastly less expensive (though still not cheap). Its very low density and high hardness make it attractive for the outer layer of bulletproof body armor and as an abrasive.

· *Typical Uses* Lightweight armor, bulletproof surfaces, abrasives, sandblasting nozzles, high temperature thermocouples.

Tungsten Carbide

Tungsten carbide (WC) is most commonly used in the form of a "cemented" carbide, or cermet: a metal carbide held by a small amount (5–20%) of metallic binder, usually cobalt. Its exceptional hardness and stability make it an attractive material when wear resistance is essential. Its properties are governed by the type of carbide, grain size and shape and the proportion of carbide to metal. Cermets are expensive but, as cutting tools, they survive cutting speeds 10 times those of the best tool steel. Shaping is usually done by pressing, sintering and then grinding; the tool bit is brazed to a shank or blade made from a cheaper steel. Tungsten carbide can be vapor-coated with Ti-nitride to improve wear resistance even further.

· *Typical Uses* Cutting tools, saw blades, dental drills, oil drilling bits, dies for wire drawing, knife edges.

Silicon Carbide

Silicon carbide (SiC, carborundum), made by fusing sand and coke at 2200 C, is the grit on high quality sandpaper. It is very hard and maintains its strength to high temperature, has good thermal shock resistance, excellent abrasion resistance and chemical stability, but, like all ceramics, it is brittle. Silicon carbide is a blue-black material. High strength SiC fibers such as Nicalon, made by CVD processes, are used as reinforcement in ceramic or metal matrix composites.

- *Typical Uses* Cutting tools, dies and molding materials, catalytic converters, engine components, mechanical seals, sliding bearings, wear protection sleeves, heat exchange tubes, furnace equipment, heating elements.

Natural silicon carbide

Textured soda lime glass

Attributes of Soda-lime Glass

Price, \$/kg	0.68–1
Density, Mg/m^3	2.44–2.5

Technical Attributes

El. modulus, GPa	68–72
Elongation, %	0
Fr. toughness, MPa·m$^{1/2}$	0.55–0.7
Vickers hardness, H_VO	440–480
Tensile strength	31–35
Service temp., C	-270–250
Specific heat, J/kg·K	850–950
Th. conduct., W/m·K	0.7–1.3
Th. expansion, 10^{-6}/K	9–9.5

Eco-Attributes

Energy content, MJ/kg	20–25
Recycle potential	High

Aesthetic Attributes

Low to high pitch, 0–10	7–8
Muffled to ringing, 0–10	8–9
Soft to hard, 0–10	7–8
Warm to cool, 0–10	5–6
Optically clear	

Features Relative to Other Glasses

✓ Corrosion resistant
✗ Ductile
✗ Impact resistant
✗ Resilient
✗ Tough
✓ Wear resistant

Glass

Soda-Lime Glass

Soda-lime is the glass of windows, bottles and light bulbs, used in vast quantities, the commonest of them all. The name suggests its composition: 13–17% NaO (the "soda"), 5–10% CaO (the "lime") and 70–75% SiO_2 (the "glass"). It has a low melting point, is easy to blow and mold, and it is cheap. It is optically clear unless impure, when it is typically green or brown. Windows, today have to be flat and that was not – until 1950 – easy to do; now the float-glass process, solidifying glass on a bed of liquid tin, makes "plate" glass cheaply and quickly.

· *Typical Uses* Windows, bottles, containers, tubing.

Borosilicate Glass

When most of the lime in soda lime glass is replaced by borax, B_2O_3, it becomes borosilicate glass ("Pyrex"). It has a higher melting point than soda lime glass and is harder to work; but it has a lower expansion coefficient and a high resistance to thermal shock, so its used for glass wear and laboratory equipment.

· *Typical Uses* Laboratory glassware, ovenware, headlights, electrical insulators, metal/glass seals, telescope mirrors, sights, gages, piping.

Silica Glass

Silica is a glass of great transparency. It is nearly pure SiO_2, it has an exceptionally high melting point and is difficult to work, but, more than any other glass, it resists temperature and thermal shock.

· *Typical Uses* Envelopes for high-temperature lamps.

Glass Ceramic

Glass ceramics are glasses that, to a greater or lesser extent, have crystallized. They are shaped while in the glassy state, using ordinary molding methods and then cooled in such a way that the additives they contain nucleate small crystals. It's sold for cooking as Pyroceram and is used for high-performance heat-resisting applications.

· *Typical Uses* Cookware, stove surfaces, high-performance heat-resisting applications.

Fibers

What are they? Fibers clothe us, carpet us, cushion us, keep us warm and protected. They are the substance of thread, of rope, of fabrics, of textiles and of fiber reinforced composites. Fibers have the unique virtue of being strong yet flexible – pull them and they resist, flex them and they comply. Spin them into rope or weave them into cloth and these inherit the same properties. Metal fibers can be made – they are used to reinforce car tires – but metals are heavy. More interesting are those that are light, stiff and strong. Here are some that, through their strength and flexibility, allow the creation of novel materials.

Glass Fibers

Glass fibers are made by drawing molten glass through a spinner, giving continuous fibers of diameter between 10 and 100 microns. Their perfection gives them exceptional strength in tension, yet they are flexible. They can be aggregated into loose felt that has very low heat conduction (it is used to insulate houses). They can be woven into a fabric, and printed or colored to give a fire-resistant substitute for curtains or covers (when the woven fabric is treated with silicone it can be used up to 250 C). As chopped strand or as continuous fibers or yarns (bundles of fibers), they form the reinforcement in glass fiber reinforced polymers, GFRPs. There are several grades of glass fiber, differing in composition and strength. E-glass is the standard reinforcement. C-glass has better corrosion resistance than E; R and S have better mechanical properties than E but are more expensive. AR-glass resists alkalis, allowing it to be used to reinforce cement.

- *Typical Uses* Thermal insulation, fire-resistant fabric, reinforcement of polymers to make GFRP.

Glass-fiber mat

Carbon Fibers

Carbon fibers are made by pyrolysing organic fibers such as viscose, rayon or polyacrylonitrile (PAN), or from petroleum pitch. The PAN type has the better mechanical properties, but those produced from pitch are cheaper. PAN fibers are first stretched for alignment, then oxidized in air at slightly elevated temperatures, then carbonized in an inert environment at very high temperatures, and finally heated under tension to convert the crystal structure to that of graphite. Carbon fibers have high strength and stiffness with low density, but they oxidize at high temperatures unless the atmosphere is reducing. They come in four grades: high modulus, high strength, ultra high modulus and ultra high strength – with cost increasing in that order. The single fibers are very thin (<10 microns in diameter); they are generally spun into tows and woven into textiles. They are primarily used as reinforcement in polymer, metal or carbon matrices.

- *Typical Uses* Reinforcement of polymers to make CFRP, and of metals and ceramics to make metal-matrix and ceramic matrix composites. Carbon fiber-reinforced carbon is used for brake pads in racing cars and aircraft.

Carbon fiber weaves

Fibers

Silicon Carbide Fibers

Carbon fibers oxidize, limiting their continuous use above 400 C. Silicon carbide fibers such as Nicalon were developed to overcome this difficulty. They are used where high strength and high modulus are needed at very high temperatures. The fibers are produced by the pyrolization of a polymer fiber or by CVD processing. They are produced in a range of sizes and forms including continuous fibers, whiskers, chopped fibers, multifilament tows and woven cloth.

· *Typical Uses* Reinforcement for metal and ceramic matrix composites.

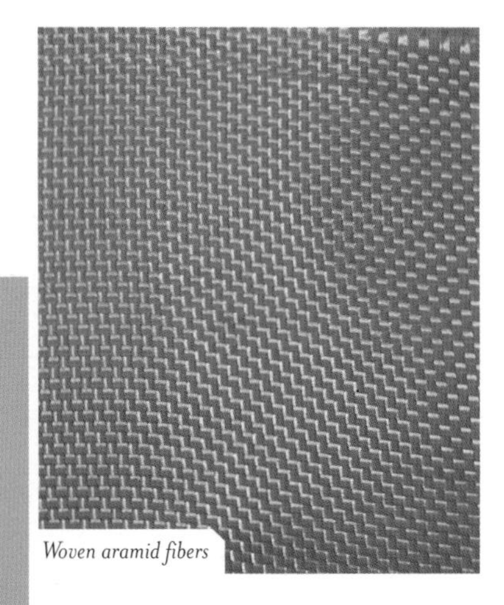

Woven aramid fibers

Aramid Fibers

Originally produced by DuPont as Kevlar, aramid fibers are processed in such a way that the polymer chains are aligned parallel to the axis of the fiber. The chemical unit is an aromatic polyamide with a ring structure that gives high stiffness; the strong covalent bonding gives high strength. They are available in low density/high strength form (Kevlar 29) and in a form suited for reinforcement (Kevlar 49). The first is used in ropes, cables and armor; the second as reinforcement in polymers for aerospace, marine and automotive components. Nomex fibers have excellent flame and abrasion resistance; they are made into a paper that is used to make honeycomb structures. These materials are exceptionally stable and have good strength, toughness and stiffness up to 170 C.

· *Typical Uses* As woven cloth: protective clothing, bomb and projectile protection, and, in combination with boron carbide ceramic, bulletproof vests. As paper: honeycomb cores for sandwich panels. As fibers and weaves: as reinforcement in polymer matrix composites.

Hemp

Hemp is a natural fiber valued for its great strength. It is used for cord, rope, sacks, packaging and, increasingly, as a reinforcement in polymers. Other rope materials, like abaca, are more water-resistant and have replaced hemp in marine applications. Hemp-reinforced composites cannot at present be recycled, but the hemp is a renewable resource and the energy overhead in producing it is low – far lower than carbon or Kevlar.

· *Typical Uses* Rope and strong fabric; reinforcement in polymer matrix composites.

Natural Materials

Bamboo

Bamboo is nature's gift to the construction industry. Think of it: a hollow tube, exceptionally strong and light, growing so fast that it can be harvested after a year, and – given a little longer – reaching a diameter of 0.3 meters and a height of 15 meters. This and its hard surface and ease of working makes it the most versatile of materials. Bamboo is used for building and scaffolding, for roofs and flooring, for pipes, buckets, baskets, walking sticks, fishing poles, window blinds, mats, arrows and furniture. Tonkin bamboo is strong and flexible (fishing poles); Tali bamboo is used for structural applications (houses or furniture); Eeta bamboo is the fastest growing and is used as a source of cellulose for the production of cellulose or Rayon.

Balsa

Balsa is another of nature's miracles: exceptionally light, buoyant and insulating, yet strong and stiff along the grain. Balsa is used for life preservers, buoys, floats, paneling, vibration isolation, insulating partitions, inside trim of aircraft and model airplanes. Its main use today is as cores for sandwich hulls of yachts and power boats; here "end-grain" balsa is used, meaning that the grain lies at right angles to the face of the hull or panel.

Parallel balsa blocks

Cork

Cork is the subcutaneous bark of the cork-oak, *Quercus Suber*. Its cellular structure is up to 80% air by volume. When dried, cork is light, porous, easily compressed and elastic. It has a low thermal conductivity, chars at 120 C, and ignites only with difficulty when in contact with flame. Cork is a natural closed-cell foam, and is waterproof and remarkably stable, surviving in the neck of a wine bottle for up to 30 years without decay or contaminating the wine. Corkboard, made by compressing granulated cork under heat, is used for wall and ceiling insulation against heat and sound. Cork is used for bottle stoppers, insulation, vibration control, floats, insulation and packaging for fruit transportation. In pressed form it is used for gaskets, oil retainers and polishing wheels and as a component of linoleum.

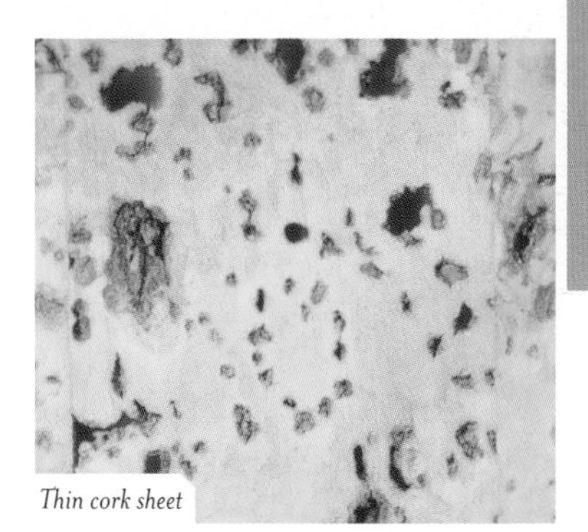

Thin cork sheet

Leather

Leather is a natural fabric. It has high tensile strength and is exceptionally tough and resilient, yet it is flexible and – as suede – is soft to touch. It is prepared by the tanning of animal hide, a smelly process in which the hide is soaked in solutions of tannins for weeks or months, making it pliable and resistant to decay. Leather is used for belts, gaskets, shoes, jackets, handbags, linings and coverings.

Wood

Wood has been used to make products since the earliest recorded time. The ancient Egyptians used it for furniture, sculpture and coffins before 2500 BC. The Greeks at the peak of their empire (700 BC) and the Romans at the peak of theirs (around 0 AD) made elaborate boats, chariots and weapons of wood, and established the craft of furniture making that is still with us today. More diversity of use appeared in Medieval times, with the use of wood for large-scale building, and mechanisms such as pumps, windmills, even clocks, so that, right up to the end of the 17th century, wood was the principal material of engineering. Since then cast iron, steel and concrete have displaced it in some of its uses, but timber continues to be used on a massive scale.

Wood offers a remarkable combination of properties. It is light, and, parallel to the grain, it is stiff, strong and tough – as good, per unit weight, as any man-made material. It is cheap, it is renewable, and the fossil-fuel energy needed to cultivate and harvest it is outweighed by the energy it captures from the sun during growth. It is easily machined, carved and joined, and – when laminated – it can be molded to complex shapes. And it is aesthetically pleasing, warm both in color and feel, and with associations of craftsmanship and quality.

Metal Foams

What are they? Metals that float in water? Doesn't sound sensible. Yet metal foams do – some have densities that are less that one tenth of that of H_2O. They are a new class of material, as yet imperfectly characterized, but with alluring properties. They are light and stiff, they have good energy-absorbing characteristics (making them good for crash-protection and packaging) and they have attractive heat-transfer properties (used to cool electronic equipment and as heat exchangers in engines). Some have open cells, very much like polymer foams but with the characteristics of metals (ductility, electrical conductivity, weldability, and so forth). Others have closed cells, like "metallic cork." And they are visually appealing, suggesting use in industrial design. Metal foams are made by casting methods that entrap gas in the semi-liquid metal, or by a replication technique using a polymer foam as a precursor. Once cast, they are as chemically stable as the metal from which they were made, have the same melting point and specific heat, but greatly reduced density, stiffness and strength. The stiffness-to-weight and strength-to-weight ratios, however, remain attractive.

At this point in time there are some 12 suppliers marketing a range of metal foams, mostly based on aluminum, but other metals – copper, nickel, stainless steel and titanium – can be foamed.

Design Notes There is little to go on here. Metal foams can be machined, and some can be cast to shape but this at present is a specialized process. They are best joined by using adhesives, which gives a strong bond. Some have a natural surface skin with an attractive texture, but this is lost if the foam is cut. The most striking characteristics of the materials are their low weight, good stiffness and the ability to absorb energy when crushed.

Potential Uses Metal foams have promise as cores for light and stiff sandwich panels; as stiffeners to inhibit buckling in light shell structures; as energy absorbing units, both inside and outside of motor vehicles and trains; and as efficient heat exchanges to cool high powered electronics (by blowing air through the open cells of the aluminum foam, attached to the heat source). Industrial designers have seen potential in exploiting the reflectivity and light filtering of open-cell foams, and the interesting textures of those with closed cells.

Competing Materials Rigid polymer foams; aluminum honeycombs; woods.

The Environment Metal foams are not flammable (unlike most other foams) and they can be recycled.

Suppliers Shinko Wire Company Ltd., 10-1 Nakahama-machi, Amagasaki-shi 660, Japan. Phone 81-6-411-1081; Fax 81-6-411-1056; ERG Materials and Aerospace Corp., 900 Stanford Ave., Oakland, CA 94608, USA. Phone 1-510-658-9785; Fax 1-510-658-7428; Fraunhofer Institute for Manufacturing and Advanced Materials (IFAM), Wiener Strasse 12, D-29359, Bremen, Germany. Phone 49-421-2246-211; Fax 49-421-2246-300

Open cell aluminum foam

Attributes of Closed-cell Aluminum Foam

Price, \$/kg	12.00–20.00
Density, Mg/m³	0.07–0.5

Technical Attributes

El. modulus, GPa	0.02–2
Fr. toughness, MPa·m$^{1/2}$	0.03–0.5
Vickers hardness, H_v	0.08–1
Yield strength, MPa	0.025–30
Service temp., C	-100–200
Th. conduct., W/m·K	0.3–10
Th. expansion, 10^{-6}/K	19–21

Eco-Attributes

Energy content, MJ/kg	250–290
Recycle potential	High

Aesthetic Attributes

Reflectivity, %	5–10
Translucent to opaque	

Features

- ✓ Flame retardant
- ✓ Impact resistant
- ✓ Light
- ✓ Stiff
- ✓ UV resistant

Amorphous Metals

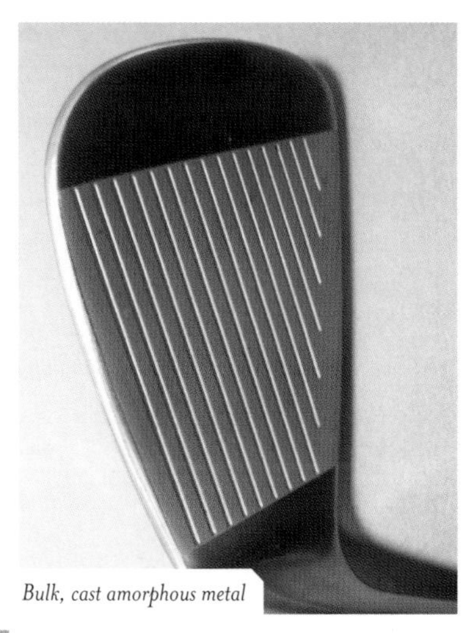
Bulk, cast amorphous metal

Amorphous Metals

Attributes of Ti-based Amorphous Metal

Price, $/kg	32.00–40.00
Density, Mg/m^3	6.2–6.4

Technical Attributes

El. modulus, GPa	91–100
Elongation, %	1.9–2.1
Fr. toughness, $MPa{\cdot}m^{1/2}$	52–57
Vickers hardness, H_V	555–664
Yield strength, MPa	1800–2000

Eco-Attributes

Energy content, MJ/kg	350–400
Recycle potential	High

Aesthetic Attributes

Reflectivity, %	95–99
Opaque	

Features

- ✓ Flame retardant
- ✓ Resilient
- ✓ Stiff
- ✓ Strong

What are they? Metals that are glasses? Amorphous, or glassy, metals are certainly not transparent, so what does this mean? Here "glass" is used in a technical sense, meaning "a material that has no recognizable structure," latinized to "a-morphous." That may not sound remarkable but it is – indeed such materials did not exist before 1960, when the first tiny sliver was made by rapidly cooling ("quenching") a complex alloy of gold at an incredible rate of 1,000,000 C/sec. That does not suggest much promise for the large-scale production of bulk materials, but such has been the pace of alloy development that you can now buy golf clubs with amorphous metal heads.

When metals are cast, crystals nucleate in the coldest part of the mold and grow inwards until they meet, giving a solid casting made up of a large number of interlocking crystals or grains. Most metals, as liquids, are very fluid – about as runny as water – meaning that atoms can rearrange easily and quickly. As a crystal grows into the liquid, atoms joining it have sufficient mobility that each seeks the spot where it fits best, and that automatically extends the ordered crystal. But if the alloy has many components of different size and affinity, it becomes much more viscous when liquid – more like honey than water – and the atoms in it have less mobility. Further, this affinity between the different components of the alloy causes clusters to form in the liquid and these do not easily rearrange to join the growing crystal. All this means that if the alloy is cooled fast, limiting the time for possible rearrangements, the liquid simply goes solid without crystallization.

Design Notes New alloys based on zirconium, titanium, iron and magnesium, alloyed with many other components, form glasses at easily achievable cooling rates, opening up commercial exploitation of amorphous metals. They are, at present, expensive (about $18/lb or more) but their properties, even at that price, are enticing. They are very hard and they have very high strength, stronger, for the same weight, than titanium alloys. They can be molded by molding techniques, giving process routes that are not available for conventional metals. They have excellent corrosion resistance. And they can be sprayed to give coatings that have the same properties.

Potential Uses Glassy metals are exceptionally hard, extremely corrosion resistant, and they have the combination of properties which makes them better than any other metals for springs, snap together parts, knife-edges, strong, light casings and other applications in which high strength in thin sections is essential. Their process ability, too, is exceptional, allowing them to be molded to complex shapes.

Competing Materials Titanium alloys, high-strength steels.

The Environment Some glassy metals contain beryllium, making disposal a problem; rapid progress in alloy development can be expected to deliver materials that have no toxic ingredients.

Suppliers Howmet Research Corporation Inc., www.howmet.com; Liquidmetal Technologies, www.liquidmetal.com

Shape-memory Alloys

What are they? Shape-memory alloys (SMAs) are a group of metals that have the remarkable ability to return to some previously defined shape when heated. Some exhibit shape memory only upon heating (one-way shape memory); others undergo a change in shape upon re-cooling (two-way shape memory). They work by having a crystal structure called thermo-elastic martensite: this allows the alloy to be deformed by a twinning mechanism below the transformation temperature. The deformation is reversed when the structure reverts to the parent phase on heating.

The principal commercial grades are the NiTi alloys and the copper-based CuZnAl and CuAlNi alloys. The NiTi alloys have greater shape memory strain (up to 8% versus 4 to 5% for the copper-based alloys), can generate stresses as high as 700 MPa, and have excellent corrosion resistance, important in biomedical applications. The copper-based alloys are less corrosion resistant and are susceptibile to stress-corrosion – cracking, but they are much less expensive. The transformation temperature can be adjusted between -200 and 110 C by tweeking the alloy composition. The maximum transformation strain is 8% for single use; for 100 cycles it is 6%; for 100,000 cycles it falls to 4%.

Design Notes In free recovery devices a SMA component is deformed while martensitic, and the only function required of the shape memory is that the component return to its previous shape (while doing minimal work) on heating. An application of this is the NiTi blood-clot filter – a wire that is shaped to anchor itself in a vein and catch passing clots. The part is chilled so it can be collapsed and inserted into the vein, then body heat is sufficient to turn the part to its functional shape.

In constrained recovery devices the SMA part is required to exert a force. An example is the hydraulic coupling manufactured as a cylindrical sleeve slightly smaller than the metal tubing it is to join. The sleeve diameter is expanded while martensitic, and, on warming to austenite, it shrinks in diameter and clamps the tube ends.

In force-actuators applications, the shape memory component is designed to exert force over a considerable range of motion, often for many cycles. An example is a fire safety valve, which incorporates a CuZnAl actuator designed to shut off toxic or flammable gas flow when fire occurs.

Proportional control applications use only a part of the shape recovery to accurately position a mechanism, exploiting the fact that the transformation occurs over a range of temperatures.

In super-elastic applications and SMA alloy is designed to be above its transformation temperature at room temperature. This allows it to be deformed heavily, yet immediately recover its original shape on unloading, so that it appears to be elastic, as in super-elastic NiTi eyeglass frames.

Potential Uses The properties of the NiTi alloys, particularly, suggest increasing use in biomedical applications. They are extremely corrosion resistant and bio-compatible, can be fabricated into the very small sizes, and have properties of elasticity and force delivery that allow applications not possible any other way.

Suppliers Shape Memory Applications, Inc., www.sma-inc.com

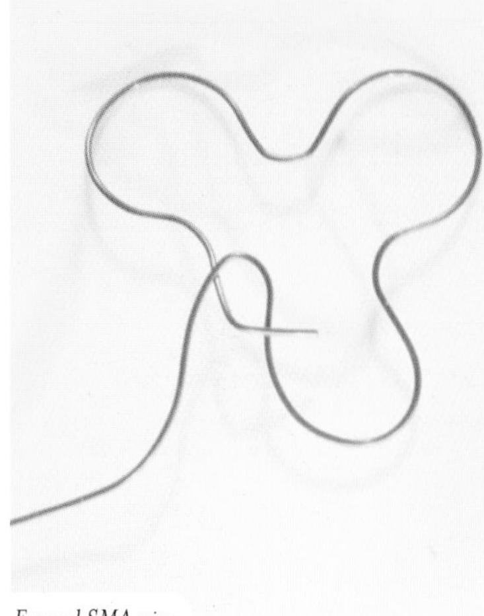

Formed SMA wire

Attributes of Nitanol, NiTi Alloy

Density, Mg/m^3	6.42–6.47

Technical Attributes

El. modulus, GPa	8–83
Elongation, %	5–25
Yield strength, MPa	195–690
Th. conduct., W/m·K	17–18
Th. expansion, 10^{-6}/K	10.5–11.5

Eco-Attributes

Energy content, MJ/kg	320–380
Recycle potential	High

Aesthetic Attributes

Reflectivity, %	80–90
Opaque	

Features

- ✓ Corrosion resistant
- ✓ Flame retardant
- ✓ FDA approval
- ✓ Resilient
- ✓ Strong

Shaping Profiles

Shaping processes give form to materials. The choice of process depends on the material, on the shape, on the required size, precision and surface finish; and the associated cost depends, most critically, on the number of components to be made – the batch size.

There are a number of broad families of shaping process, suggested by the index on the facing page. Some are economic even when the batch size is small: vacuum forming of polymers and sand-casting of metals are examples. Others become economic only when a large number is made: injection molding of polymers and die casting of metals, for instance. Variants of each process are adapted to be economic at a chosen range of batch size; and special processes have been developed that are viable for prototyping – a batch size of 1. This and certain other attributes of a process are readily quantified: the size of the components it can make; the minimum thickness of section it allows; the precision of dimensions ("allowable tolerance"); the roughness or smoothness of the surface. One, however, is not easy to quantify: that of shape. Wire drawing makes only one shape: a cylinder. Investment casting makes shapes of unlimited complexity: hollow, re-entrant, multi-connected. There have been many attempts to classify shape, none completely successful. In these profiles we will refer to a simple set of shapes; each is illustrated in the figure on the next page.

More information can be found in the sources listed under "Further Reading."

Introduction 238
Further Reading 239

Molding
Injection Molding 240
Rotational Molding 241
Blow Molding 242
Expanded Foam Molding 243
Compression Molding 244
Resin Transfer Molding 245

Casting
Die Casting 246
Sand Casting 247
Investment Casting 248
Polymer Casting 249

Bulk Forming
Shape Rolling and Die Forging 250
Extrusion 251

Sheet Forming
Press Forming, Roll Forming and Spinning 252
Thermoforming 253

Lay-up Methods 254
Powder Methods 255

Rapid Prototyping
Laser Prototyping 256
Deposition Prototyping 257

Generic Shapes
A set of possible shapes common in product design.

Prismatic	Circular		Non-Circular	
Solid	*Plain*	*Stepped*	*Plain*	*Stepped*
Hollow	*Plain*	*Stepped*	*Plain*	*Stepped*

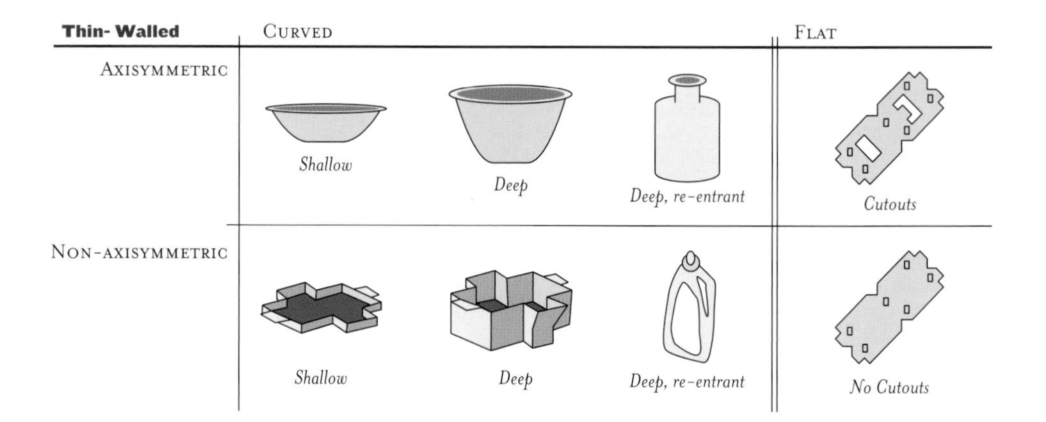

Bulk	Parallel Features		Transverse Features	
Solid	*Simple*	*Complex*	*Simple*	*Complex*
Hollow	*Simple*	*Complex*	*Simple*	*Complex*

Further Reading

ASM (1962) Casting Design Handbook, Metals Park, OH, USA. *(Old, but still a useful resource.)*

Bralla, J.G. (1998) Handbook of Product Design for Manufacture, 2nd edition, McGraw Hill, NY, USA. ISBN 0-07-007139-X. *(The bible – a massive compilation of information about manufacturing processes, authored by experts from different fields, and compiled by Bralla. A handbook rather than a teaching text – not an easy road.)*

Clegg, A.J. (1991) Precision Casting Processes, Pergamon Press, Oxford, UK. ISBN 0-08-037879-X. *(Descriptions and capabilities of the principal casting production processes.)*

DeGarmo, E.P., Black, J.T. and Kohser, R.A. (1984) Materials and Processes in Manufacturing, Macmillan Publishing Company, USA. ISBN 0-471-29769-0. *(A comprehensive text focusing on manufacturing processes, with a brief introduction to materials. The perspective is that of metal processing; the processing of polymers, ceramics and glasses gets much briefer treatment.)*

Edwards, L. and Endean, M., editors (1990) Manufacturing with Materials, Materials in Action Series, The Open University, Butterworths, London, UK. ISBN 0-408-02770-3. *(A teaching text on material shaping and joining, nicely illustrated by case studies.)*

Kalpakjian, S. (1984) Manufacturing Processes for Engineering Materials, Addison Wesley, Reading, Mass, USA. ISBN 0-201-11690-1. *(A comprehensive and widely used text on manufacturing processes for all classes of materials.)*

Lesko, J. (1999) Materials and Manufacturing Guide: Industrial Design, John Wiley and Sons, New York, NY. ISBN 0-471-29769-0. *(Brief descriptions, drawings and photographs of materials and manufacturing processes, with useful matrices of characteristics, written by a consultant with many years of experience of industrial design.)*

Mayer, R.M. (1993) "Design with Reinforced Plastics," The Design Council and Bourne Press Ltd., London, UK. ISBN 0-85072-294-2. *(A text aimed at engineers who wish to design with fiber-reinforced plastics; a useful source of information for shaping processes for polymer composites.)*

Swift, K. and Booker, J.D. (1998) Process Selection: From Design to Manufacture, John Wiley & Sons, UK. ISBN 0-340-69249-9. *(Datasheets in standard formats for 48 processes, some of which are for shaping.)*

Injection Molding

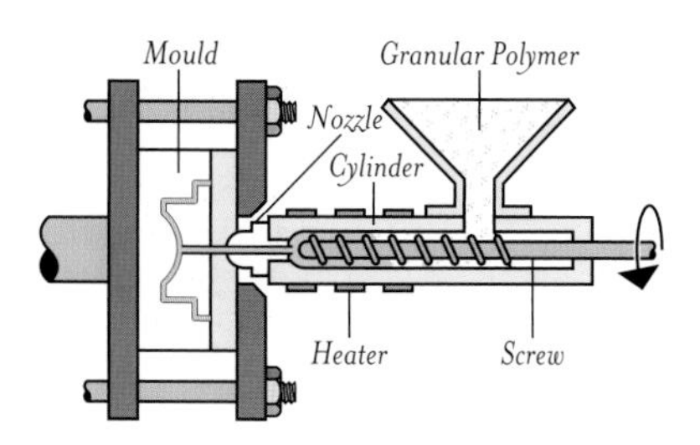

Attributes of Injection Molding

Weight range, kg	0.01–25
Minimum thickness, mm	0.3–10
Shape complexity	High
Allowable tolerance, mm	0.05–1
Surface roughness, μm	0.2–1.6
Economic batch size	10k–1,000k

What is it? No other process has changed product design more than injection molding. Injection molded products appear in every sector of product design: consumer products, business, industrial, computers, communication, medical and research products, toys, cosmetic packaging and sports equipment. The most common equipment for molding thermoplastics is the reciprocating screw machine, shown schematically in the figure. Polymer granules are fed into a spiral press where they mix and soften to a dough-like consistency that can be forced through one or more channels ("sprues") into the die. The polymer solidifies under pressure and the component is then ejected. Thermoplastics, thermosets and elastomers can all be injection molded. Co-injection allows molding of components with different materials, colors and features. Injection foam molding allows economical production of large molded components by using inert gas or chemical blowing agents to make components that have a solid skin and a cellular inner structure.

Shapes Simple and complex solid bulk shapes.

Design Notes Injection molding is the best way to mass-produce small, precise, polymer components with complex shapes. The surface finish is good; texture and pattern can be easily altered in the tool, and fine detail reproduces well. Decorative labels can be molded onto the surface of the component (see In-mold Decoration). The only finishing operation is the removal of the sprue.

Technical Notes Most thermoplastics can be injection molded, although those with high melting temperatures (e.g. PTFE) are difficult. Thermoplastic based composites (short fiber and particulate filled) can be processed providing the filler-loading is not too large. Large changes in section area are not recommended. Small re-entrant angles and complex shapes are possible, though some features (e.g. undercuts, screw threads, inserts) may result in increased tooling costs. The process may also be used with thermosets and elastomers.

The Economics Capital cost are medium to high, tooling costs are usually high – making injection molding economic only for large batch sizes. Production rate can be high particularly for small moldings. Multi-cavity molds are often used. Prototype moldings can be made using single cavity molds of cheaper materials.

Typical Products Housings, containers, covers, knobs, tool handles, plumbing fittings, lenses, toys and models.

The Environment Thermoplastic sprues can be recycled. Extraction may be required for volatile fumes. Significant dust exposures may occur in the formulation of the resins. Thermostatic controller malfunctions can be hazardous.

Competing Processes Polymer casting, die casting of metals.

Rotational Molding

What is it? The chunky, brightly-colored chairs for children are examples of what can be done with rotational molding. A pre-measured quantity of polymer powder is fed into a cold die, which is then heated in a large oven while it is rotated about two axes simultaneously. This tumbles and melts the powder, coating the inside walls of the mold to a thickness set by the initial load of powder, creating a hollow shell. The component then cools in a dwell cycle, the mold chilled by air or water spray. The process is best suited for components that are large, hollow and closed; although small, thin-walled components can be made and open shapes can be created by subsequent machining.

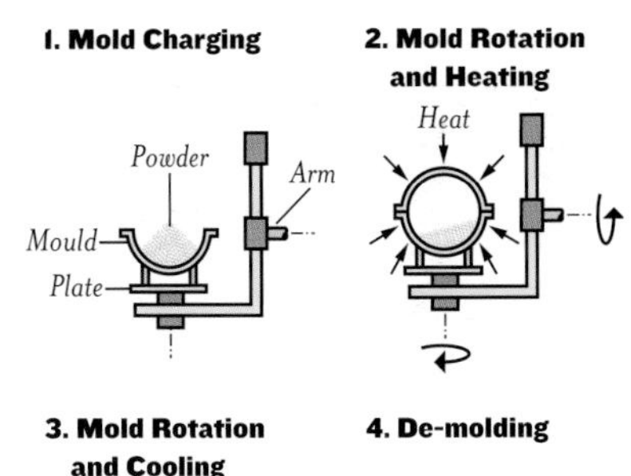

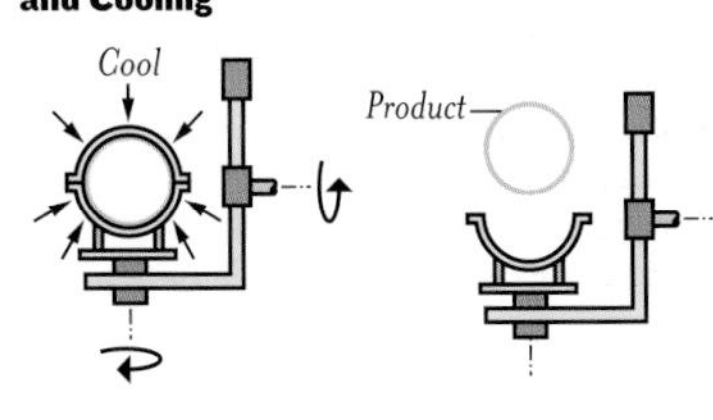

Shapes Simple, closed, hollow thin-walled shapes.

Design Notes Rotational molding is versatile, and one of the few processes able to make hollow (and thus material-efficient) shapes. The low pressure limits the possible sharpness of detail, favoring rounded forms without finely-detailed moldings. Inserts and pre-formed sections of different color or material can be molded in, allowing scope for creativity. Components with large openings – trash cans and road cones, for example – are molded in pairs and separated by cutting.

Attributes of Rotational Molding

Weight range, kg	0.1–50
Minimum thickness, mm	2.5–6
Shape complexity	Low
Allowable tolerance, mm	0.4–1
Surface roughness, µm	0.5–2
Economic batch size	100–10K

Technical notes In principle, most thermoplastics can be rotationally molded, though in normal practice the range is more restricted – the most common is polyethylene. The process makes large components, but the thickness is limited by the thermal conductivity of the polymer, since it has to melt through this thickness. One advantage: it produces virtually stress-free components. It is possible (but difficult) to vary the wall thickness; abrupt changes of section are not possible. Because of the two axes rotation, component length is restricted to less than 4 times the diameter. Thermosets can now be rotationally molded too – polyurethane is the most usual.

The Economics Equipment and tooling costs are low – much lower than for injection molding – but cycle times are longer than any other molding process, and it is labor-intensive.

Typical Products Tanks, food and shipping containers, housings, portable lavatories, traffic cones, large toys, trashcans, buckets, boat hulls, pallets.

The Environment No problems here.

Competing Processes Thermoforming; blow molding; lay-up methods.

Blow Molding

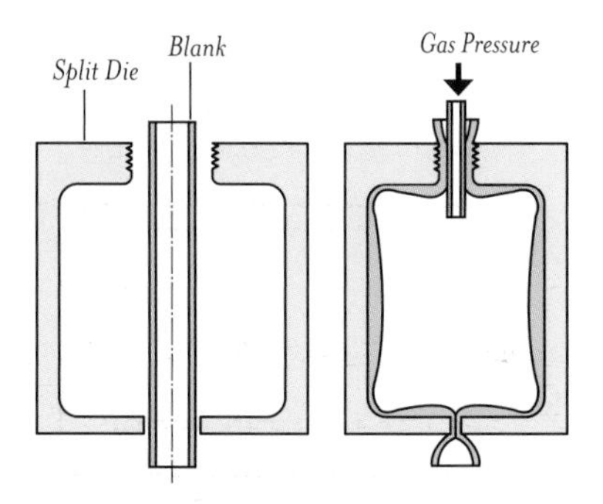

Attributes of Blow Molding

Weight range, kg	0.001–0.3
Minimum thickness, mm	0.4–3
Shape complexity	Low
Allowable tolerance, mm	0.25–1
Surface roughness, µm	0.2–1.6
Economic batch size	1K–10,000K

What is it? Blow molding is glass blowing technology adapted to polymers. In extrusion blow molding a tube or parison is extruded and clamped in a split mold with a hollow mandrel at one end. Hot air is forced under pressure through the mandrel, blowing the polymer against the mold walls where it cools and freezes. In injection blow molding, a pre-form is injection molded over a mandrel and transferred to the blowing die. Air is injected under pressure through the mandrel blowing the polymer against the mold walls where it cools and freezes, as before. The process gives better control over finished component weight and wall thickness than extrusion blow molding, with better precison in the unblown, injection molded neck areas, lending itself to screw closures, etc. Solid handles can be molded in. In stretch blow molding, an important variant, the temperature is chosen such that the polymer is drawn as it expands, orienting the molecules in the plane of the surface. It is used in the production of PET drink bottles.

Shapes Large, hollow thin-walled shapes.

Design Notes Extrusion blow molding allows a wide variety of hollow shapes to be formed, offering channels for air or liquid. Multi-layer injection blow molding is more commonly used for components that need to be gas tight and strong; here barrier layers are added to prevent gas diffusion and the outer layers to provide strength, impact resistance and good acceptance of printing and decoration. Insert molding is possible.

Technical Notes Blow molding is limited to thermoplastics, commonly PET, PC, HDPE, LDPE, PP, ABS and some PVCs. Limited levels of reinforcement are possible for composite materials. The wall thickness should be as uniform as possible to avoid distortion.

The Economics Extrusion blow molding is the cheaper of the two processes, because the die costs are less. It is competitive for large containers (capacity above 0.5 liters) and high batch sizes. Tooling costs for extrusion blow molding are much higher, limiting the process to volume production.

Typical Products Injection blow molding: bottles and containers, particularly those with threaded closures. Extrusion blow molding: containers, cases for tools and portable machinery, and large hollow structures such as car bumpers.

The Environment Waste material can be recycled. The resins can generate dust and vapors, requiring good ventilation.

Competing Processes Rotational molding.

Expanded Foam Molding

What is it? The snow-white, crisp, light protective packaging in which your computer arrived, and which you discarded, oblivious of its visual aesthetics, was made by expanded foam molding of polystyrene. It is a low temperature, low pressure, process using cheap mold materials, with uses far beyond those of throwaway packaging. There are two stages in its operation. Solid polymer granules, colored, if desired, containing a foaming agent that releases CO_2 on heating (a commodity product, widely available) are first softened and expanded by steam-heating under a small pressure. The softened particles are transferred to aluminum molds (to give good heat transfer) and steam-heated at 3 atmospheres, causing them to expand to 20 or more times their original volume and fuse together, filling the mold and taking up its shape.

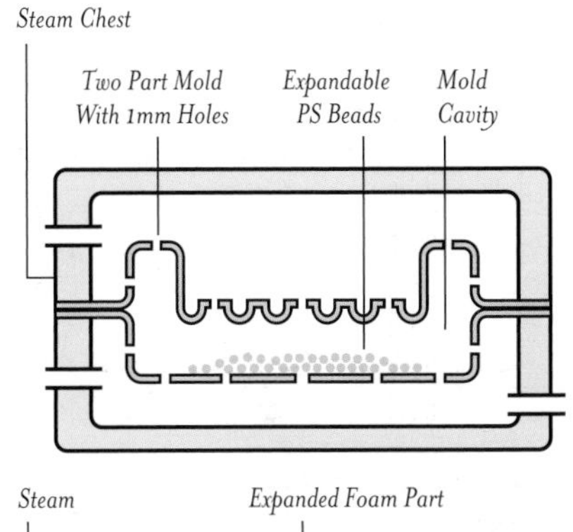

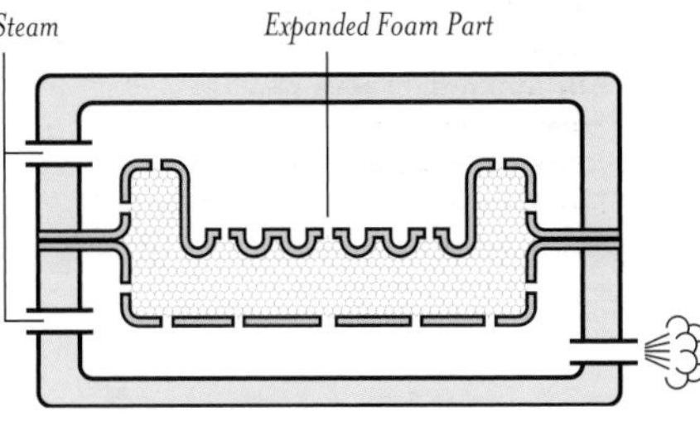

Shape. 3D solid shapes.

Design Notes Expanded foam moldings acquire a smooth skin where they contact the mold surface, and accept detail provided the radii are greater than 2 mm. The product is very cheap, gives good impact protection and can be recycled – so as packaging goes it is attractive. It is exceptionally good as a thermal insulator – hence its use in disposable cups and cooler bags and walls for kitchen fridges, refrigerated trucks and storage buildings. Increasingly, designers see other possibilities: expanded bead moldings have interesting acoustic properties, complex profiles can be molded, the material is very light, and has visual and tactile properties that are interesting. But the products are very easily damaged.

Attributes of Expanded Foam Molding

Weight range, kg	0.01–10
Minimum thickness, mm	5–100
Shape complexity	Low/Med
Allowable tolerance, mm	0.5–2
Surface roughness, µm	50–500
Economic batch size	2K–1000K

Technical Notes Thermoplastics are easy to mold, polystyrene particularly so. Expanded foam molding gives a closed cell foam with a porosity of 80–95%. The high porosity and closed cells make it a good thermal insulator; the low density (about 10% that of water) gives it outstanding buoyancy.

The Economics The low molding pressure and temperature makes this a particularly cheap process – hence its wide use in disposal, apparently valueless, objects.

Typical Products Disposable packaging for drinks and food (McDonalds has been held to ransom for this), lifesaving equipment, water-sport products like surfboards and life vests; thermal insulation in coolers, containers and carriers, core materials for sandwich structures (window and door frames, floor panels), protection of valuable products such as electronic, computer or audio systems.

The Environment At one time, CFC blowing agents were used for polystyrene expansion. These have now been replaced by gases that – in the volume in which they are used – do not damage the environment.

Competing Processes Nothing else is quite as cheap.

Compression Molding

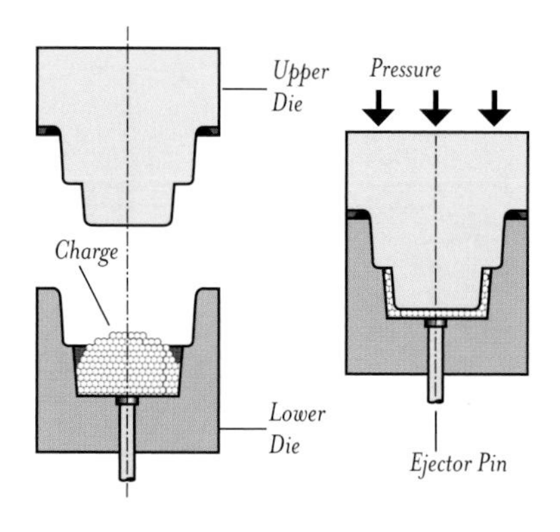

Attributes of Polymer Molding

Weight range, kg	0.2–20
Minimum thickness, mm	1.5–25
Shape complexity	Low/Med
Allowable tolerance, mm	0.1–1
Surface roughness, μm	0.2–2
Economic batch size	2K–200K

Attributes of BMC and SMC Molding

Weight range, kg	0.3–60
Minimum thickness, mm	1.5–25
Shape complexity	Low/Med
Allowable tolerance, mm	0.1–1
Surface roughness, μm	0.1–1.6
Economic batch size	5K–1,000K

What is it? In compression molding a pre-measured quantity of polymer – usually a thermoset – in the form of granules or a pre-formed tablet containing resin and hardener is placed in a heated mold. The mold is closed creating sufficient pressure to force the polymer into the mold cavity. The polymer is allowed to cure, the mold is opened and the component removed. A variant, polymer forging, is used to form thermoplastics that are difficult to mold, such as ultra high molecular weight polyethylene.Compression molding is widely used to shape the composites BMC and SMC.

Shapes Simple bulk shapes (BMC); flat, dished thin-walled shapes (SMC).

Design Notes Compression molding is limited to simple shapes without undercuts. The components generally require some finishing to remove flash. BMC and SMC moldings have good surface finish and accurate dimensioning, good enough for auto manufacturers to use them for external body components. SMC moldings yield high quality panels and casings; shapes in which the sheet thickness is more or less uniform. BMC moldings yield simple 3-dimensional shapes with changes in section.

Technical Notes BMC (Bulk Molding Compound) and SMC (Sheet Molding Compound) differ in the shape and content of reinforcement and filler. BMC has less (15–25% of glass fiber) and it is the easiest to mold to 3-dimensional shapes. SMC has more (up to 35%) of glass fiber and is limited to sheet shapes. DMC (Dough Molding Compound) is the genesis – a dough-like mix of thermosetting polyester, polyurethane or epoxy with hardener, chopped glass fiber, filler and coloring agent. Two more – GMT (Glass Mat Thermoplastics) and TSC (Thermoplastic Sheet Compounds) – are the thermoplastic equivalent, based on nylon 6 or polypropylene. The word "dough" conveys well the way in which they are shaped: squeezed between a pair of dies, like a pie crust.

The Economics Tooling costs are modest for polymer molding, higher for BMC and SMC molding, making the economic batch size larger for these two.

Typical Products Electrical and electronic components, tableware, washing machine agitators, utensil handles, container caps, appliance housings; SMC: body panels and bumpers for cars and trucks, gas and electricity meter boxes, and electrical housings. BMC: more complicated shapes.

The Environment The process itself does not damage the environment, but flash and scrap cannot be recycled for thermosets.

Competing Processes Injection molding; resin transfer molding; lay-up methods.

Resin Transfer Molding

What is it? Resin transfer molding (RTM) is an easy way of manufacturing complex shapes in fiber-reinforced composites without high tooling costs. It uses a closed mold, in two or more parts, usually made of glass-reinforced polymers or light metal alloys, with injection points and vents to allow air to escape. Reinforcement is cut to shape and placed in the mold, together with any inserts or fittings. The mold is closed and a low viscosity thermosetting resin (usually polyester) is injected under low pressure through a mixing head in which hardener is blended with the resin. The mold is allowed to cure at room temperature. The fluidity of the resin, and the low molding pressure, give tools a longer life at low cost.

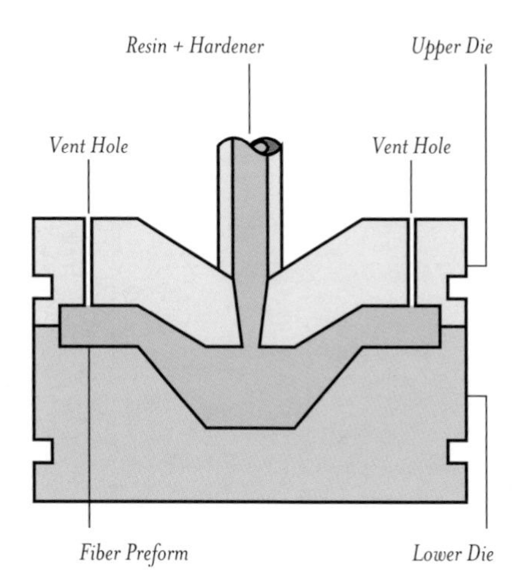

Attributes of Resin Transfer Molding

Weight range, kg	0.2–20
Minimum thickness, mm	1.5–13
Shape complexity	Med/High
Allowable tolerance, mm	0.25–1
Surface roughness, µm	0.2–1.6
Economic batch size	10K–100K

Shapes Simple bulk shapes; shaped panels.

Design Notes Resin transfer molding is increasingly used for large fiber-reinforced polymer components. Shapes can be complex. Ribs, bosses, and inserts are possible. Foam panels can be insert molded to reduce component weight.

Technical Notes A range of resin systems can be used with RTM: almost any thermosetting resin with a low viscosity is possible, for example, polyester, epoxies, vinyl esters and phenolics. The reinforcement, typically, is a 25–30% volume fraction of glass or carbon in the form of continuous fiber mat.

The Economics Tooling costs are low, and the process is not particularly labor intensive, making it attractive economically.

Typical Products Manhole covers, compress soil casings, car doors and side panels, propeller blades, boats, canoe paddles, water tanks, bath tubs, roof sections, airplane escape doors.

The Environment From the vented, closed mold – with adequate ventilation – it is easy to control emissions, so worker exposure to unpleasant chemicals is drastically reduced, when compared with other composite processes.

Competing Processes Lay-up methods; press molding; BMC and SMC molding.

Die-Casting

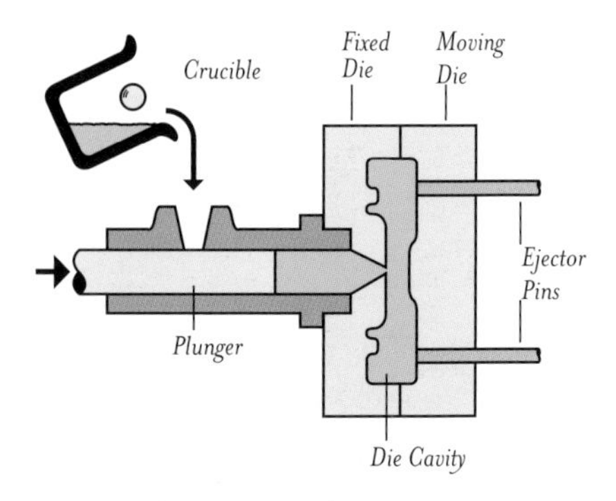

Attributes of Die-Casting

Weight range, kg	0.05–20
Minimum thickness, mm	1–8
Shape complexity	Med/High
Allowable tolerance, mm	0.15–0.5
Surface roughness, μm	0.5–1.6
Economic batch size	5K–1,000K

What is it? Most small aluminum, zinc or magnesium components with a complex shape – camera bodies, housings, the chassis of video recorders – are made by die-casting. It is to metals what injection molding is to polymers, and the two compete directly. In the process, molten metal is injected under high pressure into a metal die through a system of sprues and runners. The pressure is maintained until the component is solid, when the die is opened and the component ejected. For large batch sizes the dies are precision-machined from heat-resistant steel and are water cooled to increase life. For small batch sizes cheaper mold materials are possible.

Shapes Complex bulk shapes; hollow shapes have to be molded in sections and subsequently joined.

Design Notes Pressure die-casting allows thin-walled shapes and excellent surface detail. The integrity of the material properties is less good: turbulent filling of the mold and fast cycle time can lead to shrinkage and porosity. The process can make complex shapes, but elaborate movable cores increase the tooling costs.

Technical Notes Die-casting is widely used to shape aluminum, magnesium and zinc alloys. Two types of die-casting equipment are commonly used: cold or hot. In the "cold" process, the hot metal is held in a separate container and passed to the pressure chamber only for casting. In the "hot" process the reservoir of hot metal is held in the pressure chamber. The prolonged contact times restrict this process to zinc alloys.

The Economics High tooling costs mean that pressure die-casting becomes economic only for large batch sizes, but the process is one of the few that allows thin-walled castings. Aluminum has a small solubility for iron, limiting die-life to about 100,000 parts. Magnesium has none, giving almost unlimited die-life. Gravity die-casting has lower equipment costs but is usually less economic because the molten metal has to be more fluid – and thus hotter – to fill the mold well, this reduces the production rates.

Typical Products Record player and video player chassis, pulleys and drives, motor frames and cases, switch-gear housings, housings for small appliances and power tools, carburettor and distributor housings, housings for gear-boxes and clutches.

The Environment Aluminum, zinc and magnesium scrap can all be recycled. The process poses no particular environmental problems.

Competing Processes Investment casting, sand casting; injection molding of polymers.

Sand Casting

What is it? Sand casting probably started on beaches – every child knows how easy it is to make sand castles. Add a binder and much more complex shapes become possible. And sand is a refractory; even iron can be cast in it. In green sand casting, a mixture of sand and clay is compacted in the split mold around a pattern that has the shape of the desired casting. The pattern is removed to leave the cavity in which the metal is poured. Cheap wooden patterns, with gates and risers attached, are used when the batch size is small and the process is manual, but this is slow and labor intensive. Automated systems use aluminum patterns and automated compaction. In CO_2 /silicate sand casting, a mixture of sand and sodium silicate binder is packed around a pattern as before and flooded with CO_2 to seal the sodium silicate gel. In evaporative sand casting, the pattern is made from polystyrene foam and embedded in sand. When molten metal is poured into the mold, the polystyrene pattern vaporizes. In shell casting, a mixture of fine-grained sand and thermosetting resin is applied to a heated metal pattern (aluminum or cast iron) and cured to form a shell.

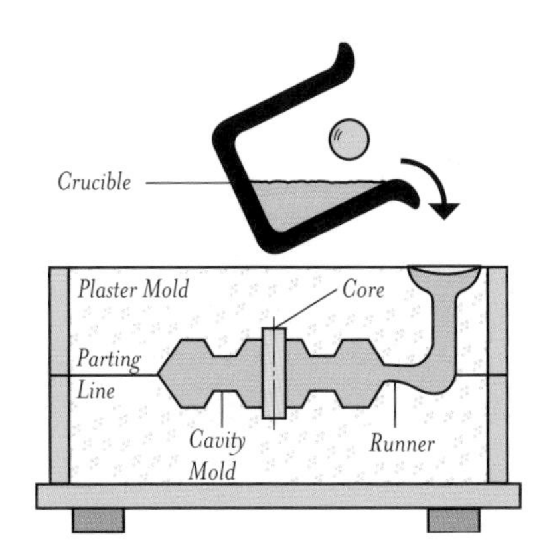

Attributes of Sand Casting

Weight range, kg	0.3–1000
Minimum thickness, mm	5–100
Shape complexity	Med/High
Allowable tolerance, mm	1–3
Surface roughness, µm	12–25
Economic batch size	1–1000K

Shapes Very complex shapes are possible – think of auto engine blocks.

Design Notes Few processes are as cheap and versatile as sand casting for shaping metals. It allows complex shapes, but the surface is rough and the surface detail is poor. The process has particular advantages for complicated castings with varying section thickness. Bosses, undercuts, inserts and hollow sections are all practical. Evaporative pattern casting leaves no parting lines, reducing finishing requirements. The minimum wall thickness is typically 3 mm for light alloys, and 6 mm for ferrous alloys.

Technical Notes In principle, any non-reactive, non-refractory metal (melting temperature less than 1800 C) can be sand cast. In particular, aluminum alloys, copper alloys, cast irons and steels are routinely shaped in this way. Lead, zinc and tin can be cast in dry sand, but melt at too low a temperature to evaporate the foam pattern in evaporative pattern casting.

The Economics Capital and tooling costs for manual sand casting are low, making the process attractive for small batches. Automated systems are expensive, but capable of producing very complex castings.

Typical Products Machine tool frames and bodies, engine blocks and cylinder heads, transmission cases, gear blanks, crank shafts, connecting rods. Evaporative pattern casting: manifolds, heat exchangers, pipe fittings, valve casings, other engine parts.

The Environment The mold materials, in many cases, can be reused.

Competing Processes Investment casting; die-casting.

Investment Casting

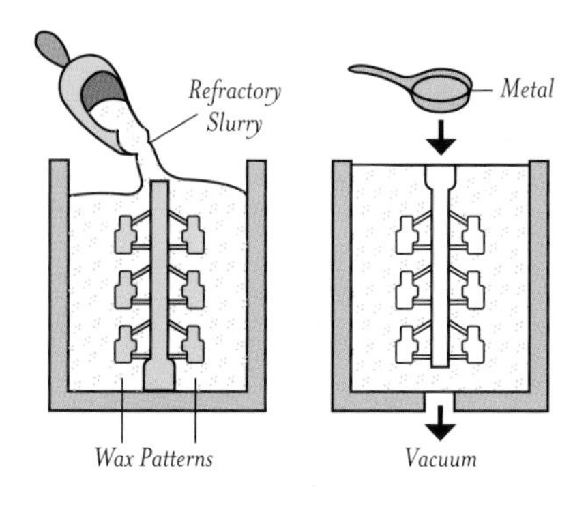

Attributes of Investment Casting

Weight range, kg	0.001–20
Minimum thickness, mm	1–30
Shape complexity	Med/High
Allowable tolerance, mm	0.1–0.4
Surface roughness, μm	1.6–3.2
Economic batch size	1–50K

What is it? If you have gold fillings in your mouth, be thankful for this process – it was used to make them. The lost wax process – the old name for investment casting – has been practiced for at least three thousand years; sophisticated jewelery, ornaments and icons were being made in Egypt and Greece well before 1500 BC. In investment casting, wax patterns are made and assembled (if small) into a tree with feeding and gating systems. The assembled pattern is dipped into refractory slurry, then covered in refractory stucco and allowed to dry. The procedure is repeated a number of times until about 8 layers have built up, creating a ceramic investment shell. The wax is then melted out and the ceramic shell fired before the molten metal is cast. Gravity casting is adequate for simple shapes, but air pressure, vacuum or centrifugal pressure is needed to give complete filling of the mold when complex, thin sections are involved. The mold is broken up to remove the castings. The process is suitable for most metals of melting temperatures below 2200 C. Because the wax pattern is melted out, the shape can be very complex, with contours, undercuts, bosses, and recesses. Hollow shapes are possible: bronze statues are hollow, and they are made by an elaboration of this process.

Shapes As complex as you like.

Design Notes The process is extremely versatile, allowing great freedom of form. It offers excellent reproduction of detail in small 3D components.

Technical Notes Investment casting is one of the few processes that can be used to cast metals with high melting temperatures to give complex shapes; it can also be used for low melting metals. The traditional uses of investment casting were for the shaping of silver, copper, gold, bronze, pewter and lead. Today the most significant engineering applications are those for nickel, cobalt and iron-based alloys to make turbine blades.

The Economics Investment casting lends itself both to small and large batch sizes. For small batch sizes the process is manual, with low capital and tooling costs, but significant labor costs. When automated, the capital costs are high, but quality control and speed are greater. The production rate is increased by the use of multiple molds.

Typical Products Jewelery, dental implants, statuary, metal sculpture and decorative objects, high temperature gas turbines and similar equipment, monumental statues.

The Environment There are the usual hazards associated with casting molten metal, but procedures for dealing with these are routine. The mold materials cannot, at present, be recycled.

Competing Processes Die-casting (for metals that melt below 800 C); sand casting.

Polymer Casting

What is it? Many resins are sufficiently fluid before polymerization that they can be cast – poured into a shaped mold without applying pressure, where they react and solidify. A few thermoplastics – notably acrylics – and most thermosets can be cast by doing so before they polymerize.

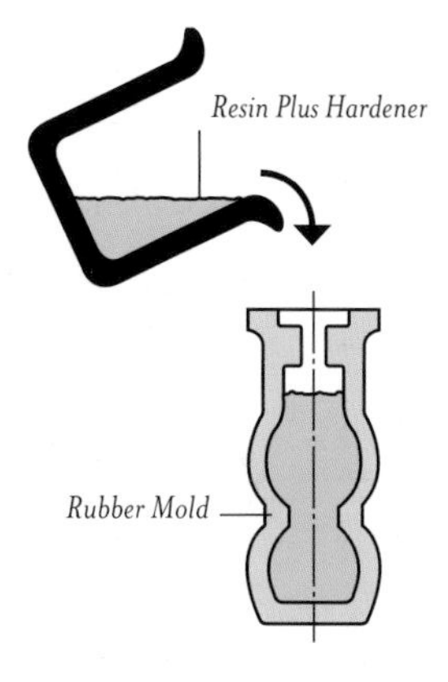

Shapes Most shapes – complex shapes require a flexible mold.

Design Notes The optical properties of cast transparent polymers like acrylics are better than if molded. Fillers can be added, but for this a similar process – centrifugal casting – is frequently used. Large parts and large section thicknesses are common, but the quality of the final part depends heavily on the skill of the operator – trapped air and gas evolution are both potential problems.

Technical Notes In casting methyl methacrylate (acrylic), a monomer-soluble initiator is used. The reaction is vigorous and liberates much heat that must be dissipated to keep the temperature within safe limits and prevent the monomer from boiling. Considerable shrinkage occurs – as much as 21% for methyl methacrylate – and must be taken into account when designing molds for casting. Using monomer-polymer syrups made by interrupted polymerization helps control both heat and shrinkage.

Attributes of Polymer Casting

Weight range, kg	0.1–700
Minimum thickness, mm	2–100
Shape complexity	High
Allowable tolerance, mm	0.8–2
Surface roughness, μm	0.5–1.6
Economic batch size	10–1000

The Economics The tooling required is cheap; stiff (metal or epoxy) or flexible (elastomer) molds are both possible. For small tooling the price is under $100, for large tooling it is a few thousand dollars.

Typical Products Elevator buckets, bearings, large gears, sheets, tubes, electronic encapsulation, rod stock, bowling balls, epoxy tooling; centrifugal casting is used for pipes, tanks and containers.

The Environment Most polymers give off vapors when curing – adequate ventilation is important.

Competing Processes Injection molding, compression molding – but in both cases the dies are much more expensive.

Shape Rolling and Die Forging

Shape Rolling

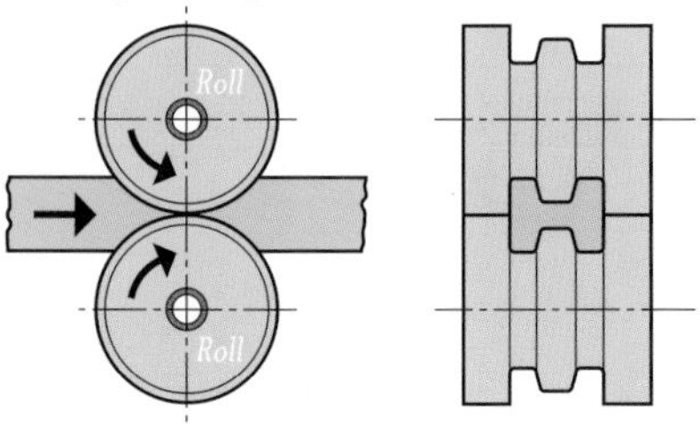

Roll

Roll

Die Forging

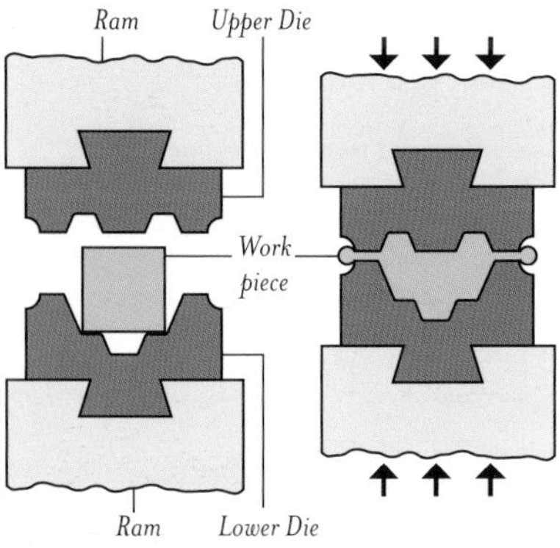

Ram Upper Die

Work piece

Ram Lower Die

Attributes of Shape Rolling and Die Forging

Weight range, kg	0.1–100
Minimum thickness, mm	2–100
Shape complexity	Low
Allowable tolerance, mm	0.3–2
Surface roughness, µm	3.2–12.5
Economic batch size	10K–1000K

What are they? In rolling and forging a metal ingot is squeezed to shape by massive rolls or dies that subject it to large plastic deformation. Nearly 90% of all steel products are rolled or forged. I-beams and other such continuous sections are made by shape rolling. In hot rolling, the ingot, heated to about 2/3 of its melting temperature, is forced through a series of rolls that progressively shape the profile. In hot die forging, by contrast, a heated blank is formed between open or closed dies in a single compressive stroke. In cold rolling and forging the metal blank is initially cold, though deformation causes some heating. Often a succession of dies is used to create the final shape.

Shapes Rolling: prismatic shapes only. Forging: Simple solid bulk shapes; re-entrant angles are not possible.

Design Notes The process produces components with particularly good mechanical properties because of the way in which the deformation refines the structure and reduces the porosity. During hot rolling and forging the metal recrystallizes, remaining relatively soft, and its surface may oxidize. Cold rolling and forging, by contrast, impart a high surface finish and cause extreme work-hardening, raising the strength of the product but limiting the extent of deformation.

Technical Notes Most metals can be rolled or forged, but the extent of deformation that is possible varies widely. Those best suited are the range of aluminum and magnesium alloys designed for deformation processing (the "wrought" grades), copper alloys and steels.

The Economics Rolls and dies for rolling and forging have to be made from exceptionally hard materials and are expensive, meaning that shape rolling and closed die forging are suitable only for large batches.

Typical Products Rolling: continuous rods, square sections, I-beams. Forging: highly stressed mechanical parts such as aircraft components, connecting rods, crank shafts, gear blanks, valve bodies, tube and hose bodies, hand and machine tools.

The Environment The lubricants used in rolling and forging generate oil mist and unpleasant vapors, requiring good ventilation.

Competing Processes Extrusion (for continuous prismatic sections); die, sand or investment casting for solid bulk shapes.

Extrusion

What is it? The process of squeezing toothpaste from its tube is one of extrusion. The gooey toothpaste – or the gooey polymer or metal in the industrial process – is forced by pressure to flow through a shaped die, taking up the profile of the die orifice. In co-extrusion two materials are extruded at the same time and bond together – a trick used in toothpaste to create colored stripes in it.

Shapes Prismatic shapes: solid or hollow.

Design Notes Metal extrusion is limited to ductile materials with melting points below 1700 C, commonly aluminum, copper, magnesium, low and medium carbon steels, low alloy steels and stainless steels. The tolerance is often lower than expected because of creep and die wear; it can be improved by cold drawing as a secondary process. Symmetric cross-sections, constant wall thickness and generous radii are the easiest to form; the aspect ratio of the section should not exceed 14:1 for steel or 20:1 for magnesium. Impact extrusion is a cold process for metals (aluminum, copper, lead, magnesium, tin, zinc, carbon steels, low alloy steels) which combines the principles of forging and extrusion.

Polymer extrusion begins with powder or pellets; pressure is built up by a rotating screw, forcing the polymer through a heating chamber and through the die. Most polymers can be extruded (including those with particulate and short fiber-reinforcement). In ceramic extrusion a ceramic powder is mixed with a polymer binder and extruded like a polymer. The extruded section is then fired, burning off the binder and sintering the ceramic powder.

Technical Notes There are two variants of material extrusion: direct – where the die is stationary and the metal is forced through it; and indirect – where the die itself compresses the metal. Indirect extrusion has less friction between the billet and die, so the extrusion forces are lower but the equipment is more complex and the product length is restricted.

The Economics For metals, rolling is more economical for simple shapes and large product runs; for polymers, the tooling costs are relatively low although capital costs are high and secondary processing is often required.

Typical Products Tubing, window frame sections, building and automotive trim, aircraft structural parts, railings, rods, channels, plastic-coated wire, seals, filaments, film, sheet, pellets, bricks.

The Environment No special problems here.

Competing Processes For metals: rolling, drawing; for polymers: extrusion blow molding.

Direct Extrusion

Billet

Ram *Die*

Extruded Product

Indirect Extrusion

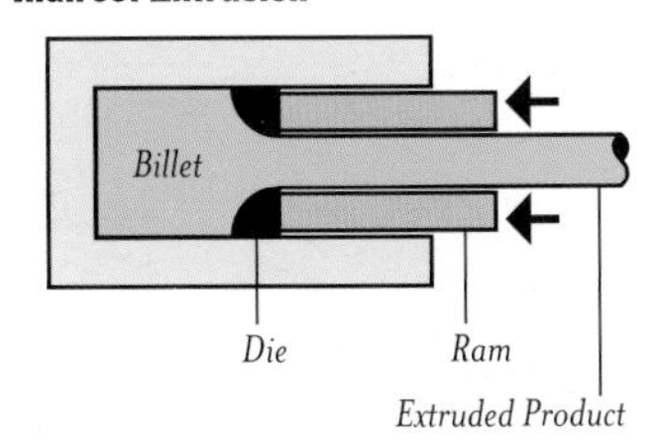

Polymer and Ceramic Extrusion

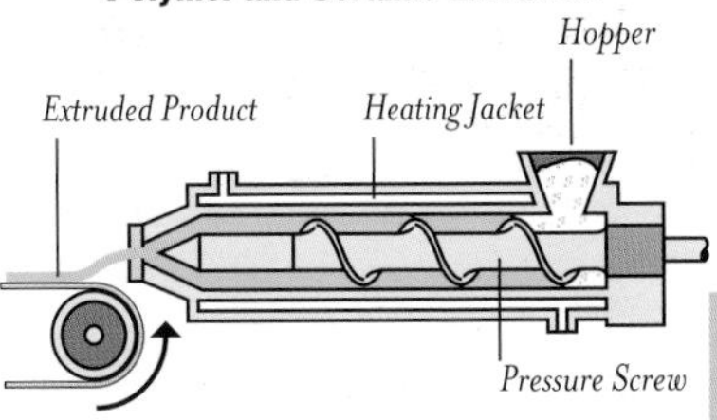

Attributes of Extrusion	
Weight range, kg	1–1000
Minimum thickness, mm	0.1–900
Shape complexity	Low
Allowable tolerance, mm	0.2–2
Surface roughness, μm	0.5–12.5
Economic batch size	1K–1,000K

Press Forming, Roll Forming and Spinning

Press Forming

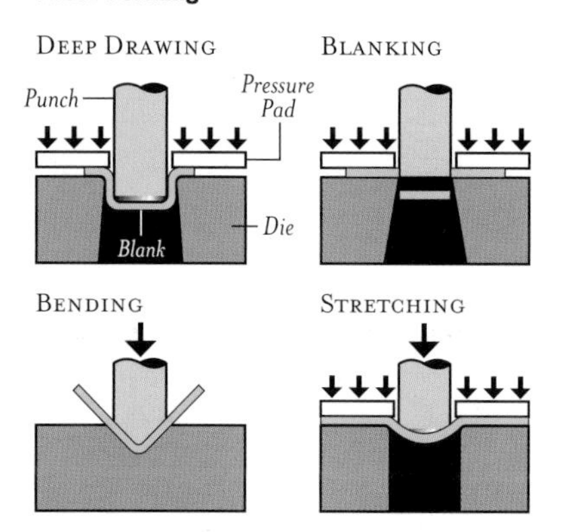

Roll Forming

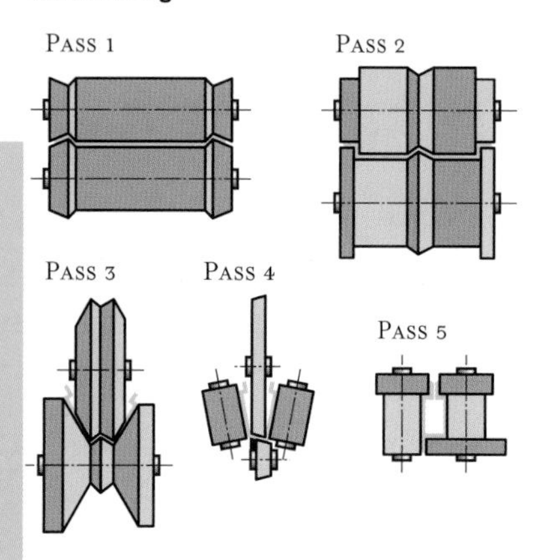

Spinning

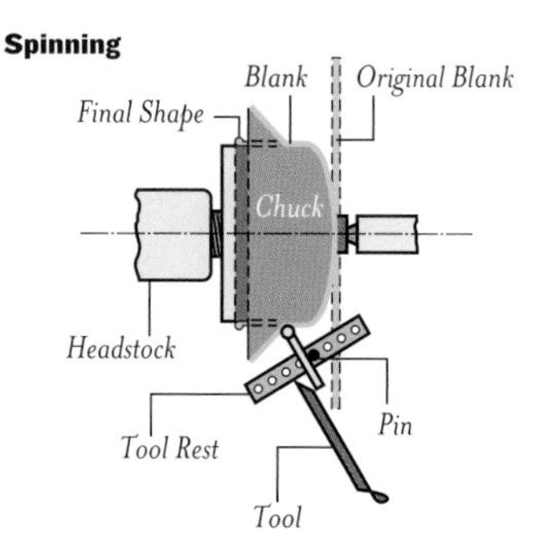

Attributes of Press Forming, Roll Forming and Spinning

Weight range, kg	0.01–30
Minimum thickness, mm	0.2–5
Shape complexity	Med
Allowable tolerance, mm	0.1–0.8
Surface roughness, µm	0.5–12.5
Economic batch size	25K–250K

What is it? Many products are made by cutting, pressing, punching, folding or spinning of sheet. Press forming covers a range of sheet forming processes that use a die and a press; these include blanking, shearing, drawing and stretching. They can be performed consecutively to form complex shapes. Tools are dedicated, so tooling costs are high. In roll forming a continuous strip of sheet metal is fed through a series of shaped rolls, gradually forming it to the desired profile. The process is suited for long lengths of constant, sometimes complex, cross sections. Hollow components are possible by incorporating seem welding into the process. High production rates, tooling and capital costs make the process economic only for large production levels. In spinning a circular blank of sheet metal is formed over a rotating mandrill or forming block, against which it is pressed by a rigid tool or roller as it spins. The tools are very simple, made of wood or metal, and are therefore cheap.

Shapes Flat and curved thin-walled shapes. Rolling: prismatic shapes only. Spinning: curved thin-walled shapes.

Design Notes Skilled metal workers can form sheet into intricate shapes using drawing, bending, and forming operations. Shapes with holes, curves, recesses, cavities and raised sections are common. Sheet forming starts with sheet stock, giving the products an almost constant cross-section unless folded. Spinning is limited to relatively simple hemispherical, conical, or cylindrical shapes, although re-entrant shapes are possible.

Technical Notes Sheet forming is most commonly applied to metals, particularly steels, aluminum, copper, nickel, zinc, magnesium and titanium. Polymer and composite sheet can be processed, though here the operations are limited to blanking and shearing.

The Economics The cost of sheet forming operations depends largely on tooling costs. Dedicated tools and dies are usually expensive, and require large batch sizes to make them economic. Processes with low-cost tooling, such as spinning, are best suited to small batch sizes, because they are labor intensive.

Typical Products Press forming: automobile body parts, casings, shells, containers. Stamping and blanking: smaller mechanical components such as washers, hinges, pans and cups. Roll forming: architectural trim, window frames, roof and wall panels, fluorescent light fixtures, curtain rods, sliding door tracks, bicycle wheels. Spinning: rocket motor casings, missile nose cones, pressure vessels, kitchen utensils, light reflectors.

The Environment Sheet metal forming carries no particular environmental hazard.

Competing Processes Die-casting; injection molding.

Thermoforming

What is it? Large thermoplastic sheet moldings are economically made by thermoforming. In vacuum thermoforming a thermoplastic sheet, heated to its softening point, is sucked against the contours of a mold, taking up its profile; it is then cooled, solidifying against the mold. Drape thermoforming relies partly on vacuum and partly on the natural sag of the hot polymer to form the shape. Plug-assisted thermoforming augments the vacuum with a compression plug. Pressure thermoforming uses a pressure of several atmospheres to force the hot polymer sheet onto the mold. Male or female molds are possible and – for vacuum thermoforming – can be machined from wood, polymer foam, or from aluminum (for larger batch sizes).

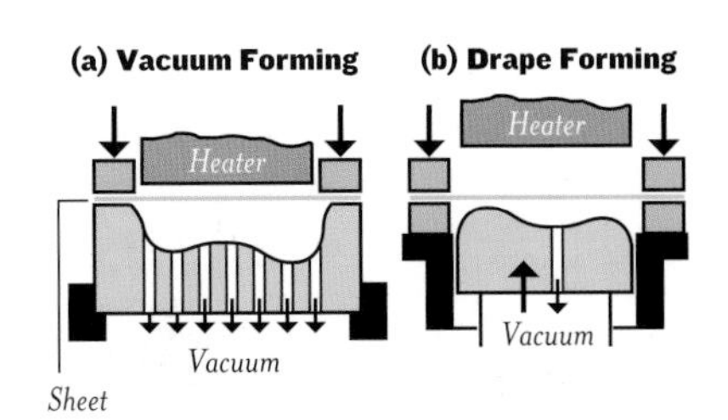

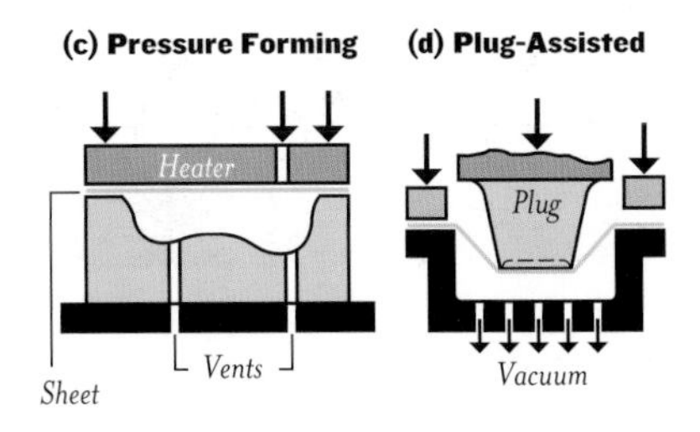

Attributes of Thermoforming

Weight range, kg	0.003–50
Minimum thickness, mm	0.25–6
Shape complexity	Low
Allowable tolerance, mm	0.5–1
Surface roughness, μm	0.3–1.6
Economic batch size	10–100k

Shapes Curved thin-walled shapes.

Design Notes The low pressure in vacuum forming gives poor reproduction of fine details; pressure forming, using higher pressures, gives sharper features but is more expensive because wooden molds cannot be used. The surface of the sheet in contact with the mold tends to mark, so the tool is usually designed with the finished side away from the mold. Colored, textured or pre-decorated sheet can be molded, reducing finishing costs.

Technical Notes Thermoforming is used to shape thermoplastic sheet, particularly ABS, PA, PC, PS, PP, PVC, Polysulphones, PBT, PET, foams, and shortfiber-reinforced thermoplastics. The maximum depth-to-width ratio of the molding is in the range 0.5 to 2. Inserts can be molded in. The process is able to cope with a very large range of sizes from products as small as disposable drink cups to those as large as boat hulls; and it is economic for both small and large batch sizes. It gives products with excellent physical properties but the starting material is more expensive (sheet rather than pellet). The product has to be trimmed after forming, and sheet scrap cannot be directly recycled.

The Economics Both the capital and the tooling costs of thermoforming equipment are low but the process can be labor intensive.

Typical Products Appliances, refrigerated liners, bath tubs, shower stalls, aircraft interior panels, trays, signs, boat hulls, drink cups.

The Environment No environmental problems here.

Competing Processes Injection molding (for large batch sizes); rotational molding.

Lay-up Methods

Hand Lay-up

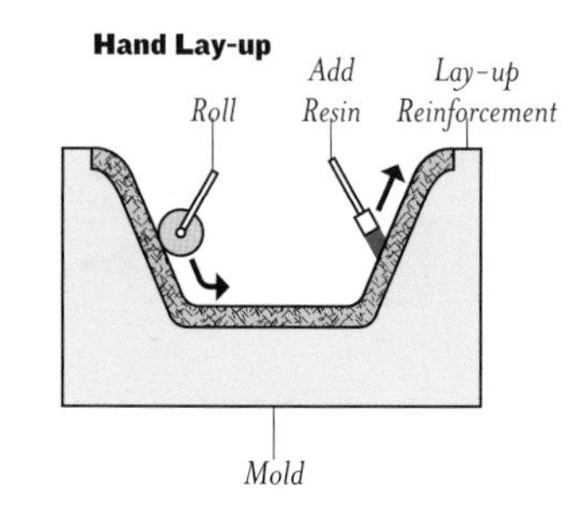

Vacuum Bag

Pump · Pump · Flexible Bag · Mold · Release Coat · Resin +Glass

Pressure Bag

Flexible Bag · Mold · Release Coat · Resin +Glass · Heater

Attributes of Lay-up Methods

Weight range, kg	1–6000
Minimum thickness, mm	2–10
Shape complexity	Low/Med
Allowable tolerance, mm	0.6–1 mm
Surface roughness, µm	1–500
Economic batch size	1–500

What are they? The low density and high resilience of polymer-matrix composites gives them an increasingly large fraction of the market for shells, casings, body panels and structural components. They can be formed in a number of ways.

In hand lay-up, an open mold (made of glass-reinforced polymers, wood, plaster, cement or light metal alloys) is coated with a resin to give a smooth surface skin. When this has cured, a layer of reinforcement (woven or knitted glass or carbon fiber) is laid on by hand, resin is applied by a brush or spray gun, and the layer is rolled to distribute the resin fully through the fibers. The process is repeated, layer by layer until the desired thickness is reached. The type of weave influences its ability to take up double curvature ("drapability"): random mat ("glass wool") and knitted fibers have good drapability; weaves with straight wefts do not. Flame retardants and inert fillers are added to reduce cost and improve properties. In spray-up a resin mixed with chopped fibers is sprayed into the mold; it is used for large components where the reinforcement fraction need not be large. In vacuum/pressure bag molding, the reinforcement and the resin are applied to the mold by conventional hand or spray-up methods. A bag is then placed over the curing composite and forced down onto it by vacuum or pressure, giving dense materials with better properties.

Shapes Curved thin-walled shapes with a single or double curvature.

Design Notes Lay-up methods give the greatest freedom to exploit the potential of fiber-reinforced polymers. They are generally limited to shapes with high surface area to thickness ratios. Ribs, bushes and foam panels inserts are all possible. Vacuum and pressure bag methods give higher quality than hand lay-up because air bubbles in the resin are removed.

Technical Notes The resin systems are all thermosetting, based on polyesters, epoxies, vinyl esters or phenolics. The reinforcement is commonly glass fiber but carbon and natural fibers, such as jute, hemp or flax can also be used.

The Economics The mold materials can be cheap (wood, cement); large moldings and small batch sizes are practical, but lay-up methods are labor intensive.

Typical Products Hand lay-up: boat hulls, building panels, vehicle bodies, ducts, tanks, sleighs, tubs, shower units. Vacuum/pressure bag molding: aerospace components, radomes.

The Environment Open molds lead to evaporation of resin, creating a health hazard: adequate ventilation is important.

Competing Processes Press molding; resin transfer molding.

Powder Methods

What are they? When you make a snowman by squeezing and modeling snow, you are compacting and sintering a powder. Many materials are most easily formed by powder methods. The technique has particular value in shaping difficult materials – ceramics and refractory metals for instance – that are too brittle to deform and have high melting temperatures and so cannot be easily cast. In powder sintering, the loose powder is packed in a shaped mold of steel, graphite or ceramic and sintered at a temperature, typically, of 2/3 of the melting temperature of the powder. In pressing and sintering, the powder is first compressed in a cold die, giving it sufficient strength to be handled, allowing it to be sintered as a free-standing body. Better densification, strength and ductility are given by hot pressing, which combines compaction and sintering into one operation. Vacuum hot pressing is used for powders that are particularly sensitive to contamination by oxygen or nitrogen (titanium is an example). In hot isostatic pressing (HIPing), the powder is placed in a deformable metal container and then subjected to high temperature and high pressure (1100 C, 200 MPa) in an argon atmosphere that acts as the pressure medium. In powder extrusion or powder injection molding (also known as metal injection molding) the powder is mixed with up to 50% of a polymer binder and extruded or injection molded (See "Extrusion"). The resulting shape is sintered in a separate operation, burning off the polymer and bonding the powder particles.

Shapes Prismatic or simple 3D shapes.

Design Notes Powder methods give great freedom of choice of materials (particularly as powders of different materials can be mixed to give composites) but they are limited in the shapes that can be made. Side walls must, in general, be parallel, undercuts at right angles to the pressing direction, and screw threads, lateral holes and re-entrant grooves cannot be molded in easily. There is considerable shrinkage (up to 35%) on sintering – the dimensional changes must be allowed for.

Technical Notes Powder methods are routinely used for brass, bronze, iron-based alloys, stainless steels, cobalt, molybdenum, titanium, tungsten, beryllium, metal matrix composites and ceramics.

The Economics Powder methods are fast and use material efficiently, but powders are expensive.

Typical Products Powder sintering: filters, porous bearings and low density ceramics. Pressing and sintering: small gears and cams, small bearings, electrical and magnetic components. Hot isostatic pressing: tool steel and tungsten carbide cutting tools, super alloy components for aerospace. Powder injection molding: spark plugs, heat engine components, medical and dental devices.

The Environment Burn-off of binder generates toxic fumes.

Competing Processes None that directly compete.

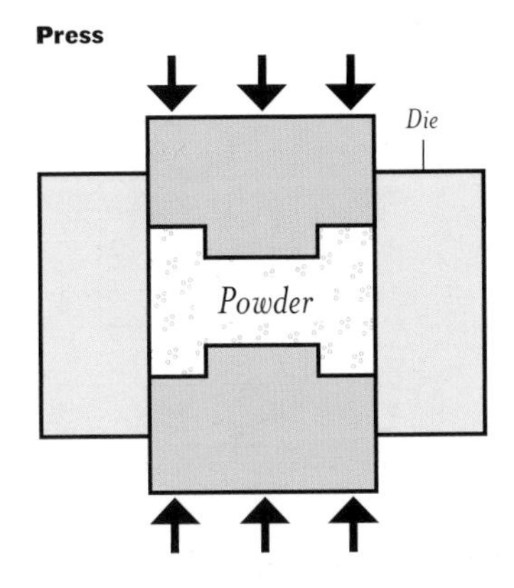

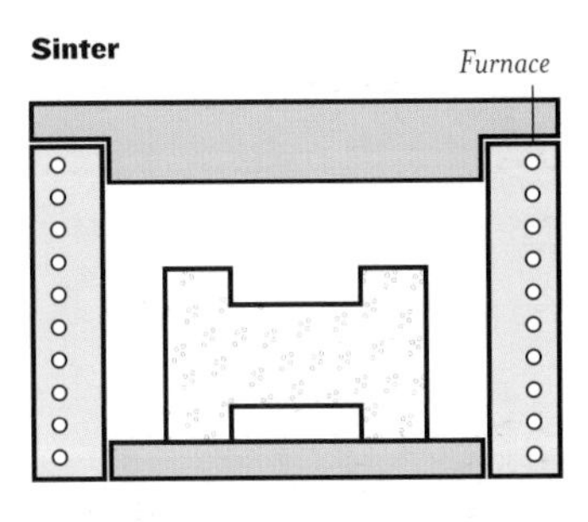

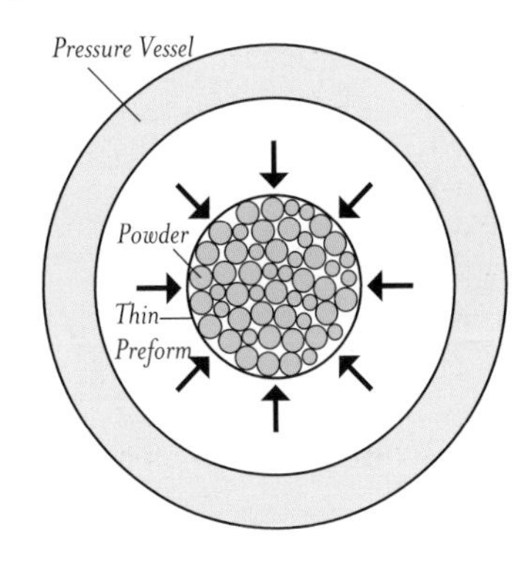

Attributes of Powder Methods

Weight range, kg	0.01–5
Minimum thickness, mm	1.5–8
Shape complexity	Low/Med
Allowable tolerance, mm	0.1–1
Surface roughness, μm	1.6–6.3
Economic batch size	1K–1,000K

Powder Methods

Laser Prototyping

Stereolithography

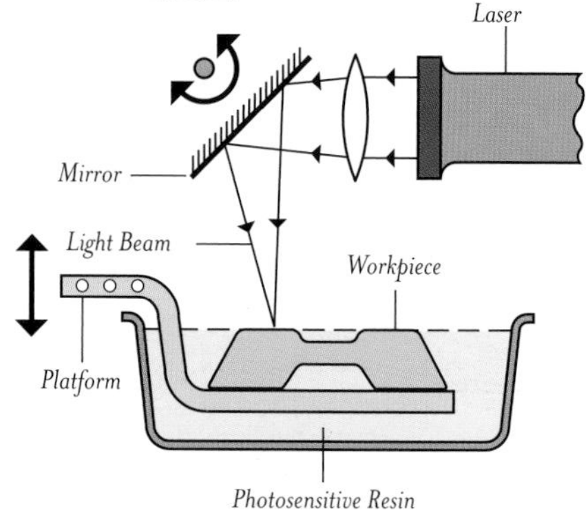

Attributes of Laser Prototyping

Weight range, kg	0.1–20
Minimum thickness, mm	0.5–100
Shape complexity	High
Allowable tolerance, mm	0.2–2
Surface roughness, µm	100–125
Economic batch size	1–100

What are they? All designers dream of effortlessly transforming their visions into real three-dimensional objects, of arbitrary complexity, that can be handled, viewed, and tried out in the context in which they are to be used. Methods for rapid prototyping – one of the fastest-growing sectors of manufacturing technology – allow this. The input is a CAD file describing the form of the component; the output is a single realization of the component in polymer, metal or compacted laminated paper. In stereolithography (SL) a photo sensitive resin is polymerized by a UV laser beam that scans the surface of a bath containing liquid resin. Each scan creates a thin slice of the object; when one scan is complete the platform is lowered slightly (typically 0.1 mm) flooding the surface with fresh liquid that is polymerized by the next scan. The object is thus built layer-by-layer from the bottom up. In selective laser sintering (SLS) the liquid resin is replaced by a fine heat-fusible powder (a thermoplastic or wax) that is melted by the scanning laser beam, creating a single slice of the object. The table is then lowered and a new layer of powder is spread across the surface by a wiper, ready for the next scan. The method has now been extended to sinter metals. In laminated object manufacture (LOM) the slices are created by laser cutting a sheet of paper coated with a thermoplastic. The paper is fed from a roller and the profile of one slice is cut by the laser; a heated roller then presses the slice onto the stack of previously-cut slices, bonding it in place. The process is faster than the other two because the laser only traces the outline of the slice. It is best for large components with thick wall-sections. The finished material resembles wood.

Shapes Solid or hollow complex bulk shapes.

Design Notes Rapid prototyping in ABS or nylon offers the designer the ability to make prototypes of almost unlimited complexity. All the processes are slow: the time to make an object scales with volume for SL and SLS, and with surface area for LOM – imposing a practical size limit of about 300 mm.

Technical Notes Rapid prototyping systems are evolving very rapidly, and are already an essential part of the model-building capability of designers. Their speed and precision will increase and their cost will decrease in the future.

The Economics The cost of making an object depends on size and process – \$500 to \$5000 gives an idea. This will decrease as faster systems become available.

Typical Products Rapid prototyping is widely used in product design to realize a 3D model of an evolving concept.

The Environment Liquid resins are volatile, fine powders can be explosive, and laser beams can damage eyes – all require appropriate precautions.

Competing Processes Deposition prototyping (next page).

Deposition Prototyping

What are they? Think of the way in which the top of a wedding cake is built up: the icing is squeezed, like toothpaste, from a gun. Now reduce the scale by a factor of 100, and you have the basis of deposition-based rapid prototyping. The input – as with laser-based prototyping – is a CAD file describing in the form of the component; the output is a single realization of the component in polymer or wax. The figure shows one variant of the technology. In fused deposition modeling (FDM) a fine stream of thermoplastic or wax is deposited by a two-axis heated extrusion head. The semi-liquid polymer or wax is extruded from the heated head and deposited in layers, typically 0.1 mm thick, one layer at a time, starting at the base. Successive layers bond by thermal fusion. The process is well adapted to form ABS and nylon and requires no post-curing (making it faster). Overhangs can be created without support because the polymer sets very quickly. In ballistic particle manufacture (BPM) particles of molten thermoplastic are shot by a piezo-electric jet and freeze when they hit the object. The jet-head scans the surface, creating single layers at each scan, a range of thermoplastics and waxes can be used in the jet. In true three-dimensional (3D) printing the thermoplastic is jetted from a print-head with up to 96 jets in a linear array, like an ink jet printer. The resolution is like that of the printer – about 300 dpi. The process, which is rapidly evolving, has the potential to build objects both accurately and quickly. All three processes can be adapted to deposit fine powders of metals or ceramics, though commercially available systems that do this are not yet available.

Fused Deposition Modeling

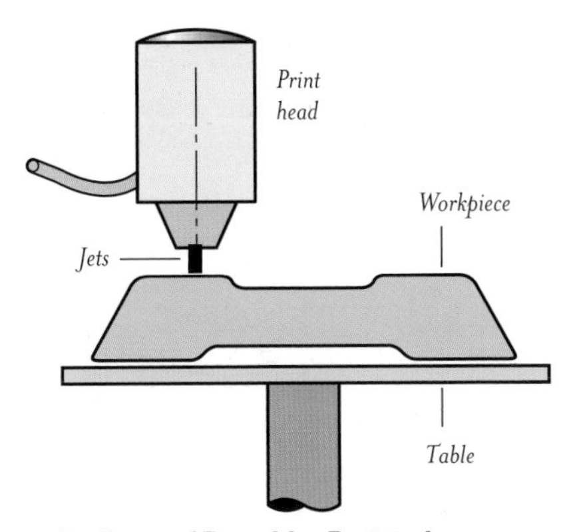

Attributes of Deposition Prototyping

Weight range, kg	0.1–10
Minimum thickness, mm	1.2–100
Shape complexity	High
Allowable tolerance, mm	0.3–2
Surface roughness, µm	75–100
Economic batch size	1–100

Shapes Solid or hollow complex bulk shapes.

Design Notes Rapid prototyping in ABS or nylon offers the designer the ability to make prototypes of strong materials and almost unlimited complexity.

Technical Notes Rapid prototyping systems are evolving very rapidly, and are already an essential part of the model-building capability of designers. Their speed and precision will increase and their cost will decrease in the future.

The Economics The cost of making an object depends on size and process – \$500 to \$5000 gives an idea. This will decrease as faster systems become available.

Typical Products Rapid prototyping is widely used in product design to realize a three-dimensional model of a evolving concept.

The Environment No particular hazards.

Competing Processes Laser-based prototyping.

Joining Profiles

The manufacture of a product involves many steps, one of which is that of assembly – and it is one that adds cost. Two techniques contribute to economical assembly: that of design to minimize the number of components and to enable ease of assembly, and that of selecting the joining process that best suits the materials, joint geometry and the required performance of the joint during its life. The figure on the next page shows joint geometries with their common names.

In this section we have assembled information for representative joining processes listed on the facing page. Three broad families can be distinguished: adhesives, mechanical fasteners, welding. The pages that follow give information about these. Choice of process also depends on the materials to be joined and the geometry of the joint. The processes used to weld polymers differ in obvious ways from those used to weld metals; less obviously, adhesives and mechanical fastening, too, are material-specific. Many processes can only join components made of the same material: steel to steel, polyethylene to polyethylene. Others can bond dissimilar materials: metal to glass, or polymer to ceramic, or composite to metal.

More information can be found in the sources listed under "Further Reading."

Introduction 260
Further Reading 261

Adhesives 262

Fasteners
Sewing 266
Rivets and Staples 267
Threaded Fasteners 268
Snap Fits 269

Welding
Hot Gas Welding 270
Hot Bar Welding 271
Hot Plate Welding 272
Ultrasonic Welding 273
Power-beam Welding 274
Brazing 276
Soldering 277
Torch Welding 278
MIG Welding 279
TIG Welding 280
Resistance Welding 281
Friction Welding 282
Diffusion and
Glaze Bonding 283

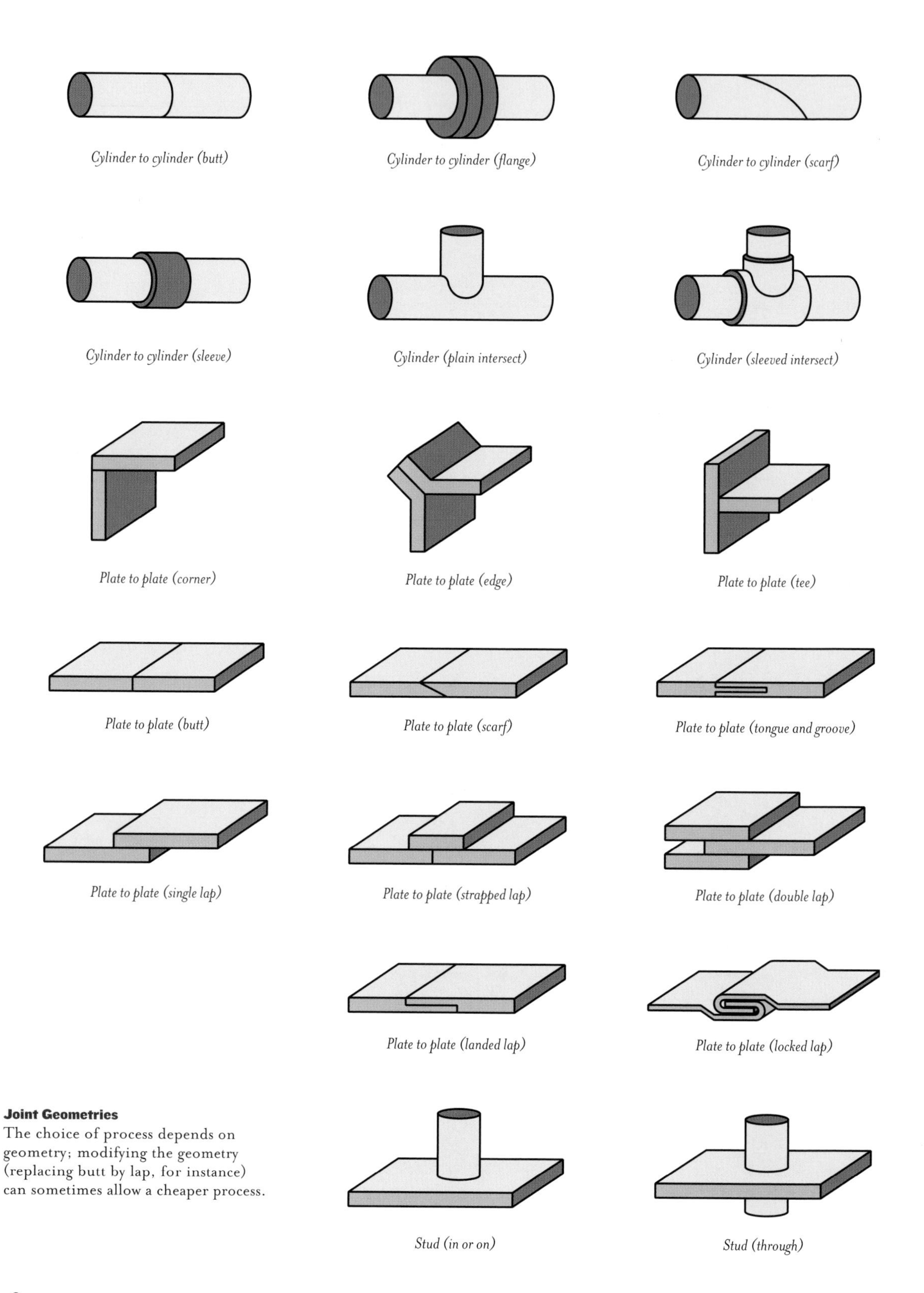

Joint Geometries
The choice of process depends on geometry; modifying the geometry (replacing butt by lap, for instance) can sometimes allow a cheaper process.

Further Reading

Bralla, J.G. (1998) Handbook of Product Design for Manufacture, 2nd edition, McGraw Hill, NY, USA. ISBN 0-07-007139-X. *(The bible – a massive compilation of information about manufacturing processes, authored by experts from different fields, and compiled by Bralla. A handbook rather than a teaching text.)*

DeGarmo, E.P., Black, J.T., and Kohser, R. A. (1984) Materials and Processes in Manufacturing, Macmillan Publishing Company, USA. ISBN 0-471-29769-0. *(A comprehensive text focusing on manufacturing processes, with a brief introduction to materials. The perspective is that of metal processing; the processing of polymers, ceramics and glasses gets much briefer treatment.)*

Houldcroft, P. (1990) Which Process?, Abington Publishing, Cambridge, UK. ISBN 1085573-008-1. *(Brief profiles of 28 welding and other processes for joining metals, largely based on the expertise of TWI International in the UK. Good, but like Wise (below), difficult to get.)*

Hussey, R. and Wilson, J. (1996) "Structural Adhesives: Directory and Data Book" Chapman & Hall, London. *(A comprehensive compilation of data for structural adhesives from many different suppliers.)*

Kalpakjian, S. (1984) Manufacturing Processes for Engineering Materials, Addison Wesley, Reading, MA, USA. ISBN 0-201-11690-1. *(A comprehensive and widely used text on manufacturing processes for all classes of materials.)*

Le Bacq, C., Jeggy, T. and Brechet, Y. (1999). "Logiciel d'aide a la preselection des procedes d'assemblage," INPG/CETIM, St Etienne, France.

Lesko, J. (1999) Materials and Manufacturing Guide: Industrial Design, John Wiley and Sons, NY. ISBN 0-471-29769-0. *(Brief descriptions, drawings and photographs of materials and manufacturing processes, with useful matrices of characteristics, written by a consultant with many years of experience of industrial design.)*

Swift, K. and Booker, J.D. (1998) Process Selection: From Design to Manufacture, John Wiley & Sons, UK. ISBN 0-340-69249-9. *(Datasheets in standard formats, for 48 processes, 10 of which are for joining.)*

Wise, R.J. (1999) Thermal Welding of Polymers, Abington Publishing, Cambridge, UK. *(A compilation of information on processes for welding polymers, largely based on the expertise of TWI International in the UK. Good, but difficult to get.)*

Adhesives

Caulking Gum Adhesion

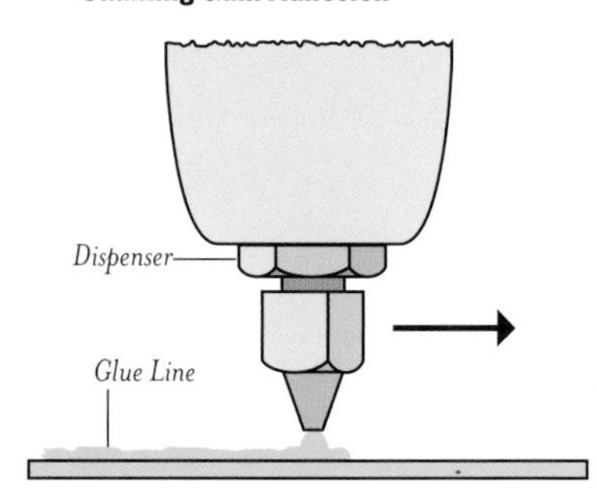

Spray Adhesion

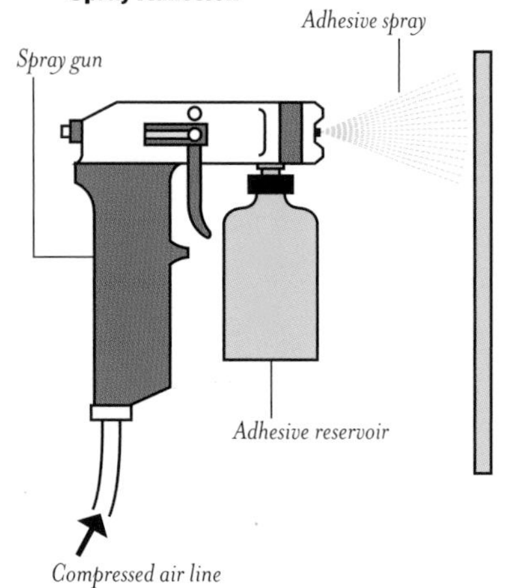

Attributes of Adhesives

Size of joint	Unrestricted
Maximum thickness, mm	Unrestricted
Unequal thickness?	Yes
Join dissimilar materials?	Yes
Impermeable?	Yes
Processing temp., C	16–180

What are they? Adhesives – glues – are so ancient a technology that they have become part of the language: "...to cement a friendship," "...the bonds of family," "...adhering to policy." And ancient the technology is: all early civilizations seem to have known how to make starch paste and glues from animal gelatines, and how to join materials with bitumen and gum arabic. Synthetic adhesives appeared around 1900, and since then advances in polymer chemistry have greatly expanded their range and performance. Synthetic structural adhesives are now used in every industrial sector. Structural adhesives are those that are used to perform some mechanical function, though they may have a secondary role as a sealant. Many are rigid, giving a stiff bond, but flexible adhesives also play an important role in design. Within each group, it is usual to use a classification based on chemistry.

Joint Type All geometries, particularly lap joints.

Design Notes Adhesives have a number of features that allow great design freedom: almost any material or combination of materials can be adhesively bonded; they can be of very different thickness (thin foils can be bonded to massive sections); the processing temperatures are low, seldom exceeding 180 C; the flexibility of some adhesives tolerates differential thermal expansion on either side of the joint; adhesive joints are usually lighter than the equivalent mechanical fasteners; and adhesives can give joints that are impermeable to water and air. The main disadvantages are the limited service temperatures (most adhesives are unstable above 190 C, though some are usable up to 260 C), the uncertain long-term stability and the unpleasant solvents that some contain.

Technical Notes Adhesive joints resist shear, tension and compression better than tear or peel – these last two should be avoided. Typical lap shear strengths are in the range 8–45 MPa. For joints loaded in shear, width "normal to the direction of shear" is more important than length "parallel to the shear direction." Butt joints are practical only when the area is large. Thin bond lines (typically 25 microns) are best, except when high impact strength is required. The essential equipment for adhesive bonding includes hot glue-guns and caulking-guns, both of which are used to apply adhesives in a paste or semi-liquid form. Spray guns are used to apply liquid adhesives and can be automated. Brushes and sprays are used for manual application.

The Environment Good ventilation is essential wherever adhesives are used.

Competing Processes Mechanical fasteners – although these generally require holes with associated stress concentrations.

Acrylic Adhesives

Acrylic adhesives are two-part systems which, when mixed or activated by UV radiation cure to form a strong, impact resistant bonding layer. Mostly used for wood-to-metal bonds.

- *Features:* durable, tough, water resistant, optically clear, able to bond a wide range of materials.
- *Typical Products:* aerospace, automotive, computer components.

Cyanoacrylate Adhesives

Cyanoacrylate adhesives are familiar as "super glue." They are single part systems that cure in seconds when exposed to damp air to give a strong, rather brittle bond.

- *Features:* cyanoacrylate adhesives bond to practically anything, and do so "instantly" requiring no heat or clamping.
- *Typical Products:* rapid joining of small parts, general domestic repairs.

Epoxies and Epoxy-phenolics

Epoxies and epoxy-phenolics are thermosetting adhesives with high tensile strengths (up to 45 MPa) and low peel strengths (1.8 kg/mm), resistant to solvents, acids, bases and salts. Nylon-epoxies have the highest strengths and are used primarily to bond aluminum, magnesium and steel. Epoxy-phenolics retain their strength up to 150–250 C and are used to bond

Matrix for Adhesives

Bonding of different materials.

	Metals	Wood	Polymers	Elastomers	Ceramics	Fiber-Composites	Textiles
Metals	Acrylic, CA, Epoxy, PU, Phenolic, Silicone						
Wood	Acrylic, Epoxy, Phenolic, Hot-melt	Epoxy, Phenolic, PVA					
Polymers	Acrylic, CA, Epoxy, Phenolic	Epoxy, PU, Phenolic, PVA	Acrylic, CA, Epoxy, Phenolic				
Elastomers	CA, Epoxy, Silicone	Acrylic, Phenolic, Silicone	CA, Epoxy, Phenolic, Silicone	PU, Silicone			
Ceramics	Acrylic, CA, Epoxy, Ceram	CA, Epoxy, PVA,Ceram	Acrylic, Epoxy, PU, PVA, Ceram	Acrylic, Epoxy, PU, PVA, Silicone	Acrylic, CA, Epoxy, Ceram		
Fiber-Composites	Acrylic, CA, Epoxy, Imide	Acrylic, CA, Epoxy, PVA	Acrylic, Epoxy, PVA, Silicone	Epoxy, PU, Silicone	CA, Epoxy, Silicone	Epoxy, Imide, PES, Phenolic	
Textiles	PU, Hot-melt	Acrylic, PVA, Hot-melt	Acrylic, PVA	Acrylic, PU, PVA	Acrylic, PU, PVA, Hot-melt	Acrylic, PU, PVA	PVA, Hot-melt

Key

Polyurethane = PU
Thermoplastic = Hot-melt
Cyanoacrylate = CA
Polyester = PES
Polyvinylacetate = PVA
Ceramic-based = Ceram

metals, glass and phenolic resins. Most are two-part systems, curing at temperatures between 20–175 C, depending on grade. They are used in relatively small quantities because of price: but, together with imide-based adhesives, they dominate the high-performance adhesives markets.

- *Features:* good adhesion to a wide range of substrates, with low shrinkage and good mechanical properties from -250 to 260 C.
- *Typical Products:* widely used in the aerospace, automotive and boat building industries.

Imide-based Adhesives (Dismaleimides, BMI and Polyimides, PI)

Imide-based adhesives, like epoxies, are widely used as a matrix in fibre-reinforced polymers. They have better elevated temperature performance than most other organic adhesives. They are used to bond ceramics and metals.

- *Features:* excellent dielectric properties (hence their use in radomes); BMIs can be used at temperatures as extreme as -250 C and 230 C; PIs up to 300 C.
- *Typical Products:* radomes.

Phenolic Adhesives

Phenolic adhesives are as old as Bakelite (1905), which they resemble chemically. Phenolics are sometimes blended with epoxies and neoprenes.

- *Features:* high strength, good water resistance and heat resistance, flame retardant.
- *Typical Products:* manufacture of plywood, chipboard and laminated wood structures; abrasive wheels and brake pads; foundry use for sand binding.

Polyurethanes

Polyurethanes include both polyurethane and isocyanate-based adhesives, with lap shear strengths of about 8 MPa.

- *Features:* bond well to a wide range of materials, are tough and flexible, have good resistance to water and solvents, and perform well from -250–80 C.
- *Typical Products:* widely used in the automotive, construction, furniture and footwear industries.

Silicones (SIL)

Silicones are synthetic polymers in which silicon replaces carbon as the major chain element. Most are two-part systems. Their chemistry gives them exceptional flexibility and chemical stability.

- *Features:* flexible, useful properties from -115–260 C; good resistance to water and UV and IR radiation.
- *Typical Products:* adhesives and seals in automotive, construction and marine industries.

Hot-melt Adhesives

Hot-melt adhesives are based on thermoplastics such as PE, nylons (PA), EVA and polyesters. These melt on heating to 100–150 C. By modifying their chemistry, they can be made to stick to other materials. Used to bond woods, textiles, polymers, elastomers and metals.

- *Features:* fast and easy to use; no solvent fumes; a wide range of chemistries that can be adapted to specific tasks.
- *Typical Products:* prototyping, crafts.

Anaerobic Adhesives

Anaerobic adhesives cure at room temperature when deprived of oxygen.

- *Features:* strong bond with low shrinkage.
- *Typical Products:* thread-locking of mechanical fasteners; fixing of gears and bearings, large flat panels.

Ceramic-based Cements

Ceramic-based cements are inorganic two-part systems that set to give a bond that is stable to high temperatures; used to bond thermal insulation to ceramics, metals, glass or wood.

- *Features:* chemically stable bond, usable up to 1600 C.
- *Typical Products:* bonding of refractory bricks in kilns and furnaces; foundry applications.

Sewing

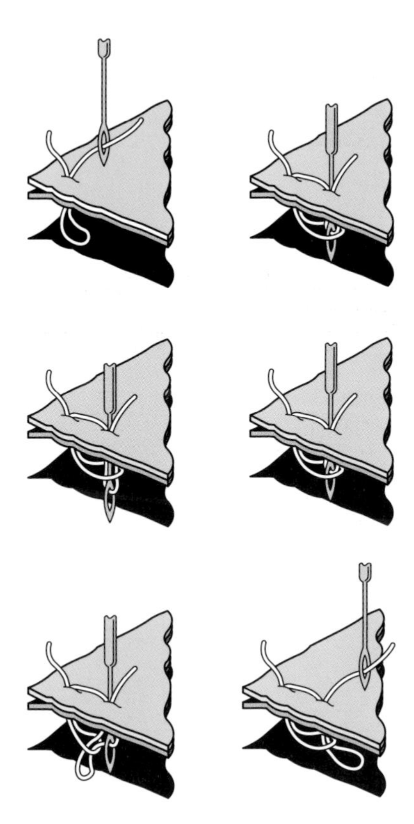

What is it? Sewing, like investment casting and enameling, has been practiced for at least 5000 years – indeed, there is evidence that it is much older still. In sewing, a series of stitches are applied, using one or more continuous threads of fibers, to join two or more thicknesses of material. There are many types of stitching, broadly described as "lock stitch" in which loops of one thread pass through the material and are locked on the other side by a second thread; or "chain stitch" in which a single thread zigzags through the material. Sewing machines automate the process. The figure shows a chain stitch of a single lap joint.

Joint Type Lap and butt.

Design Notes Sewing is an exceptionally flexible process, both in the range of materials it can join and in the shapes to which it can be applied. It provides not only mechanical fastening, but also decoration, often of a very elaborate kind.

Technical Notes Dissimilar materials can be joined. The threads used for conventional sewing are the natural fibers cotton, silk and flax, the cellulose derivative (rayon or viscose), and drawn polymer fibers made from polyethylene, polyester, polyamide (nylons) or aramids (such as Kevlar). It is also possible to sew with metal threads. The joining thread or fibre must be sufficiently strong and flexible to tolerate the tensions and curvatures involved in making a stitch. All fabrics, paper, cardboard, leather, and polymer films can be sewn. Sewing can also be used to join fabric and film to metal, glass or composite if eye-holes for the threads are molded or cut into these. There are many possible joint configurations, some involving simple through-stitching of a single lap or zigzag stitching across a simple butt, others requiring folding to give locked-lap and butt configurations.

The Economics Hand or small electric sewing machines are inexpensive; large automated machines can be expensive. Sewing is fast, cheap, and very flexible, both in the materials it can join, and in the joint configurations.

Typical Products Sewing is the principal joining process used by the clothing industry. It is important in tent and sail making, shoe construction, and book-binding. It is used to join polymer sheet to make wallets, pockets, cases and travel gear.

The Environment The thimble was devised to protect the finger from puncture when sewing. This minor risk aside, sewing offers no threat to health or safety, and is environmentally benign.

Competing Processes Adhesives; threaded fasteners.

Attributes of Sewing

Size of joint	Unrestricted
Maximum thickness, mm	1–10
Unequal thickness?	Limited
Join dissimilar materials?	Yes
Impermeable?	No
Processing temp., C	16–30

Rivets and Staples

What are they? Mechanical fasteners have three special attractions: they do not require heat; they can join dissimilar materials; and these can be of very different thickness. Rivets are widely used in aircraft design – a testament to their strength, permanence and reliability. Riveting is done by inserting a stud with a head on one end through pre-drilled holes in the mating components, and clinching (squashing) the other end where it sticks out by hitting it with a shaped hammer. Staples – familiar as a way of binding paper and attaching leather and cloth to frames in furniture – are fast and cheap; they can also be used to assemble sheet metal. Staples are applied with a spring-loaded, electric or pneumatic jig that forces the staple through the materials and onto a grooved anvil, bending the legs inwards and pinching the materials together. In blind stapling there is no anvil – the legs of the staple are simply driven through one material into the other, where they stick in the same way that nails do.

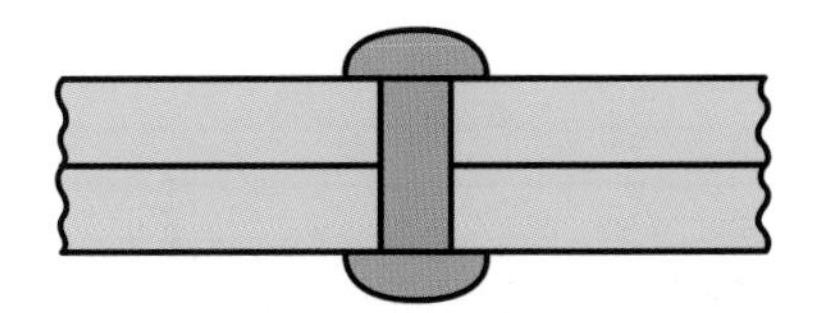

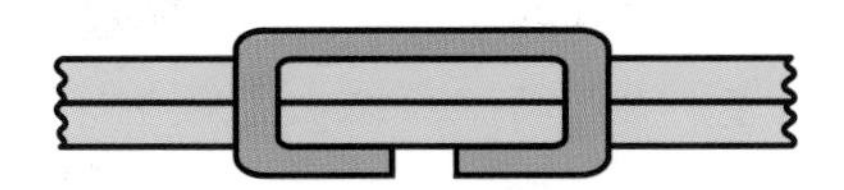

Attributes of Rivets and Staples

Size of joint	Unrestricted
Maximum thickness, mm	0.01–10
Unequal thickness?	Limited
Join dissimilar materials?	Yes
Impermeable?	With sealant
Processing temp., C	16–30

Joint Type Lap.

Design Notes Both rivets and staples can be used to join similar materials, but they can also be used to join one material to another even when there is a large difference in their strengths – leather or polymer to steel or aluminum for instance. Both allow great flexibility of design although a stress concentration where the fastener penetrates the material should be allowed for. Rivets should have heads that are 2.5–3 times the shank diameter; when one material is soft, it is best to put a washer under the head on that side to avoid pull-out. Staples are good when materials are thin; when metal is stapled, the maximum thickness is about 1mm, when non-metallic it can be up to 10 mm.

Technical Notes Rivets and staples are usually metallic: steel, aluminum and copper are common. Polymeric rivets and staples are possible: they are clinched by using heat as well as pressure. Almost any material, in the form of sheet, mesh or weave, can be joined by these methods; stapling also allows wire to be joined to sheet.

The Economics Both riveting and stapling are cheap, fast and economic even for very low production runs. Equipment, tooling and labor costs are all low. The processes can be automated.

Typical Products Stapling: joining of paper, leather, cloth, fiberboard. Rivets are extensively used in aerospace, automotive and marine applications, but have much wider potential: think of the riveting of the leather label to the denim of jeans.

The Environment The sound of the shipyard is that of riveting – it can be very loud. Over-enthusiastic staplers have been known to staple themselves. These aside, both processes are environmentally benign.

Competing Processes Adhesives; sewing; threaded fasteners.

Attributes of Threaded Fasteners

Size of joint	Unrestricted
Maximum thickness, mm	Unrestricted
Unequal thickness?	Yes
Join dissimilar materials?	Yes
Impermeable?	With sealant
Processing temp., C	16–30

Threaded Fasteners

What are they? When we get "down to the nuts and bolts" we are getting down to basics. Screws are as old as engineering – the olive press of Roman times relied on a gigantic wooden screw. Threaded fasteners are the most versatile of mechanical fasteners, with all the advantages they offer: they do not involve heat, they can join dissimilar materials of very different thickness and they can be disassembled. Ordinary screws require a pre-threaded hole or a nut; self-tapping screws cut their own thread.

Joint Type Most commonly, lap; but almost all joints can be adapted for threaded attachment.

Design Notes Mechanical fasteners allow great freedom of design, while allowing replacement of components or access because of the ease of which they can be disassembled. They can be used up to high temperatures (700 C) and – with proper location – allow high precision assembly.

Technical Notes Threaded fasteners are commonly made of carbon steel, stainless steel, nylon or other rigid polymers. Stainless steel and nickel alloy screws can be used at high temperatures and in corrosive environments. Tightening is critical: too little, and the fastener will loosen; too much, and both the fastener and the components it fastens may be damaged – torque wrenches overcome the problem. Locking washers or adhesives are used to prevent loosening.

The Economics Threaded fasteners are cheap, as is the equipment to insert them when this is done by hand. But the insertion is difficult to automate, making other methods (welding, riveting, adhesives) more attractive for a permanent bond.

Typical Products Threaded fasteners are universal in engineering design. But increasingly their use is becoming limited to products in which disassembly or the ability to have access is essential because other joining methods are cheaper, less likely to loosen, lighter and easier to automate.

The Environment Threaded fasteners have impeccable environmental credentials.

Competing Processes Snap fits, rivets and stapes, adhesives, sewing.

Snap Fits

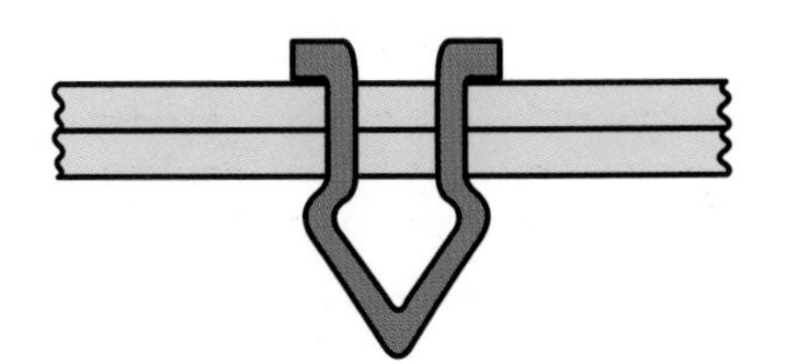

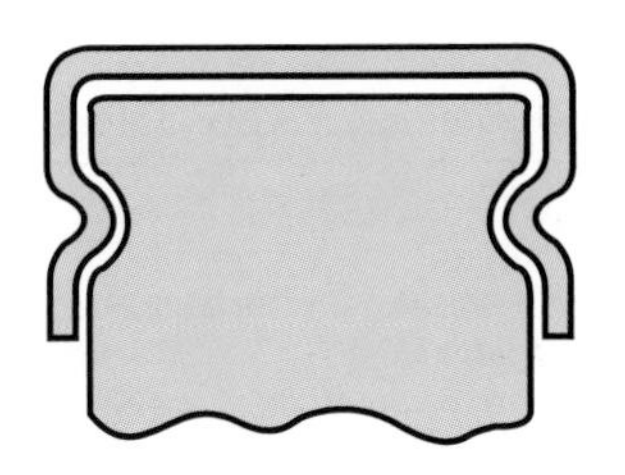

What are they? Snap fits, like other mechanical fastenings, involve no heat, they join dissimilar materials, they are fast and cheap and – if designed to do so – they can be disassembled. It is essential that the snap can tolerate the relatively large elastic deflection required for assembly or disassembly. Polymers, particularly, meet this requirement, though springy metals, too, make good snap fits.

Joint Type All geometries can be adapted for snap fitting, provided the material is chosen appropriately.

Design Notes Snap fits allow components of every different shape, material, color and texture to be locked together, or to be attached while allowing rotation in one or more direction (snap hinges). The snap fit can be permanent or allow disassembly, depending on the detailed shape of the mating components. The process allows great flexibility in design and aesthetic variety.

Technical Notes The best choices are materials with large yield strains (yield strain = yield strength/elastic modulus) and with moduli that are high enough to ensure good registration and positive locking. Polymers (particularly SAN, nylons, polyethylenes and polypropylenes) have much larger values of yield strain than metals. Elastomers have the largest of all materials, but their low modulus means that the assembly will be too flexible and pop apart easily. Among metals, those used to make springs (spring steel, copper beryllium alloys and cold worked brass) are the best choices, for the same reasons.

The Economics Snap fits are fast and cheap, and they reduce assembly time and cost, both in production and in use. Hand assembly requires no special equipment. Automated assembly requires equipment that can be expensive, but it is very fast.

Typical Products Snap fits are increasingly used because of the freedom of material and shape that they allow. Typically they are used to join small or medium sized polymer parts, metal casings, sheet parts etc.

The Environment Snap fits disassemble easily, making recycling easier. In this and every other way they are environmentally benign.

Competing Processes Adhesives; staples; sewing; threaded fasteners.

Attributes of Snap Fits

Size of joint	Restricted
Maximum thickness, mm	Unrestricted
Unequal thickness?	Yes
Join dissimilar materials?	Yes
Impermeable?	Possible
Processing temp., C	16–30

Hot Gas Welding

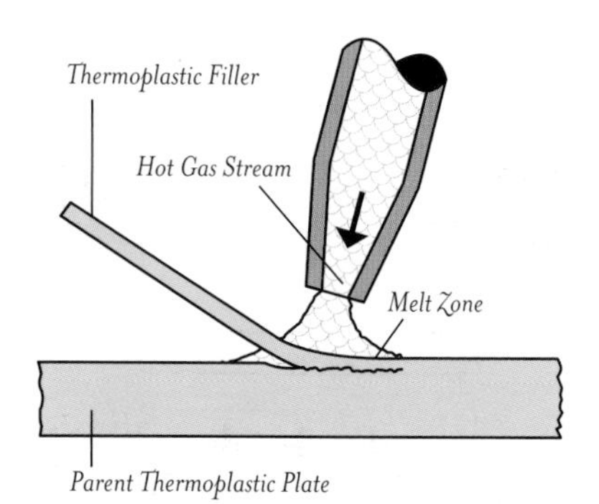

Attributes of Hot Gas Welding

Size of joint	Unrestricted
Maximum thickness, mm	2.5–10
Unequal thickness?	Limited
Join dissimilar materials?	If similar T_m
Impermeable?	Yes
Processing temp., C	200–300

What is it? Hot gas welding is a way of joining thermoplastics that is similar, in many ways, to the gas welding of metals. The weld torch is just a souped-up hairdryer that directs a stream of hot gas at scalp-searing temperatures (200–300 C) at the joint area and at a rod of the same thermoplastic as the substrate. Polymers are very viscous when heated, so they do not flow in the way that metals do – that is why filler material is needed; the filler is pressed into the softened joint to form the bond. The process is slow and poorly adapted for mass production, but it is easily portable, making it the best way to assemble and repair large polymer components.

Joint Type All geometries, particularly plate to butt and lap.

Design Notes Hot gas welding is limited to thermoplastics. It is a manual process, best suited to on-site assembly of large structures – pipes, liners and roof membranes for example. It cannot be used with very thin materials – 2.5 mm is the minimum.

Technical Notes Hot gas welding is commonly used to join polyethylene, PVC, polypropylene, acrylic, some blends of ABS and other thermoplastics. The welding tool contains an electric heater to heat the gas, air, nitrogen or CO_2, and a nozzle to direct it onto the work piece. If air is used, it is provided by a compressor or blower; nitrogen or CO_2 require a more complicated supply of gas.

The Economics The equipment cost and set up time are both low, making this an economic process that can be performed on site; but it is slow, and not suitable for mass production.

Typical Products Installation and repair of thermoplastic pipe and duct work, the manufacture of large chemical-resistant linings, the assembly of storage bins and architecture uses for joining polymer roofing like that of the Millennium Dome in London.

The Environment No problems here – just a little hot air.

Competing Processes Laser welding; threaded fasteners; adhesives.

Hot Bar Welding

What is it? The simplest of all welding processes for polymers is the one that you can use in your kitchen to seal food into freezer-bags: hot bar welding. In this process, overlapping thermoplastic polymer films are pinched between electrically heated, PTFE coated, bars. One bar is hinged to allow the films to be inserted and removed. Mechanical or pneumatic actuators close the bar and exert the pinch force. The joint itself is heated by conduction through the film, limiting the film thickness to less than 0.5 mm. A typical weld in a 100 micron thermoplastic film takes 1–3 seconds. Impulse welding is a modification of hot bar welding in which the bars are pulse-heated, giving an additional control of quality.

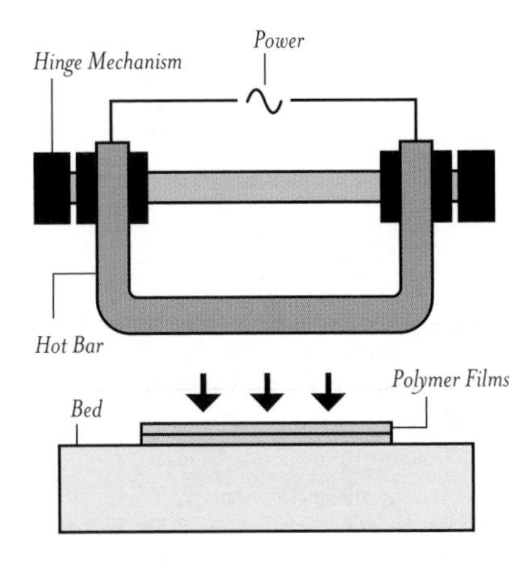

Joint Type Lap.

Design Notes The process is limited to thermoplastics and to lap joints between sheets, both of them thin. But it is fast, cheap and clean, lending itself to the sealing of food, packaging, drugs and medical equipment. The shape of the weld is defined by the shape of the bar, which can have a complex surface profile.

Attributes of Hot Bar Welding

Size of joint	Restricted
Maximum thickness, mm	0.05–0.5
Unequal thickness?	Limited
Join dissimilar materials?	Yes
Impermeable?	Yes
Processing temp., C	180–280

Technical Notes Control of time, temperature and pinch-pressure are essential for a good joint – impulse welding is one way of controlling the first two of these. The time, t, to form a bond varies inversely with the thermal diffusivity, a, of the polymer, and is proportional to the square of the thickness, x, of the sheet ($t = x^2/2a$), so adjustments are needed when material or thickness are changed.

The Economics The process is cheap, fast, and uses low-cost equipment and tooling.

Typical Products Hot bar and impulse welding are widely used for sealing of polymer packaging for the food industry, for shrink-wrapped packs, for medical and general packaging. They are used to bond polymer films to polymer moldings, to make transparent file pockets, and may other similar applications.

The Environment The process is clean, involves no chemicals, generates no fumes, and consumes little energy; it is – in a word – eco-benign.

Competing Processes Laser welding; adhesives; sewing.

Hot Plate Welding

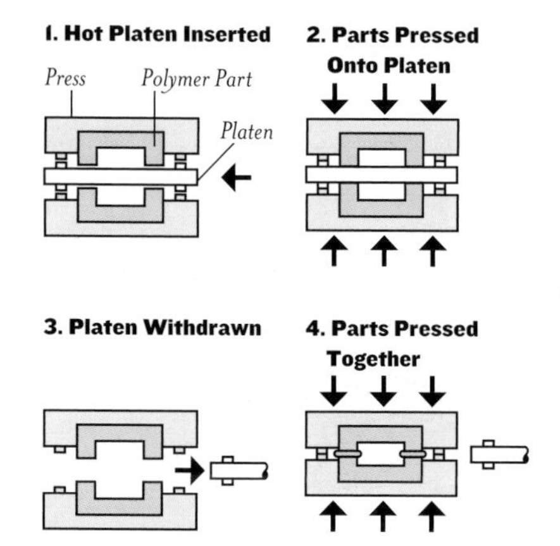

Attributes of Hot Plate Welding

Size of joint	Restricted
Maximum thickness, mm	1–30
Unequal thickness?	Limited
Join dissimilar materials?	No
Impermeable?	Yes
Processing temp., C	190–290

What is it? Hot plate welding makes butt joints between bulk thermoplastic components, in a kind of inversion of the way that hot bar welding makes lap joints. The components to be joined are held in fixtures that press them against an electrically heated, PTFE coated, platen, melting the surface and softening the material beneath it. The pressure is lifted, the tool withdrawn and the hot polymer surfaces are pressed together and held there until they have cooled. Hot plate welding can be used to form joints of large area – for example, the joining of large polyethylene gas and water pipes. The process is relatively slow, requiring weld times between 10 seconds for small components and 1 hour for very large.

Joint Type Plate to plate (butt).

Design Notes Hot plate welding can join almost all thermoplastics except nylons, where problems of oxidation lead to poor weld quality. The joint strength is usually equal to that of the parent material but joint design is limited to butt configurations. If the joint has a curved or angled profile, shaped heating tools can be used.

Technical Notes Precise temperature control of the hot plate is important for good joints. The plate temperature is generally between 190–290 C depending on the polymer that is to be joined. Pressure is applied hydraulically or pneumatically. Most thermoplastic components can be welded by the hot-plate method, but it is most effective for joining large components made from polyethylene, polypropylene or highly plasticized PVC. It creates a strong bond, impermeable to gas or water.

The Economics Equipment and tooling are moderately cheap, but the process can be slow.

Typical Products The process is used to make automotive hydraulic reservoirs and battery cases; to join unplasticized PVC door and window frames; to join thermoplastic pipes for gas and water distribution; and for sewers and outflows up to 150 mm in diameter. It is also used for smaller things: tail-light assemblies, water-pumps, refrigerator doors.

The Environment The process is clean and involves no unpleasant chemicals, and is generally eco-friendly.

Competing Processes Hot gas welding; friction welding.

Ultrasonic Welding

What is it? If you want to weld using ultrasound, you need an ultrasonic generator, a converter (transducer), a booster and a welding tool, bizarrely called a sonotrode. The generator for ultrasonic welding converts 50 Hz mains voltage into a 20 kHz signal. The converter uses the reverse piezo-electric effect to transform this into mechanical oscillations. The booster and the sonotrode transmit this to the weld zone in such a way as to create an oscillating displacement of 10 to 30 microns. A static pressure of 2–15 MPa is applied across the surfaces to be joined and the power switched on to make them slide, heat, roughen and bond.

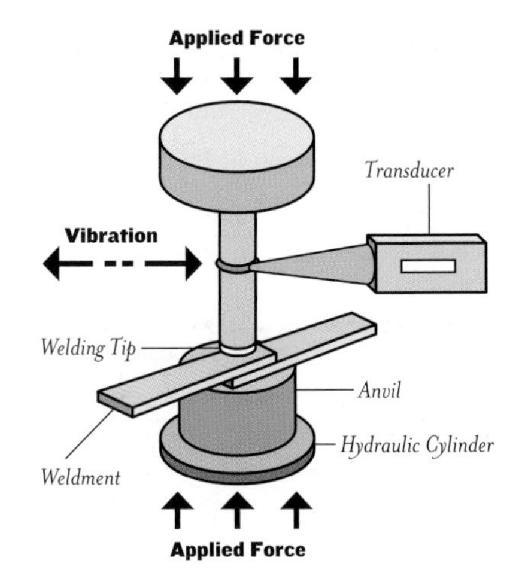

Joint Type Plate to plate (butt and lap).

Design Notes Ultrasonic welding produces fast, strong, clean and reliable welds in both polymers and metals. For polymers, the process is principally used for thermoplastic film and sheet. It can join dissimilar materials of differing thickness, applying either spot or seam welds. Most metals can also be ultrasonically welded. There is some increase in temperature at the sliding surface, but it is well below the melting temperature of the material. Instead, it appears that the rapid reversal of stress breaks up surface films and contaminants, and local plasticity, coupled with some diffusion, creates the bond. The process offers short process times (e.g. 3 seconds per weld) and relatively low temperatures, minimizing damage to the material adjacent to the weld.

Technical Notes Ultrasonic welding is generally limited to the joining of thin materials – sheet, film or wire – or the attachment of thin sheet to heavier structural members. The maximum thickness, for metals, is about 1 mm. The process is particularly useful for joining dissimilar metals. Because the temperatures are low and no arcing or current flow is involved, the process can be applied to sensitive electronic components. Joint shear strengths up to 20 MPa are possible.

The Economics Ultrasonic welding is fast and clean. Up to 1000 pieces an hour can be joined.

Typical Products Polymers: the process is widely used in the automotive, appliance, medical, textile and toy industry. Typical assemblies include sealed tail light assemblies with clear lenses and opaque bodies, double-layer insulated drinking cups, decorative multi-color name plate panels and toys. Metals: spot or line welding of thin sheet, joining of dissimilar metal in bimetallics, micro-circuit electrical contacts, encapsulating explosives or chemicals.

The Environment The process has enviable environmental credentials. It is totally clean – no fumes no chemicals, no electrical or other hazards. The input energy is almost entirely transmitted to the joint, giving energy efficiency far exceeding that of thermal processes.

Competing Processes Threaded fasteners, laser welding.

Attributes of Ultrasonic Welding

Size of joint	Restricted
Maximum thickness, mm	
(metals)	0.01–1
(polymers)	0.1–3
Unequal thickness?	Limited
Join dissimilar materials?	Yes
Impermeable?	Yes
Processing temp., C	
(metals)	300–600
(polymers)	100–250

Ultrasonic Welding

Power-beam Welding

Power-beam Welding: Metals

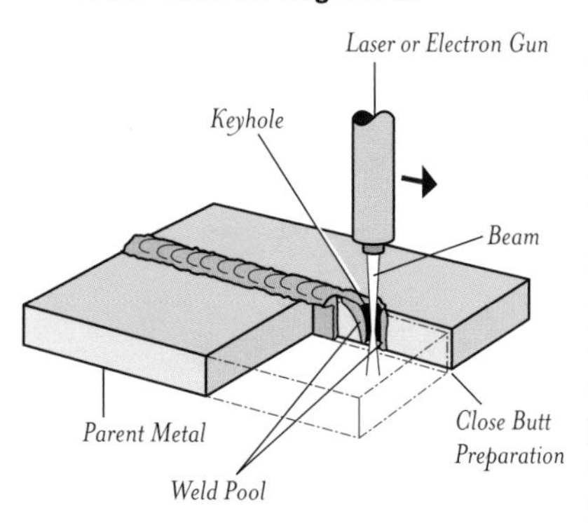

Power-beam Welding: Polymers

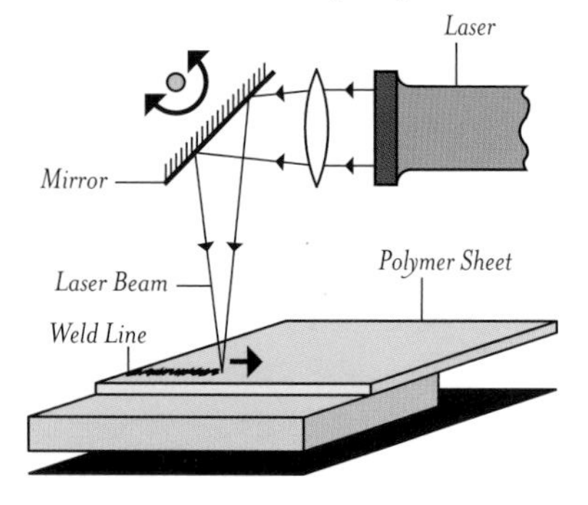

Attributes of Power-beam Welding: Metals

Size of joint	Restricted
Maximum thickness, mm	
(e-beam)	1–200
(laser beam)	0.2–20
Unequal thickness?	Limited
Join dissimilar materials?	Yes
Impermeable?	Yes
Processing temp., C	600–2K

Attributes of Power-beam Welding: Polymers

Size of joint	Restricted
Maximum thickness, mm	
(direct heat)	0.1–1
(dye absorption)	1–10
Unequal thickness?	Limited
Join dissimilar materials?	Yes
Impermeable?	Yes
Processing temp., C	200–300

What is it? Electron and laser beams are the stuff of death-rays – invisible, deadly accurate, and clean – no smoke, no fumes, no mess. Both are used to weld metals; lasers can weld polymers too. In electron beam welding, melting is produced by the heat of a focused beam of high velocity electrons. The kinetic energy of the electrons is converted into heat when it hits the work piece, which has to be contained in a vacuum chamber, and it must be a conductor. No filler metal is used. The process is more energy-efficient than laser-beam welding, allowing higher power densities and the ability to weld thicker plates. In laser beam welding, the heat source is a narrow beam of coherent monochromatic light. The process is more precise than e-beam welding and takes place in air, allowing greater design freedom, but the penetration depth is less (maximum 20 mm in metals). Shielding gas is blown though a surrounding nozzle to protect the weld. Again, no filler metal is used.

It is possible to laser weld thin, semi-transparent or opaque polymer films by simply scanning the beam across them, melting them right through, but this is not the best way to use lasers. The trick in welding polymers is to arrange that the beam is absorbed where it is most useful – at the joint interface. For transparent polymers this can be achieved by spraying a thin film of IR or UV-absorbing dye, invisible to the human eye, onto the surface where the weld is wanted; the laser beam passes through the transparent upper sheet (which can be thick – up to 10 mm) without much energy loss. But when it hits the dye at the interface it is strongly absorbed, melting the polymer there and creating a weld whilst leaving most of the rest of the material cold. Scanning the beam or tracking the work piece gives a line weld up to 10 mm wide.

Joint Type Metals: All joint geometries are possible. Polymers: lap joints.

Design Notes The high power-density in the focused beam gives narrow welds in metals with minimal penetration of heat into the work piece, allowing high weld speeds with low distortion. Accurate machining of mating components is needed, since no filler metal is used. The processes allow products to be made that could not be fabricated in any other way.

Welding polymers (particularly transparent polymers) without an interfacial dye is limited to the joining of thin thermoplastic film and sheet. Welding with a dye give more control and allows thicker sections. Dissimilar materials can be joined, although their melting temperatures must be comparable. The main feature of the process is that it is non-contact, and exceptionally clean and fast.

Technical Notes Electron beams have a power of 1–100 kW, allowing welding of metal plate from 1–200 mm thick, but the process requires a vacuum chamber, limiting the size of the assembly. Laser beam equipment has lower beam power – typically 500 W–5 kW, although lasers up to 25 kW exist. Laser beam welding is preferred in the microelectronics industry because it is clean and requires no vacuum chamber. Electron-beam welding – because of the vacuum – is partic-ularly suited for refractory metals such as tantalum, niobium, molybdenum and tungsten. Laser welding can be used for these too, but it is essential that the metal should not have a shiny surface since this reflects the beam. Coatings are available to increase absorption. The strength of a well-designed laser-welded lap joint in polymers often exceeds that of the parent film or plate. For thin films the weld speed can be up to 30 m/min, but for films as thick as 1 mm the speed is less: about 1 m/min. Control of the laser profile allows simultaneous welding and cutting (the "cut-seal process"). If one of the two sheets being joined is colored, and the other is transparent, the colored sheet should be placed furthest from the beam.

The Economics The cost of power beam equipment is high, and – for e-beam welding – cycle times are long because of the need to evacuate the vacuum chamber. But despite this, the speed and controllability compensate, making both attractive in large-scale manufacturing. When production volumes are large as in the automotive industry and the microelectronics field, the high investment can be amortized; and the high welding speed, single pass welding and freedom from the need for secondary operations make the process economic.

Typical Products Electron beam welding is used extensively to assemble gears and transmissions for automobiles, aircraft engines and aerospace products. High capacity electron beam equipment is used for pressure vessels, nuclear and process plant and chemical plant. Laser beam welding is used where precise control is important: joining of microelectronic components and thin gauged parts like bellows and watch springs. Increasingly, the process has been adapted for automobile components – gears, transmission assemblies, aerospace and domestic products. Laser welding of polymers: food packaging and the sealing of biomedical materials and equipment. There is growing interest in its use for welding of PET and other polymers for structural use.

The Environment Electron beams generate x-rays, infrared and ultra-violet radiation. Laser beams are damaging – particularly to the eyes. Proper protection is essential for both, requiring strict safety procedures. But apart from this, the processes are attractive from an environmental standpoint: they are clean, involve no chemicals, and produce no waste.

Competing Processes Metals: resistance welding, threaded fasteners. Polymers: hot bar welding.

Brazing

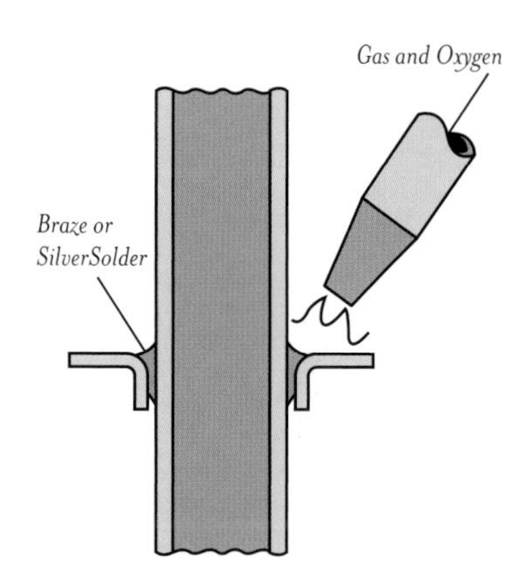

Attributes of Brazing

Size of joint	Unrestricted
Maximum thickness, mm	1–100
Unequal thickness?	Yes
Join dissimilar materials?	Yes
Impermeable?	Yes
Processing temp., C	450–600

What is it? When the components to be joined cannot tolerate the temperatures required for welding, the alternatives are brazing, soldering, mechanical fasteners or adhesives. Brazing is the hottest of these. In brazing a low melting temperature metal is melted, drawn into the space between the two solid surfaces by capilliary action, and allowed to solidify. Most brazing alloys melt above 450 C but below the melting temperature of the metals being joined. The braze is applied to the heated joint as wire, foil or powder, coated or mixed with flux, where it is melted by a gas-air torch, by induction heating or by insertion of the components into a furnace; the components are subsequently cooled in air.

Joint Type All joint geometries, particularly butt, scarf and lap.

Design Notes Almost all metals can be joined by some variant of brazing, provided they have melting temperatures above 650 C. The process can join dissimilar metals even when they have different melting temperatures. Brazing is easily adapted to mass production (cheap bicycles are brazed) and the joint is strong, permanent and durable. A large joint area is good – it compensates for the relatively low strength of the brazing metal itself. Joints need a clearance of 0.02–0.2 mm to allow a good, strong bond to form. Ceramics can be brazed if the mating surfaces are first metalized with copper or nickel.

Technical Notes Brazing alloys are designed to melt at modest temperatures (450–600 C), to wet the surfaces being brazed (often by forming an alloy with the surfaces) and to be very fluid. The most common are those based on brass-like compositions (hence the name) and on silver. Brazing brass (0.6 Cu, 0.4 Zn) and bronzes (the same with 0.0025 Mn, 0.01 Fe and 0.01 Sn) are used to join steel, stainless steel, copper and nickel; silver solders (0.3–0.6 Ag, 0.15–0.5 Cu + a little Zn) are even better for the same metals. Fluxes play an important role in removing surface contamination, particularly oxides, and in improving wetability by lowering surface tension.

The Economics The equipment and tooling costs for brazing are low. Torch brazing requires a certain skill; furnace brazing can be automated, requiring no skilled labor. The process is economical for small runs, yet allows high production rates when automated.

Typical Products Brazing is widely used to join pipe work, bicycle frames, fittings and to repair castings and assemble machine parts.

The Environment Brazing generates fumes, and some fluxes are toxic – good ventilation and clean-up are important. Otherwise, brazing has a low environmental impact; the metals involved are non-toxic.

Competing Processes Adhesives, soldering, threaded fasteners.

Soldering

What is it? Soldering is low-temperature brazing; or – another analogy – it is gluing with metal. It uses alloys that melt below 450 C. Soldered joints are less strong than brazed joints – more like an adhesive – but the equipment needed to make them is simpler and the temperatures reached by the component are much lower, an essential for the assembly of electronic equipment. Soldering can be applied in the same way as braze, or – more like an adhesive – by pre-coating ("tinning") the metal surfaces to be joined before assembling and simply heating them with a torch, an electric soldering iron, or an array of infrared lamps. In re-flow soldering, the components are heated by vapor from a boiling fluorinated hydrocarbon that condenses on components, releasing latent heat, and giving rapid, uniform heating.

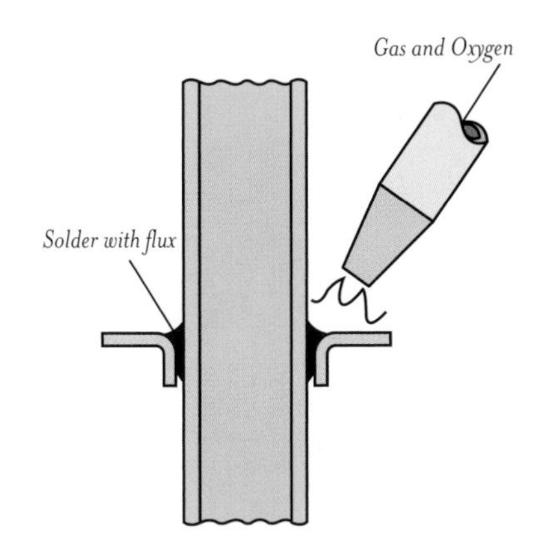

Joint Type All joint geometries, particularly lap, scarf and sleeve.

Attributes of Soldering

Size of joint	Unrestricted
Maximum thickness, mm	1–10
Unequal thickness?	Yes
Join dissimilar materials?	Yes
Impermeable?	Yes
Processing temp., C	150–450

Design Notes Solders are electrical conductors – it is this that gives them their prominence in the electronics industry. From a mechanical viewpoint they are soft metals; it is the thinness of the soldered joint that gives it high tensile strength, but the shear strength is low (below 2 MPa) thus soldered joints should have an area large compared with the load-bearing section (as in lap or sleeve joints), or be given additional mechanical strength (as in locked lap joints). Solders can join dissimilar materials of very different thickness and size.

Technical Notes Solders are alloys of low melting temperature metals – lead, tin, zinc, bismuth, cadmium, indium. The lowest melting combination of any two metals is the "eutectic" composition; lead-tin solders for electrical connections have this composition (0.6 Sn, 0.4 Pb). Lead-tin solders for other applications range in composition from 0.2–0.6 Sn. Tin-antimony solders are used for electrical connection. Indium-tin solders are used for glass to metal joints. Soldering requires fluxes. Some are merely solvents for grease, cleaning the surface; others are acids and must be removed after the joint is made or they cause trouble.

Typical Products Solders are widely used to make electrical connections, to create printed circuit boards and attach logic chips to them, for domestic pipe-work, automobile radiators, precision parts in jewelery.

The Economics Like brazing, soldering is a flexible process, allowing high production rates, yet is also economic for small batches. The costs of equipment and tooling are low, and no great skills are required.

The Environment Now we come to it. Heavy metals – lead and cadmium in particular – have a bad eco-reputation. There is mounting pressure to replace them. Alternatives are emerging, but they cost more and are harder to use.

Competing Processes Brazing; adhesives; threaded fasteners.

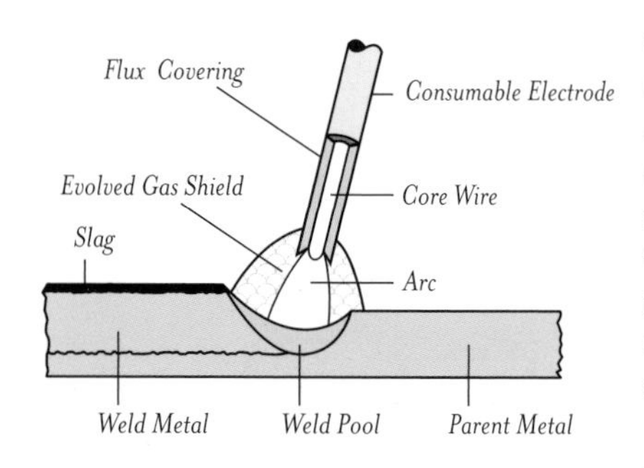

Attributes of Torch Welding

Size of joint	Unrestricted
Maximum thickness, mm	1–100
Unequal thickness?	Limited
Join dissimilar materials?	No
Impermeable?	Yes
Processing temp., C	1500–1700

Torch Welding (MMA or SMA)

What is it? Torch welding, also known as manual or shielded metal arc (MMA or SMA) welding, is the most important general purpose welding and surfacing method using low-cost equipment. In it, an electric arc is established between a flux-coated consumable rod (an electrode) and the component. The flux coating decomposes to give a gas shield; the slag that forms over the weld-pool prevents the metal from oxidizing. Appropriate choice of metal and flux allow the process to be used for a wide variety of applications, though they are almost exclusively limited to ferrous alloys.

Joint Type All joint geometries.

Design Notes MMA is easy to use and very flexible, making it a prime choice for one-off or low volume production; but – because it cannot be automated – it is not an appropriate mass production process. Distortion caused by thermal expansion is minimized by designing symmetry into the weld lines, and balancing the welds around the neutral axis of the structure. Weld lines are best designed to be straight or have simple contours, and the joint must be designed to allow access to the weld torch.

Technical Notes Carbon, low alloy and stainless steels, cast irons and certain nickel alloys can be joined by MMA welding. Dissimilar materials are difficult to weld. The weld rod is consumed, providing filler metal to the weld; but when it is used up it has to be replaced, disrupting the welding process and causing random stresses in the joints. The flux must be removed in a cleaning operation between each weld pass.

The Economics MMA welding is a versatile, low-cost process, easily transported, but it cannot easily be automated because of the limited length of each electrode.

Typical Products MMA is used to join pressure vessels, structural steel work, pipe-work, to attach stiffeners to structures in ship building, and in general engineering.

The Environment Welding fumes can present health hazards, and radiation from the weld can damage eyesight. Good ventilation and the use of welding masks and tinted goggles overcome these problems.

Competing Processes MIG and TIG welding.

MIG Welding

What is it? Gas metal arc (MIG) welding is one of the Big Three heavy-duty welding processes (together with Torch and TIG). The electrode here is a bare wire, with no flux. The flux is replaced by a stream of inert gas, which surrounds the arc formed between the consumable wire electrode and the component; the wire is advanced from a coil as the electrode is consumed. The real advantages over torch welding are that there is no flux or slag, giving a cleaner weld – and that it can be automated. But there is a penalty: because the process needs gas, it is more expensive and less portable. None the less, MIG is considered to be the most versatile of all arc welding processes.

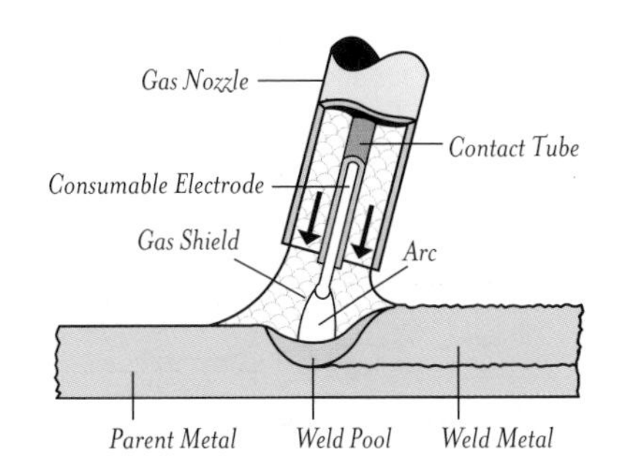

Joint Type Butt, lap.

Attributes of MIG Welding

Size of joint	Unrestricted
Maximum thickness, mm	1–100
Unequal thickness?	Limited
Join dissimilar materials?	No
Impermeable?	Yes
Processing temp., C	600–2000

Design Notes If you want high-quality welds in aluminum, magnesium, titanium, stainless steel or even mild steel, MIG welding is the process to choose. It is best for fillet welds through an adaptation – MIG spot welding – and lends itself well to lap joints. Distortion by thermal expansion is minimized by designing symmetry into the weld lines, and balancing the welds around the neutral access of the structure. Weld lines are best designed to be straight or have simple contours, and the joints are designed to allow access for the weld torch.

Technical Notes Most common metals and alloys except zinc can be welded using MIG; electrode wire (the filler material) is available for welding all of these. The shielding gas is usually argon, helium, carbon dioxide or a mixture of these – it is chosen to suit the material being welded. The process produces uniform weld beads that do not require de-slagging, making it suitable for mechanization and operation by welding robots. MIG can make most joint geometries and can be done in most orientations, but is most efficient when flat and horizontal.

The Economics Equipment costs are moderate, tooling costs are low. MIG welding is more expensive than Torch welding because of the cost of the inert gas, but it is rapid and requires less labor.

Typical Products MIG welding is used throughout the industry in both manual and automatic versions, particularly ship building, structural engineering, process plant and electrical engineering, domestic equipment and the automobile industry. It is indispensable for welding difficult, non-ferrous metals such as aluminum, magnesium and titanium.

The Environment Health hazards depend on the composition of the electrode and component; both appear as airborne fume during welding. Radiation from the weld can be harmful to the eye, requiring that a welding helmet and tinted safety goggles must be worn.

Competing Processes Torch welding, TIG welding.

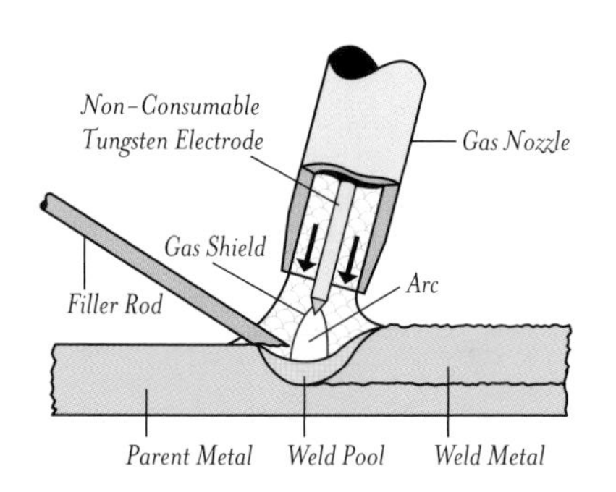

Attributes of TIG Welding

Size of joint	Unrestricted
Maximum thickness, mm	0.2– 10
Unequal thickness?	Limited
Join dissimilar materials?	No
Impermeable?	Yes
Processing temp., C	600–2000

TIG Welding

What is it? Tungsten inert-gas (TIG) welding, the third of the Big Three (the others are Torch and MIG) is the cleanest and most precise, but also the most expensive. In one regard it is very like MIG welding: an arc is struck between a non-consumable tungsten electrode and the work piece, shielded by inert gas (argon, helium, carbon dioxide) to protect the molten metal from contamination. But, in this case, the tungsten electrode is not consumed because of its extremely high melting temperature. Filler material is supplied separately as wire or rod. TIG welding works well with thin sheet and can be used manually, but is easily automated. Both penetration and deposition rates are much less than those of MIG welding, but precise control of the weld is easier.

Joint Type All joint geometries.

Design Notes Because the heating is de-coupled from the filler supply, greater control of weld conditions is possible. Thus, TIG welding is used for thin plate and for precision assemblies, made of almost any metal. Clean surfaces and well-prepared joints are important. It is principally used for thin sections and precisely made joints.

Technical Notes TIG welding produces very high quality welds on metals such as aluminum, magnesium, titanium, stainless steel, and nickel; cast iron and mild steel are also easily welded. The arc is started by a high frequency AC discharge to avoid contaminating the tungsten electrode; it is subsequently maintained by a DC current or a square wave AC current, which gives greater control of penetration.

The Economics The equipment is more expensive and less portable than Torch, and a higher skill level is required of the operator. But the greater precision, the wide choice of metals that can be welded and the quality of the weld frequently justify the expense.

Typical Products TIG welding is one of the most commonly used processes for dedicated automatic welding in the automobile, aerospace, nuclear, power generation, process plant, electrical and domestic equipment markets.

The Environment TIG welding requires the same precautions as any other arc welding process: ventilation to prevent inhalation of fumes from the weld pool, and visors or colored goggles to protect the operator from radiation.

Competing Processes Torch and MIG welding.

Resistance Welding

What is it? The tungsten filament of a light bulb is heated, by resistance, to about 2000 C. That is more than enough to melt most metals. Resistance welding relies on localizing the electric current I (and thus the I^2R heating) where heating is desired: at the interface. In resistance spot welding, the overlapping sheets are pressed between water-cooled electrodes. The current-pulse generates heat; the cooled electrodes chill the surfaces, localizing the heat in the interface between the sheets where the metal melts and welds. Projection spot welding uses an additional trick – a pre-formed pimple on the face of the joint – to confine the current path and further localize the heating; a clever idea, because it extends the use of the method to forgings, castings and machined components and makes it faster because several welds can be made at the same time. In seam resistance welding the electrodes are water-cooled wheels, between which overlapping sheets are rolled to give a seam weld.

Joint Type Butt and lap.

Design Notes Resistance welding has many advantages over rivets or screw fasteners: it is faster, easily automated, requires no drilling or punching, and no fluxes or fillers; and it gives products that are lighter. Joint design must allow for access to electrodes.

Technical Notes Resistance welding is commonly used for steels of all grades, aluminum, magnesium, brass and cast iron. Electrodes are made of low resistance copper alloys and are hollow to allow water cooling. Resistance spot welding rates are typically 12–180 welds/minute; projection spot welding rates are greater.

The Economics The capital cost of equipment is medium; high if automated. But the process is fast, reliable, lends itself to automation, and requires no post-weld treatments.

Typical Products The really big-time users of resistance welding are the automobile and domestic appliance industries. In aircraft structures it is used for assembling doors, fuselage stringers, outer skins, decking, ribs and seat frames. The electronics industry makes extensive use of miniaturized spot welding for assembling circuitry. Seam resistance welding is used for water or gas-tight assemblies: fuel tanks for vehicles, duct work, drums and cans.

The Environment The heat and the flash of welding require normal precautions – face-shield, goggles – but otherwise the environmental burden is minimal.

Competing Processes Threaded fasteners, adhesives, Power-beam welding.

Resistance Spot Welding

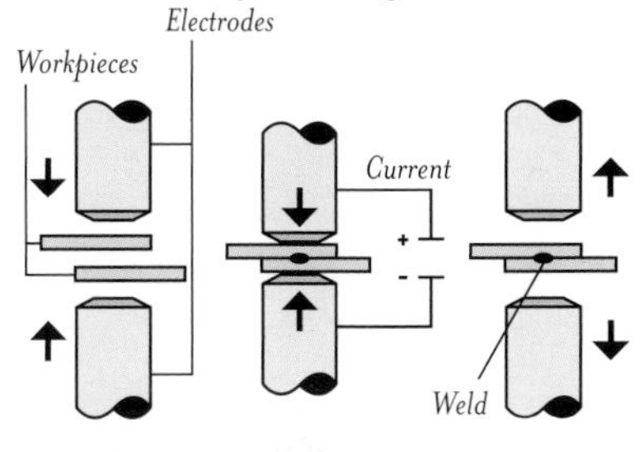

Projection Spot Welding

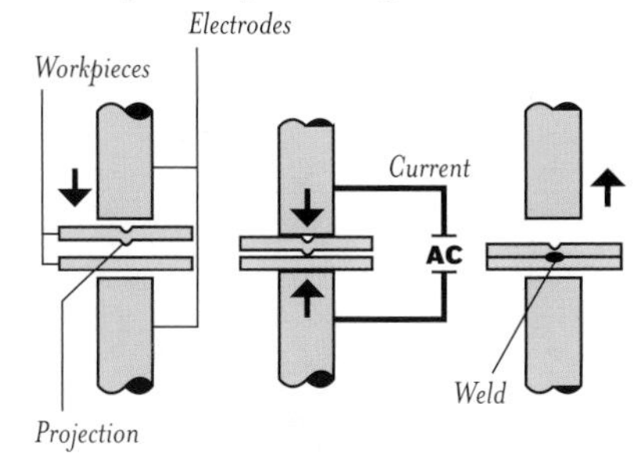

Seam Resistance Welding

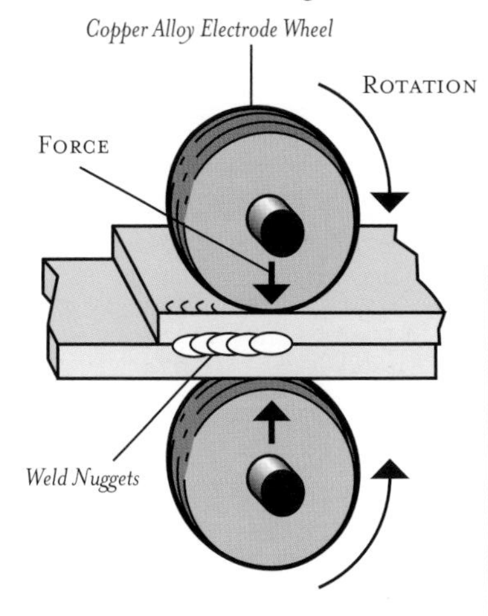

Attributes of Resistance Welding

Size of joint	Unrestricted
Maximum thickness, mm	0.1–10
Unequal thickness?	Limited
Join dissimilar materials?	Yes
Impermeable?	If seam weld
Processing temp., C	600–2000

Friction Welding

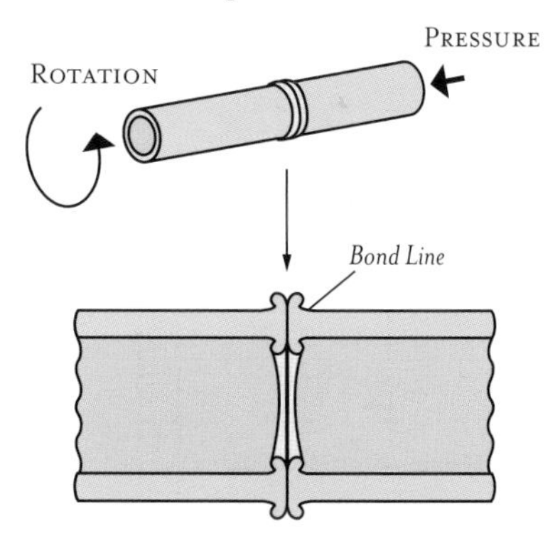

Friction-stir Welding

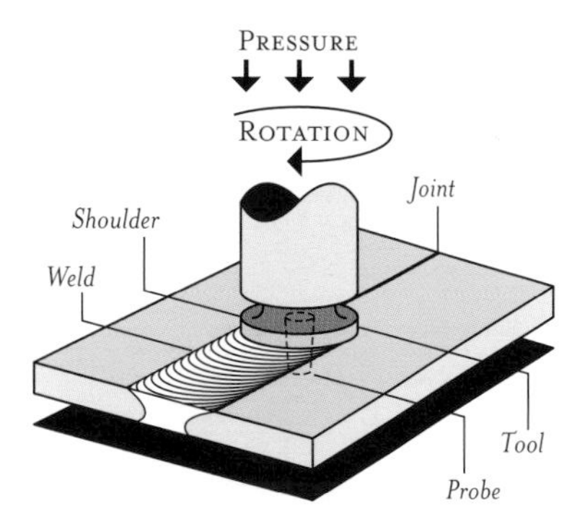

Attributes of Friction Welding

Size of joint	Restricted
Maximum thickness, mm	1–100
Unequal thickness?	Yes
Join dissimilar materials?	Yes
Impermeable?	Yes
Processing temp., C	400–1200

Friction Welding

What is it? When there is friction between two people, heat may be generated but they rarely bond. Materials are different: rub them together and there's a good chance they'll stick. In friction welding, one component is rotated or vibrated at high speed, forced into contact with the other, generating frictional heat at the interface, and – when hot – the two are forged together. In direct drive friction welding, the motor is connected to the work piece and starts and stops with each operation. In inertial friction welding, the motor drives a fly wheel that is disconnected from the motor to make the weld. In friction stir welding a non-consumable rotating tool is pushed onto the materials to be welded. The central pin and shoulder compact the two parts to be joined, heating and plasticizing the materials. As the tool moves along the joint line, material from the front of the tool is swept around to the rear, eliminating the interface. The weld quality is excellent (as good as the best fusion welds), and the process is environmentally friendly.

Joint Type Butt.

Design Notes Friction welding gives clean, high-quality joints between a wide variety of metals – particularly metal matrix composites can be joined in this way. Dissimilar materials can be joined: stainless steel and aluminum for instance. No melting is involved so no protective flux or gas are needed. Linear or orbital motion allows non-circular shapes to be joined.

Technical Notes For small components, rotational speeds of up to 80,000 rpm and a few kilograms load are used; for very large components the rotational speed is as slow as 40 rpm with thousands of kilograms forging load. Weld times range from 1–250 seconds.

The Economics The capital cost of equipment is high, but tooling costs are low. The process is fast and can be fully automated.

Typical Products Friction welding is widely used to join automobile components, agricultural machinery, and for welding high speed steel ends to twist drills. It is the best way of attaching forgings to shafts or bars, or hub bearings to axles.

The Environment Friction welding is clean and energy-efficient.

Competing Processes Torch welding, MIG welding, TIG welding, ultrasonic welding.

Diffusion and Glaze Bonding

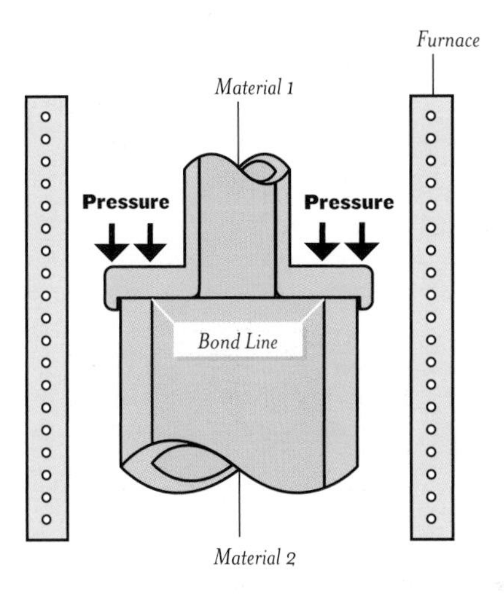

What are they? Joining ceramics poses problems because they are hard and brittle. Mechanical fasteners generally require holes, and create high clamping forces and stress concentrations. Conventional welding is impractical: the melting temperatures of ceramics are very high and residual stresses lead to cracking. Adhesives stick to ceramics, but few are reliable above 300 C, where ceramics are often used. Two processes by-pass these difficulties: diffusion and glaze bonding. In diffusion bonding the surfaces to be joined are cleaned, pressed into close contact and heated in vacuum or a controlled atmosphere. Solid-state diffusion creates the bond, which is of high quality, but the process is slow and the temperatures are high. Diffusion bonding is used both for ceramics and metals – among the metals: titanium, metal matrix composites, and certain steels and copper-based alloys are particular candidates. Glaze bonding depends on the fact that molten glass wets and bonds to almost anything. To exploit this, the surfaces to be joined are first coated with a thin layer of finely-ground glass slurry, with the composition chosen to melt at a temperature well below that for diffusion bonding, and with maximum compatibility between the two surfaces to be joined. A small pressure is applied across the interface and the assembly is heated, melting the glass and forming a thin, but strong bonding layer.

Attributes of Diffusion and Glaze Bonding

Size of joint	Restricted
Maximum thickness, mm	1–100
Unequal thickness?	Yes
Join dissimilar materials?	Yes
Impermeable?	Yes
Processing temp., C	600–1800

Joint Type All joint types.

Design Notes The processes allow differing materials to be joined: ceramic to ceramic, ceramic to glass, metal to ceramic, metal to metal. Glaze bonding is particularly versatile: tailoring the glass to match the requirements of melting temperature and thermal expansion allows bonds between metal and ceramic.

Technical Notes To make materials bond by diffusion requires temperatures above 3/4 of their melting temperature, and a modest pressure to keep the surfaces together. Ceramics melt at very high temperatures – so the processing temperatures are high. Glaze bonding overcomes this by creating an interfacial layer that bonds closely to both surfaces but melts at a much lower temperature.

The Economics These processes are energy-intensive and slow. But for some material combinations this is the only choice.

Typical Products For metals: assembling titanium aircraft panels, and smaller items where dissimilar metals must be joined. For ceramics and glasses: bonding these to each other and to metals.

The Environment The environmental impact is low except in one regard: energy demand is high.

Competing Processes For ceramics: threaded fasteners, adhesives. For metals: threaded fasteners, beam welding.

Surface Profiles

Almost every component of a product has some sort of surface process applied to it. Surface processes enhance the thermal, fatigue, friction, wear, corrosion or aesthetic qualities of the surface, leaving the bulk properties unchanged. Economically they are of great importance, extending life, allowing materials to be used in increasingly harsh service conditions, imparting multi-functionality (for example – by creating a corrosion resistant surface on a tough, but chemically reactive substrate, or a thermally-insulating surface on a strong, but metallic substrate) and, of course, creating much of the visual and tactile features of a product.

The choice of a surface process depends on the material to which it will be applied and the function it is to perform. The processes used to etch the surface of glass differ in obvious ways from those used to texture polymers; less obviously, paints and electro-coatings, too, are often material-specific. The pages that follow profile ways of treating surfaces, describe the processes and their functions, and list the materials to which they can be applied. They cover the most common processes used to create the visual and tactile qualities of products.

More information can be found in the sources listed under "Further Reading."

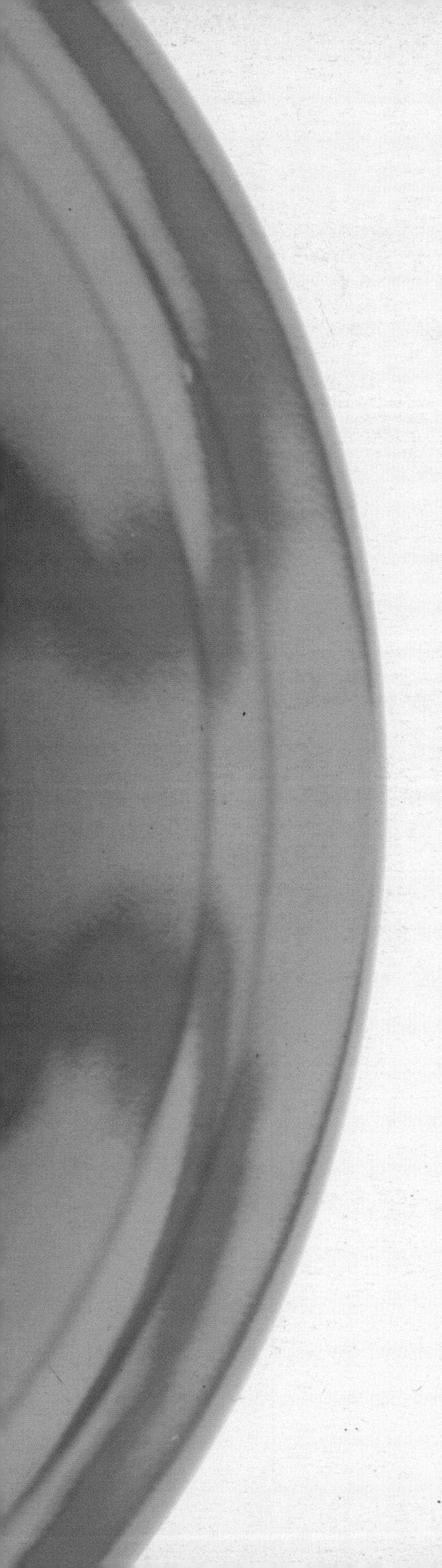

Further Reading 286

Printing
Screen Printing 288
Pad Printing 289
Cubic Printing 290
Hot Stamping 291
In-mold Decoration 292

Plating
Vapor Metalizing 293
Electro-plating 294
Electro-less Plating 297
Anodizing 298

Polishing
Mechanical Polishing 299
Electro-polishing 300
Chemical Polishing 301

Coating
Solvent-based Painting 302
Water-based Painting 303
Electro-painting 304
Powder Coating 305
Enameling 307

Etching 308
Texturing 309

Further Reading

There is a confusing diversity of surface processes. Some (such as painting) are applicable to many materials; others (like electro-less plating) are limited to one or a few. Selection is not easy, particularly as there is, at present, no single information source that allows comparison of the full range. The major contributions to systematic surface process selection derive from sources that, generally speaking, focus on one class of process or one class of material: painting, for instance, (Roodol, 1997); coating technology (Gabe, 1983; Rickerby and Matthews, 1991; Granger and Blunt, 1998); of heat treatment (Morton, 1991; Sudarshan, 1989); of corrosion (Strafford et al., 1984; Gabe, 1983); or electro-deposition (Canning, 1978). A lesser number attempt a broader review (ASM, 1982; Poeton, 1999; CES, 2002), but are far from comprehensive. Some of the sources we have found helpful are listed below. Much more information can be found on the internet, though it is uneven in content and quality. We have made extensive use of the internet, of information from companies and of the books listed below in compiling the profiles contained in this section.

ASM (1982) "ASM Metals Handbook Volume 5: Surface Cleaning, Finishing, and Coating," Ninth edition, American Society for Metals, Metals Park, OH, USA. *(The volume of this distinguished handbook-set devoted to surface engineering.)*

Bralla, J.G. (1998) "Handbook of Product Design for Manufacturing," Second edition, McGraw Hill, NY, USA. ISBN 0-07-007139. *(The bible – a massive compilation of data for manufacturing processes, authored by experts from different fields and edited by Bralla.)*

Canning, W. (1978) "The Canning Handbook on Electroplating," W. Canning Limited, Gt. Hampton St., Birmingham, UK.

CES 4 (2002) "The Cambridge Engineering Selector," Version 4, Granta Design, Cambridge, UK, www.grantadesign.com.

Gabe, D.R. (1983) "Coatings for Protection," The Institution of Production Engineering, London, UK.

Grainger, S. and Blunt, J. (1998) Engineering Coatings, Design and Application, 2nd edition, Abington Publishing, UK. ISBN 1-85573-008-1. *(A monograph aimed at technical engineers detailing processes for enhancing wear and corrosion resistance of surfaces.)*

Morton, P.H. (1991) "Surface Engineering and Heat Treatment," The Institute of Metals, London, UK.

Poeton (1999) "The Poeton Guide to Surface Engineering," www.poeton.co.uk/pands/surface/.

Surface Profiles

Rickerby, D.S. and Matthews, A. (1991) "Advanced Surface Coatings: a Handbook of Surface Engineering," Chapman and Hall, NY, USA.

Roobol, N.R. (1997) "Industrial Painting, Principles and Practice," 2nd edition, Hanser Gardner Publications, Cincinnati, OH, USA. ISBN 1-56990-215-1. *(A comprehensive guide to painting and resin coating.)*

Strafford, K.N., Dalta P.K. and Googan, C.G. (1984) "Coatings and Surface Treatment for Corrosion and Wear Resistance," Ellis Horwood Limited.

Sudarshan, T.S. (1989) "Surface Modification Technologies," Marcel Dekker, Inc. NY, USA.

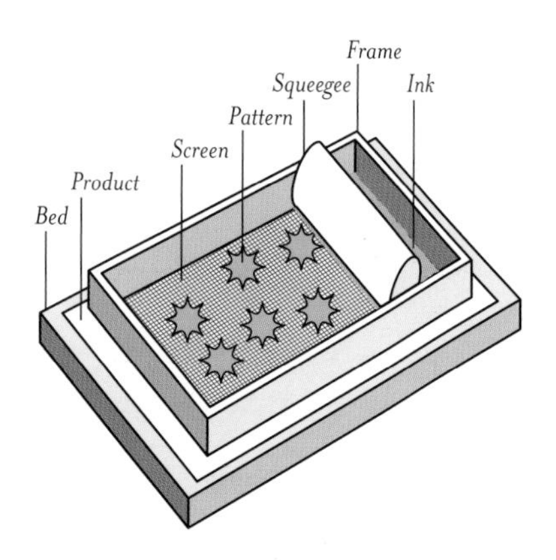

Attributes of Screen Printing

Surface hardness, Vickers	5–10
Coating thickness, μm	10–100
Curved surface coverage	Poor
Processing temp., C	15–25

Screen Printing

What is it? Walk down any street and you will see examples of screen printing: displays and posters advertising products in shops; buses with ads on their sides; badges and control panels on computers and stereo equipment; sports bags and T-shirts; and even the Pop Art creations of Andy Warhol. Screen printing has its origins in Japanese stenciling, but the process that we know today stems from patents taken out by Samuel Simon of Manchester at the turn of the last century. He used silk stretched on frames to support hand painted stencils, a process also used by William Morris to create his famous wallpapers and textile prints. During the First World War in America screen printing took off as an industrial printing process; but it was the invention of the photographic stencil in the 1930s that revolutionized the process. It is now a $5 billion per year industry.

Functions Color, reflectivity, texture.

Design Notes The process can be applied to polymers, glass, metals, wood, textiles and of course paper and board. Flat and cylindrical objects can be printed. Multiple colors can be printed, but each requires a separate screen.

Technical Notes The screen is a wooden or aluminum frame with a fine silk or nylon mesh stretched over it. The mesh is coated with a light-sensitive emulsion or film, which – when dry – blocks the holes in the mesh. The film is exposed to the image or pattern to be printed using ultra-violet light, hardening it where exposed. The screen is washed, removing the emulsion where it was not hardened, leaving an open stencil of the image that was printed onto it. The screen is fitted on the press, the product to be printed is placed under the screen, and the topside of the screen is flooded with ink. A rubber blade gripped in a wooden or metal handle called a squeegee (not unlike a giant windscreen wiper) is pulled across the top of the screen; it pushes the ink through the mesh onto the surface of the product. To repeat the process the squeegee floods the screen again with a return stroke before printing the next impression. Epoxy inks give protection against scratching and can be used with products that attack standard enamel inks.

The Economics The capital cost of the equipment is low. The process is economic for small runs and one-color printing. Each additional color adds to the cost because it is applied separately, and requires registration.

Typical Products Posters, stickers, ticketing, shelf strips, banners, exhibition panels, ring binders, mouse-mats, site boards, signs, T-shirts, control panels and badges for computers.

The Environment The chemicals, particularly the cleaning fluids, can be volatile and toxic, requiring good ventilation and operator protection.

Competing Processes Solvent and water-based painting; pad printing; cubic printing.

Pad Printing

What is it? Pad printing is a little like printing with rubber stamps; instead of rubber, the image is cut into a steel or copper plate, from which it is lifted by a rubber pad. It is used as a decorating process for irregular shapes, and for those shapes which cannot be easily printed by silk screen. The steps are as follows. (1) The image to be transferred is etched into a printing plate commonly referred to as a cliche. The cliche is flooded with ink and wiped, leaving ink only in the image area. As solvents evaporate from the image area the ink's ability to adhere to the silicone transfer pad increases. (2) The pad is pressed onto the cliche lifting the ink. Solvents evaporate from the outer ink layer where it is exposed to the atmosphere, making it tackier and more viscous. (3) The pad is pressed onto the component to be decorated, conforming to its shape and depositing the ink. Even though it compresses considerably during this step, the contoured pad is designed to roll away from the substrate surface without smudging the image by sliding. (4) When the pad is pressed onto the product, the adhesion between the ink and substrate is greater than the adhesion between the ink and pad, resulting in complete transfer of the ink, leaving the pad clean and ready for the next cycle. Automatic methods, with several pads in series (transfer pad printing), allow cheap multi-color decoration. Manual machines are used for low production volumes; for these, multi-color prints require separate steps and set-up time.

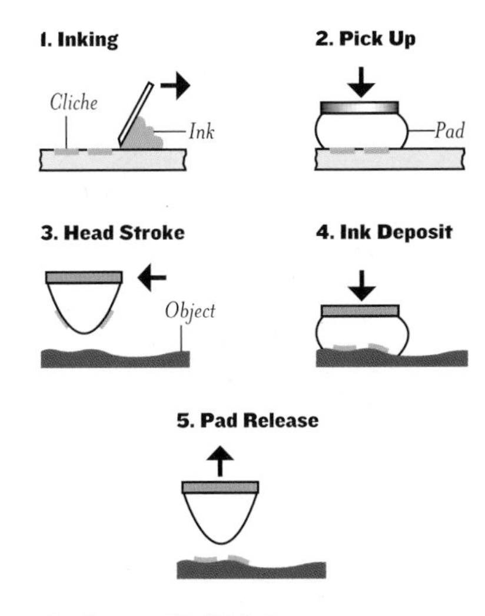

Attributes of Pad Printing

Surface hardness, Vickers	5–10
Coating thickness, μm	6–10
Curved surface coverage	Good
Processing temp., C	15–30

Functions Color, reflectivity, texture.

Design Notes The advantages of pad printing are the ability to print on irregular surfaces (such as a golf ball) and to print wet on wet multi-colors. Its excellent quality of detail makes pad printing attractive for flat items. Pad printing is limited to relatively small images compared to screen printing – usually less than 0.1 m^2. Large opened areas can be difficult to cover, requiring special, screened cliches.

Technical Notes Polymer, glass and metal products can all be printed. 3D, irregular shapes are possible. The inks are pigmented resins suspended in organic thinners.

The Economics Capital and tooling costs are low. The process is fast, and the product can be handled immediately after printing.

Typical Products Cups, pens, glass frames, lighters, golf balls.

The Environment Thinners can give toxic fumes, requiring ventilation.

Competing Processes Solvent and water-based painting; screen printing; cubic printing.

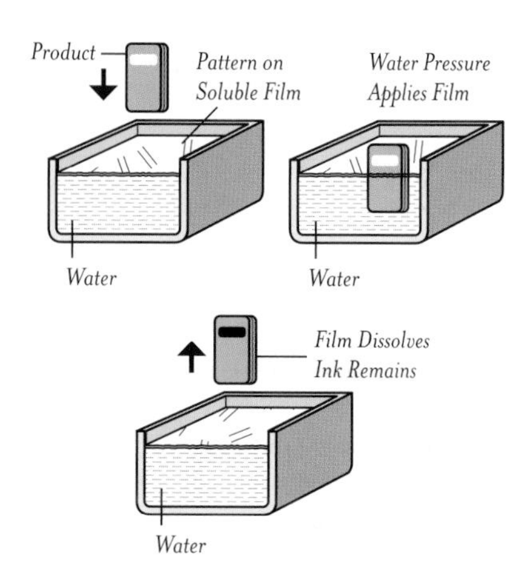

Attributes of Cubic Printing

Surface hardness, Vickers	5–10
Coating thickness, μm	6–10
Curved surface coverage	Good
Processing temp., C	15–30

Cubic Printing

What is it? Transfers are familiar to every child – a painless way of applying images to paper, to products and even to yourself. The image is printed on a thin, water-soluble film; when floated on water, the image floats free and can be transferred to the product to which it sticks. Cubic printing uses the same principle, scaled-up and made commercially viable. The film and its image are floated on the surface of a water tank. The product is immersed in the tank, where the pressure of the water presses the image onto the surface of the component.

Functions Color, reflectivity, texture.

Design Notes Cubic printing is very versatile; complex, curved shapes can be printed and the image can be multi-colored. Registration is difficult, so the image is usually applied in a single operation, and designed so that precise registry is not necessary.

Technical Notes The pattern or design is printed on a water soluble film. The film, specially treated to activate the inks, is floated onto the water, where it dissolves, leaving the decoration inks on the water surface. The component to be decorated is pressed down into the water to effect the transfer. The inks flow on the surface of the object, decorating it in three dimensions. A transparent topcoat is applied to protect the printed surface. Polymer, ceramic, glass, metal and wood products can all be printed. Printing on 3D and irregular shapes is possible.

The Economics Capital and tooling costs are low. The component must be dried and given a finishing coat, making it relatively slow.

Typical Products Automotive: dash panels, console, steering wheel, shift knobs, spoilers, bumpers, visor. Sports equipment: tennis racket, fishing reels, gun stocks, helmets, golf clubs, sunglasses, bikes. Computers products: monitor panel, keyboards, mice, PC notebooks, digital cameras, printers. Appliances: TV, refrigerators, washer-dryer, coffee maker, microwave, stereo equipment. Kitchen Products: canisters, silverware, glass stems, bowls, pitchers, salt and pepper shakers. Furniture: end tables, coffee tables, door knobs, picture frames, tissue holders, clocks.

The Environment Cubic printing presents no significant environmental problems.

Competing Processes Solvent and water-based painting; silk screen printing; pad printing.

Hot Stamping

What is it? Next time you see a book with imprinted gold lettering on its cover, think of hot stamping. It is a dry process for permanently applying a colored design, logo, text or image. It is best known as a method of applying metallic gold or silver lettering or decoration. A heated metal die (250–300 C) is pressed against a colored carrier foil and the component being printed. The hot stamp is created when the raised surface of the die contacts the foil and transfers the colored film on the face of the foil to the product being printed. The pressure from the die creates a recess for marking, which protects the stamped letter from abrasion, and heat makes the marking medium adhere to the product.

Functions Color, reflectivity, texture.

Design Notes Block lettering can be hot stamped on either raised or flat areas. For irregular surfaces, a silicone plate is used to transmit heat and pressure to the foil. A range of decorating finishes is available, including metallic foils, wood grain finishes and multi-colored graphics. The process can be applied to polymers, wood, leather, paper, vinyl, mylar, and textiles such as polyester and acetates, and – less easily – to painted metal. Decoration is permanent and resists peeling, scratching and abrasion.

Technical Notes Hot stamping dies are made from magnesium, brass, or steel. Inexpensive photo-etched dies work well for prototype or short runs on flat areas; dies for long production runs are deep-cut from either brass or steel. Foils are made in four plies: a thin film carrier, generally polyester, a release coating, a decorative coating – pigment or metallic, and an adhesive specific to the substrate material.

The Economics The equipment costs are low. Hot stamping with polymer or aluminum foil is inexpensive; gold foil can be expensive. Components can be handled and packed immediately. Minimal set-up time allows users to change graphics and colors as needed by simply changing a dry roll of foil or transfers.

Typical Products Retail and cosmetics packaging, book binding, automotive finishing, commercial printing, consumer goods, computer casings and tape cassettes.

The Environment No ink mixing or clean-up involving volatile organic solvents (VOCs) is required. The die is hot, but otherwise the process is pollutant-free.

Competing Processes Solvent and water-based painting; pad printing; cubic printing.

1. Set-up

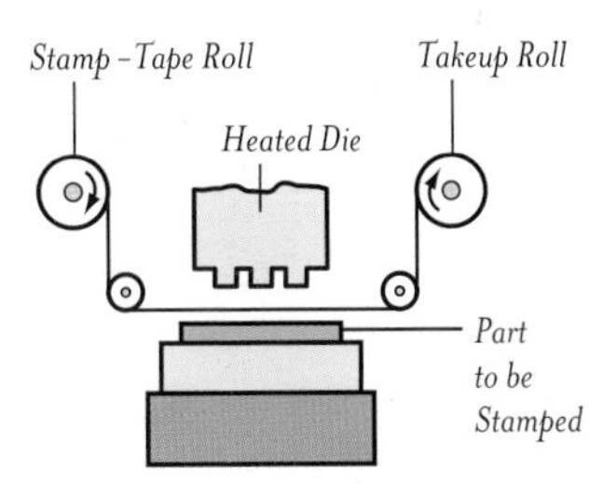

2. Stamping

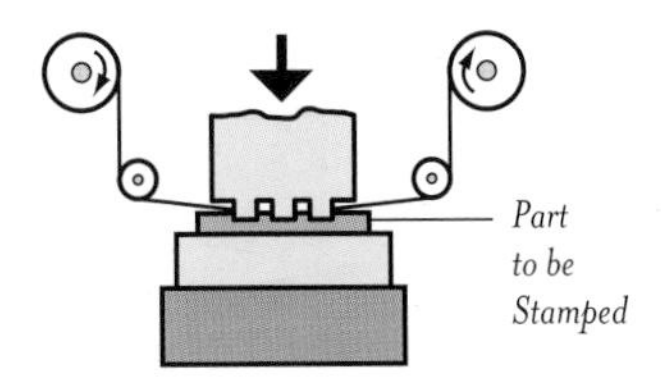

3. Finish

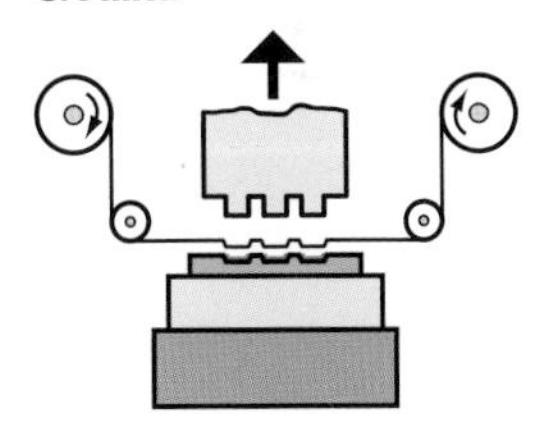

Attributes of Hot Stamping

Surface hardness, Vickers	5–50
Coating thickness, µm	1–50
Curved surface coverage	Poor
Processing temp., C	150–300

In-mold Decoration

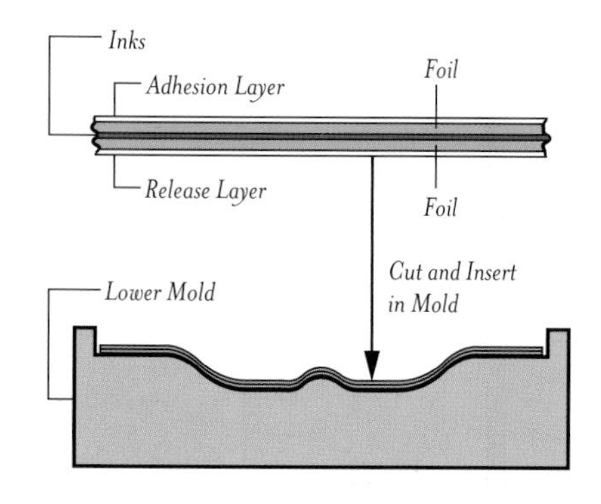

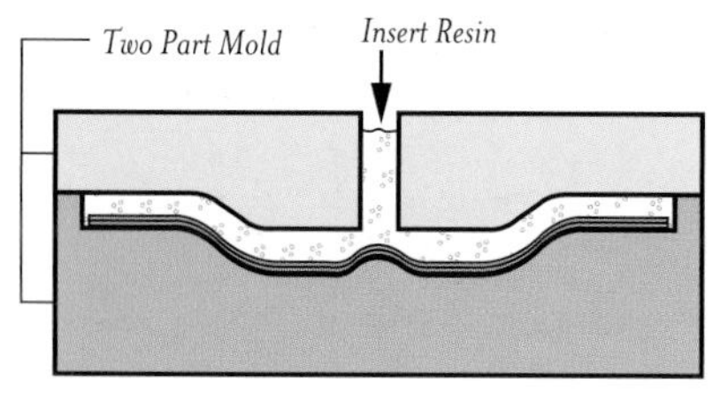

Attributes of In-mold Decoration

Surface hardness, Vickers	5–15
Coating thickness, µm	10–500
Curved surface coverage	Avg. to Good
Processing temp., C	125–200

What is it? In-mold decoration (IMD) allows accurately registered colors, designs and print to be applied to injection molded components without any secondary processing. The image, which can be full-color, is printed onto a polyester or polycarbonate film called the "foil." If the product is flat or mildly curved, the foil is fed as a continuous strip, or cut and placed directly in the mold. If true 3D components are to be made, the foil is first thermally molded to the shape of the component, then placed in the mold cavity. Hot liquid resin is injected behind the foil, bonding its surface to the molding resin and forming an integrally decorated component.

Functions Color, texture, patterning, printing; resistance to abrasion.

Design Notes The figure shows the build-up of a more complex foil. The decorative coatings are applied to the foil in reverse order: the release coating directly onto the foil, then the various color coatings, and finally the adhesive layer. By printing the graphics on the second surface, the design is held within two thin polymer layers, protecting it from wear and tear. Change of decoration simply requires a change of foil, without changing the mold or the material.

Technical Notes In the continuous IMD process modified hot-stamping foils are passed from a roller into the injection mold. On injection, the pressure and temperature of the melted polymer releases the carrier film, bonds it to the molding and presses it into the die-cavity, thereby molding the two-dimensional film into a three-dimensional geometry; however, the pattern can sometimes become distorted. In the discrete process, the foil is cut and placed in the mold. Complex shapes require that the foil is first thermally molded to the approximate shape of the mold cavity before insertion. Most thermoplastics – including polycarbonate, but not, however, polyethylene – can be decorated by IMD without problems.

The Economics IMD is simple and fast, and it reduces the number of steps in the manufacturing process, allowing the low-cost decoration of moldings that previously required a separate, costly, hot-stamping process, and it offers a greater range of graphics and colors.

The Environment IMD complies with the stringent environmental standards of the automotive industry. Recycled polymers can be used, as long as they have no gross contamination. The mold resin and the film can both be of the same material type, aiding recyclability.

Typical Products Mobile phone front and back covers, keypads, lenses, computer terminals, automotive heater controls, dashboards, toggle switches, credit cards.

Competing Processes Hot stamping, silk screen printing; cubic printing; pad printing.

Vapor Metalizing

What is it? Mirrors used to be made by a complex process involving silver dissolved in mercury. Today they are made by PVD metalizing, a process in which a thin coating of metal – usually aluminum – is deposited from a vapor onto a component. The vapor is created in a vacuum chamber by direct heating or electron beam heating of the metal, from which it condenses onto the cold component, much like steam from a hot bath condensing on a bathroom mirror. In PVD metalizing there is no potential difference between vapor source and component. In ion plating the vapor is ionized and accelerated by an electric field (the component is the cathode, and the metalizing source material is the anode). In sputtering, argon ions are accelerated by the electric field onto a metal target, ejecting metal ions onto the component surface. By introducing a reactive gas, compounds can be formed (Ti sputtering in an atmosphere of N_2, to give a hard coating of TiN, for instance).

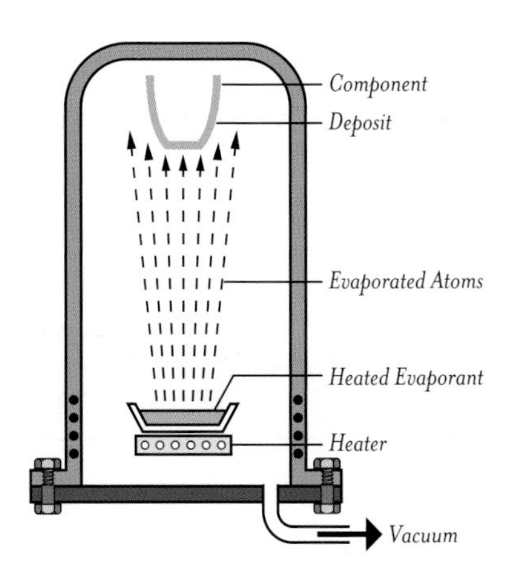

Attributes of Vapor Metalizing

Surface hardness, Vickers	10–40
Coating thickness, µm	1–80
Curved surface coverage	Avg. to Good
Processing temp., C	18–120

Functions Color, reflectivity, texture; hardness, wear; electrical conduction; protection against aqueous corrosion and by organic solvents.

Design Notes Metalizing is widely used to give a reflective metallic finish on bulk and film polymers, metal, glass and ceramic.

Technical Notes Clean surfaces are essential. Aluminum, copper, nickel, zirconium and other metals can be deposited.

The Economics The capital cost is high, the tooling cost low. For aesthetics, PVD metalizing is preferred to electroplating for polymers and ceramics because of its speed, quality and absence of unpleasant chemicals.

Typical Products Automotive trim, household appliances and kitchenware, door and window hardware, bathroom fixtures, printed circuit boards.

The Environment High volume production, good quality and cleanliness – particularly the absence of unpleasant chemicals – makes this an attractive process from an environmental standpoint.

Competing Processes Electro-plating; electro-less plating.

Electro-plating

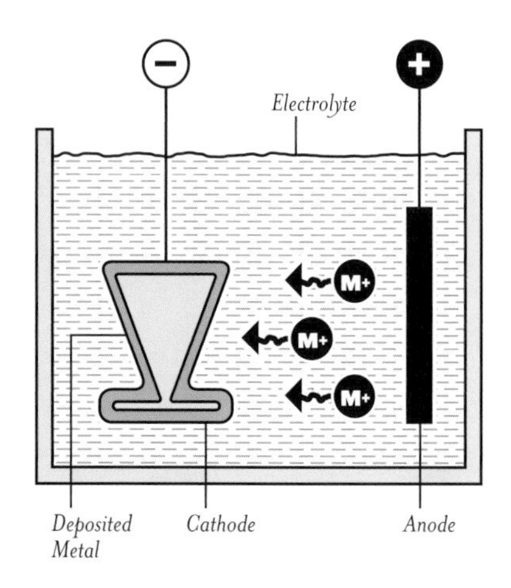

Attributes of Electro-plating

Surface hardness, Vickers	Depends on coating material
Coating thickness, µm	1–1000
Curved surface coverage	Depends on the bath
Processing temp., C	5–80

What is it? When, in 1800, Alessandro Volta discovered how to make an electric battery, he unleashed the development of electro-chemistry. By 1806 Humphrey Davy, prolific inventor, had developed ways of plating metals from their salts and solutions – electro-plating. The component (cathode) and the metalizing source material (anode) are submerged in the aqueous electrolyte where a direct electrical current drives metal ions from the source to the component, creating a thin metal coating.

Functions Color, texture, reflectivity; corrosion resistance, wear resistance, providing electrical conductivity, and good electrical contact.

Design Notes Plating for aesthetic purposes has its origins in the silver and gold plating of tableware and cutlery ("Sheffield Plate"), and in chrome and nickel plating for shiny, durable surfaces. Most polymers can now be plated, but care must be taken to prevent the coating breaking or peeling in use. The ability to plate convoluted shapes depends on what is called the "surface coverage" of the plating bath. Some have very poor surface coverage, meaning that only flat or gently curved surfaces can be plated; others – often assisted by mysterious, proprietary additives to the plating bath – can plate complex shapes with re-entrant features.

Technical Notes Almost any metal can be electroplated. Polymers (ABS, PET, ABS/polycarbonate, polyphenylene oxide, polysulfone, polypropylene, nylon, polyester, polyacetal, polystyrene, polycarbonate, epoxy/glass) and other non-electrically conductive materials must be first coated with an electrically conductive material. Cleaning and surface preparation are essential. The usual range of coating thickness is 1 to 50 microns, though thicknesses up to 1 mm are routine. Thicker coatings can generally be produced more cheaply or conveniently by other processes: thermal spraying, hot dipping or cladding. The processing temperature is in the range 5–80 C. Coatability – the ease with which an electro-plating can be applied – increases in this order: aluminum, mild steel, brass, copper. Many electro-platings have internal stresses; these can be reduced by heat treatment.

The Economics The equipment cost for electroplating is relatively high, but tooling costs are low. For thin deposits, particularly on small components or areas, large batch processing can make this a very competitive technique.

Typical Products A flavor for the immense range of applications of electroplating, technical and aesthetic, can be tasted by scanning the list of materials that completes this profile.

The Environment Many electro-plating baths pose environmental and health hazards. Some contain disagreeable chemicals – those with cyanogens are downright nasty. Protection from chemical pollutants and toxic vapor requires special precautions, as does the disposal of the plating medium. Cadmium is toxic, and now banded as plating in many European countries. Alternatives to chromium plating, one of the more unpleasant of processes, are sought but not yet found. Nickel can cause allergies and should be kept away from skin contact. Such gloomy news is not, however, universal: electro-plating is widely used to coat with copper, gold, silver, tin and zinc.

Competing Processes Solvent and water-based painting; mechanical polishing; electro-polishing; electro-less plating, anodizing, vapor metalizing.

Aluminum
Aluminum electro-plating can substitute for hot dipped aluminum, but is infrequently used.

Brass
Brass lamps and trays, low cost trim, interior automotive hardware, tubular furniture, household goods, toys, casket hardware, novelties, promote adhesion of rubber to steel.

Bronze
Bronze inexpensive jewelery, door plates, hardware, trophies, handbag frames, undercoat for nickel and chromium, bearing surfaces, tableware, household fixtures.

Cobalt
Cobalt alloy electroplates, mirrors, reflectors, applications where high hardness is required.

Co-Ni
Co-Ni magnetic recording, permanent magnet coating on memory drums in digital computers, electro-forming.

Copper
Copper undercoat (improved adhesion, prevention of hydrogen embrittlement), wire coatings, stop-off coatings during heat treatment and chemical milling, lubricant during drawing, thermally conductive coatings on cooking utensils, electro-forming.

Gold
Gold pen points, jewellery, watch and vanity cases, musical instruments, reflectors, name-plates, eyeglass frames, bracelets, trophies, novelties, electrical contacts, springs, electronic parts, laboratory apparatus.

Nickel
Nickel base for thin chromium electroplates, trim for automobiles, appliances, business machines and consumer goods, electro-forming, build-up of worn and mis-machined parts.

Rhodium
Rhodium resistant finish for costume jewellery, insignia, emblems, musical instruments, medical and surgical parts, laboratory equipment and optical goods, electrical contacts, reflectors and mirrors.

Silver
Silver tableware, candlesticks, cigarette lighters and musical instruments, bearings, surgical instruments, chemical equipment, electrical contacts.

Tin
Tin food and beverage containers, refrigerator evaporators, food and dairy equipment, hardware, appliance and electronic parts, copper wire, bearings.

Tin-Nickel
Tin-Nickel cooking utensils, analytical weights, surgical instruments, watch parts, chemical pumps, valves and flow control devices, resistance to marine corrosion.

Tin-Zinc
Tin-Zinc radio and television parts, cable connectors, relay assemblies, galvanic protection of steel parts contacting aluminum.

Zinc
Zinc appliances and automotive parts, finishing small parts (pipe couplings, bolts, nuts, rivets, washers, nails, hinges, hangers, hooks), electrical conduit pipes, silo and tie rods, screening, telephone exchange equipment.

Electro-less Plating

What is it? Electro-less plating is electro-plating without electricity. It relies instead on a difference in what is called electropotential when a metal is placed in a solution containing ions of another metal. Metal ions from the solution are deposited on the surface of the component by the action of a chemical reducing agent present in a metallic salt solution. In the case of electro-less plating of nickel (the most significant commercial application of the process), the salt is nickel chloride and the reducing agent is sodium hypophosphate. Once started the reaction can continue and there is no theoretical limit to the thickness of the coating. Electro-less plating is used when it is impossible or impractical to use normal electro-plating – when complex internal surfaces are to be plated, or when dimensional accuracy of the component is critical, for instance.

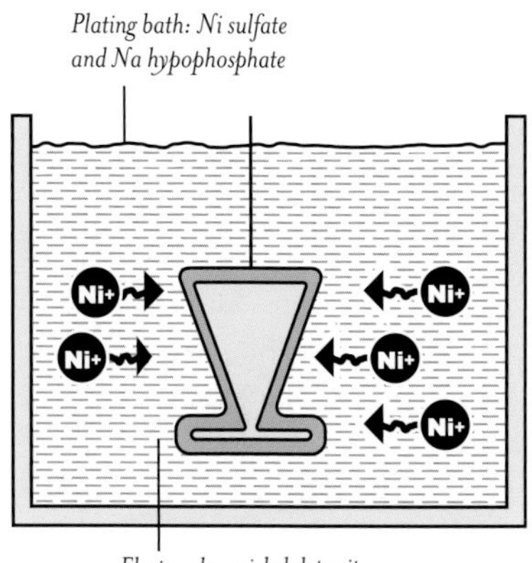

Attributes of Elctro-less Plating

Surface hardness, Vickers	600–1100
Coating thickness, µm	20–120
Curved surface coverage	Very good
Processing temp., C	20–50

Functions Surface hardness and wear resistance; corrosion protection; electrical conductivity; magnetic permeability; aesthetics.

Design Notes Most metals and polymers can be coated. The coating is typically applied to non-metallics as a conductive base for subsequent electro-plating. The uniformity of the coating is good, even on components of very complex shapes.

Technical Notes The main electro-less platings in current use are nickel (Ni, NiP, NiB, and nickel composite coatings containing SiC or PTFE) and copper. Ferromagnetic alloys Co-P, Ni-Co-P, Ni-Fe-P can be plated on magnetic tapes and discs. Special processing steps are needed for polymers because they are non-conductive and do not catalyze the chemical reduction of nickel; in this case adhesion results only from mechanical bonding of the coating to the substrate surface. To improve this, polymers are typically etched in acidic solutions or organic solvents to roughen their surfaces and to provide more bonding sites.

The Economics Electro-less plating is more costly than electro-plating: about 50% more for nickel plating. Deposition rates are much slower and chemical costs are higher; equipment and energy costs, however, are less. Despite these cost differences, the choice between the two processes is determined by physical rather than economical factors or production quantities.

Typical Products Heat sinks, bearing journals, piston heads, landing gear components, loom ratchets, radar wave guides, computer drive mechanisms, computer memory drums and discs, chassis, connectors, rotor blades, stator rings, compressor blades and impellers, chain saw engines.

The Environment The usual problems with disposal of chemical waste have to be faced, but otherwise the process is non-toxic.

Competing Processes Electro-plating, vapor metallizing.

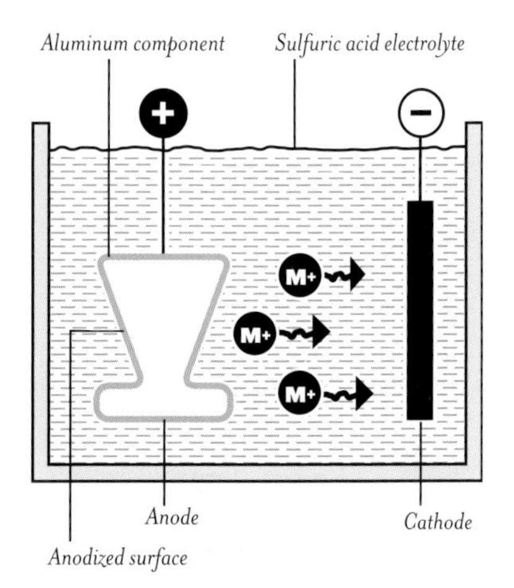

Attributes of Anodizing

Surface hardness, Vickers	600–1000
Coating thickness, µm	1–500
Curved surface coverage	Good
Processing temp., C	0–40

Anodizing

What is it? Aluminum is a reactive metal, yet in everyday objects it does not corrode or discolor. That is because a thin oxide film -Al2O3- forms spontaneously on its surface, and this film, though invisible, is highly protective. The film can be thickened and its structure controlled by anodizing. The process is electrolytic; the electrolyte, typically, is dilute (15%) sulfuric acid. The object to be anodized is made the anode (+) of the bath, with a potential difference of a few volts between it and the inert cathode. This sets up an enormous potential gradient across the oxide film, causing it to grow in thickness. The thicker film gives greater protection, and can be colored or patterned.

Functions Color, reflectivity; protection against corrosion and wear.

Design Notes Anodizing is most generally applied to aluminum, but magnesium, titanium, zirconium and zinc can all be treated in this way. The oxide formed by anodizing is hard, abrasion resistant and resists corrosion well. The oxide is micro-porous, allowing it to absorb dyes, giving metallic reflectivity with an attractive metallic, colored sheen; and it can be patterned.

Technical Notes There are three main types of electrolytes: "Weak dissolving" baths are based on boric acid; they form a thin (0.1 to 1 micron), non-porous oxide layer. "Medium dissolving" baths are based on a solution of chromic acid 3–10%, sulfuric acid 20%; oxalic acid 5%. They give a two-layer oxide: a thin, non-porous initial barrier layer with a porous layer on top of it. "Strong dissolving" baths contain phosphoric acid based electrolytes. The ability of the film to accept dyes depends on alloy compo-sition: pure aluminum and Al-5 Mg alloys are well-suited.

The Economics Equipment costs are moderately high, but tooling costs are low. Alternative treatments are chromating and phosphating: the ranked costs are: anodizing > chromating > phosphating.

Typical Products Anodizing is routinely used to protect and color aluminum; barrier layer anodic films on aluminum, titanium or tantalum are the basis of thin-film resistor and capacitor components for the electronic industry; the anodic film is exploited in the printing industry for photo-lithographic printing plates.

The Environment The chemicals involved here are aggressive but manageable. Disposal of spent anodizing fluids requires a recycling loop.

Competing Processes Solvent and water-based painting; electro-plating; electro-less plating; vapor metalization; chemical polishing; electro-polishing.

Mechanical Polishing

What is it? "Spit and polish" is a metaphor for hard work, and one that is relevant here. Mechanical polishing is slow and expensive, and should be used only when absolutely necessary. There are many variants, among them: lapping, honing and polishing. All make use of a fine abrasive, suspended in wax, oil or some other fluid, that is rubbed against the surface to be polished by a polishing disk, a belt or a shaped former.

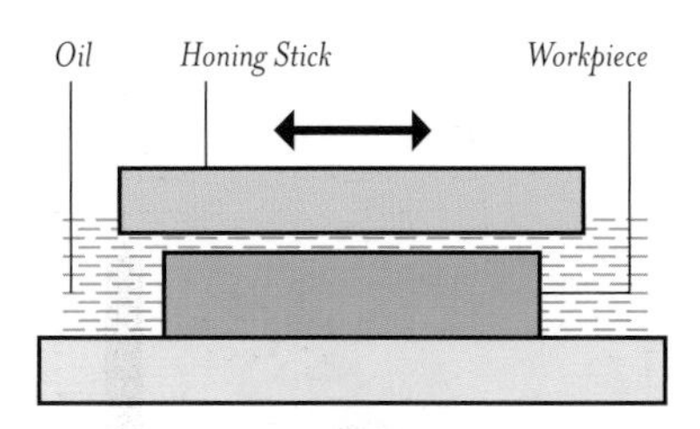

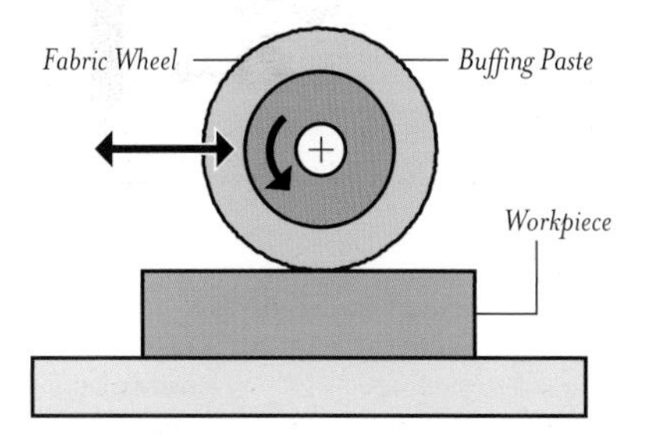

Functions Color, reflectivity, texture; wear resistance; reduced friction; improved fatigue resistance.

Design Notes Almost any metal or ceramic can be polished using suitable tools and techniques. Polishing is usually applied to conical, cylindrical, flat or spherical surfaces, but the parabolic surface of large mirrors are also polished. There are other ways of polishing – electro-polishing, chemical polishing – that give a highly reflective surface, but if precision is essential, as it is in accurate optical systems or in precision machine components, then mechanical polishing is the only choice.

Technical Notes Polishing uses rotating wheels or belts and abrasives with or without lubricants. Less material is removed than with grinding and it produces a smoother surface. Polishing can be divided into four steps: roughening, greasing, buffing, color buffing. From the roughing step to the coloring step finer and softer abrasives are used and the pressure is reduced. The polishing compound can be in the form of a solid bar or liquid. While solid compounds are used in manual operations, liquid compounds are best when used on automatic machines: production time is reduced, as no time is needed to change the bars or sticks.

Attributes of Mechanical Polishing

Surface hardness, Vickers	Same as substrate
Coating thickness	Not relevant
Curved surface coverage	Poor
Processing temp., C	0–30

The Economics The equipment and tooling cost for manual polishing are low (capital cost $100–$1000; tooling cost $10–$200) but the production rate is low. Those for automatic polishing are greater (capital cost $20,000–$1,000,000, tooling $1000–$10,000) but with a higher production rate. Mechanical polishing is expensive; over-specification should be avoided.

Typical Products Pistons, pins, gears, shafts, rivets, valves and pipe fittings, lenses and mirrors for precision optical equipment (ground and polished to a precision of better than 0.1 micron).

The Environment Hazards depend on the specific operation, the component, its surface coating and type of abrasive system in use.

Competing Processes Electro-polishing; chemical polishing.

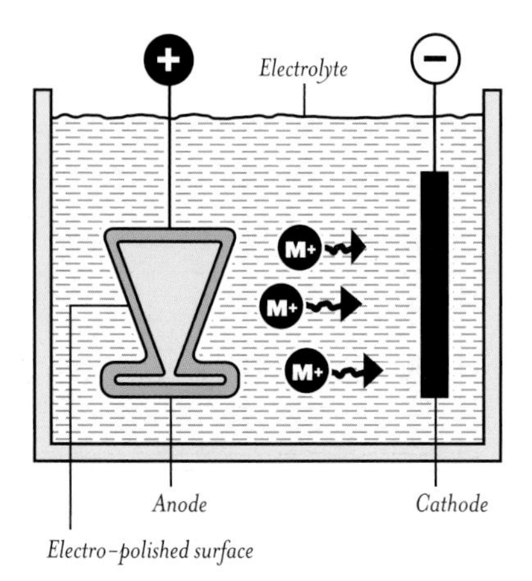

Attributes of Electro-polishing

Surface hardness, Vickers	Same as subtrate
Coating thickness	Not relevant
Curved surface coverage	Average
Processing temp., C	0–90

Electro-polishing

What is it? Electro-polishing is electro-plating in reverse. It offers an escape from the high cost of mechanical polishing, but at the penalty of much lower precision. The component is connected to the positive side of a low voltage direct current power source while a cathode is connected to the negative side, both immersed in a conductive solution. During the process, a layer of oxides or salts forms on the surface of the component; the layer controls the dissolution so that high points of a rough surface are dissolved faster than lower points. The final result is a metal surface with a smooth glossy appearance.

Functions Color, reflectivity, texture.

Design Notes Most metals can be electro-polished. Electro-polished surfaces are brilliant and have superior heat and light reflectance. The surface, however, has less mirror-like reflectivity than conventionally polished surfaces. More complex shapes can be polished than are possible by mechanical polishing, but less polishing occurs in holes, recesses and slots than on more prominent surfaces.

Technical Notes Electro-polishing is suitable for stainless steels, mild steels and low alloy steels, aluminum, brass, zinc die-casting alloys, beryllium copper, nickel silver, molybdenum, and tungsten. No special pre or post-treatments are needed. The most economical surface finish attainable by the process is 0.4 microns; 0.05 microns can be achieved – but at a higher cost.

The Economics The process does not normally require any special tooling, and so is economical for all levels of production. It requires less labour and less expensive equipment than mechanical polishing, but the initial capital cost can be high. Process time for any one component ranges from 3–10 minutes plus the time required for cleaning, rinsing and other preliminary and subsequent steps.

Typical Products Used for components as small as 20 mm^2 in area: miniature electrical contacts, rivets, screws; also used for components as large as 1 m^2: large metal panels.

The Environment Many electro-polishing baths pose environmental and health hazards. Some contain disagreeable chemicals – those with cyanogens require particular care. Protection from chemical pollutants and toxic vapor requires special precautions, as does the disposal of the polishing medium.

Competing Processes Mechanical polishing; chemical polishing; electro-plating; electro-less plating; vapor metallizing.

Chemical Polishing

What is it? If a bar of soap with a rough surface is immersed in water, it emerges smoother than it went in. That is because the soap dissolves faster at the peaks of the roughness than at the troughs because the local concentration gradient is steeper there. Chemical polishing relies on a similar principle. It is a process whereby a surface is smoothed by controlled chemical dissolution in a bath of acid containing additives that, by creating a surface boundary-layer, cause protrusions to dissolve faster than flat or intruded parts of the surface – there is no external power supply. The process is normally carried out at high temperatures to increase the polishing rate. Chemical polishing, like electro-polishing, is a way of brightening a surface, enhancing reflectivity, but without the dimensional precision of mechanical polishing. Much cheaper, of course.

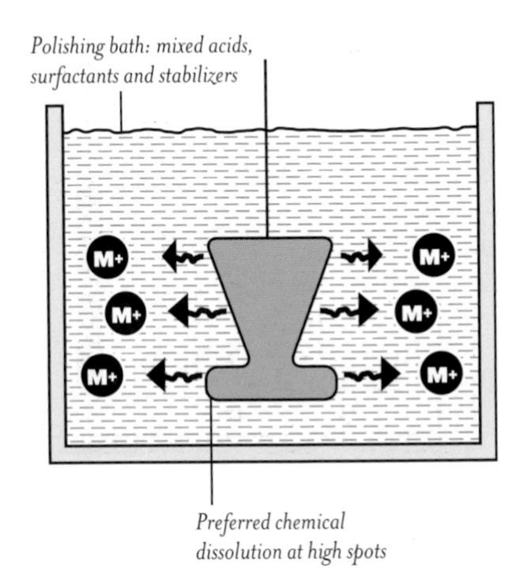

Functions Color, reflectivity; texture.

Design Notes Chemical polishing offers great freedom in polishing items with blind holes and other recessed areas. It can treat components of different shapes at the same time. The reflectivity of the surface is less than that given by electro-polishing; the chemically polished surface is often anodized subsequently to produce a clear, colorless protective oxide coating.

Attributes of Chemical polishing

Surface hardness, Vickers	Depends on substrate
Coating thickness	Not relevant
Curved surface coverage	Good
Processing temp., C	55–140

Technical Notes Aluminum, copper and stainless steel are commonly treated by chemical polishing. Chemical polishing baths are generally based on combinations of phosphoric acid, nitric acid, sulfuric acid, hydrochloric acid, organic acids and special surfactants and stabilizers; some are based on peroxides. The process is controlled by the chemistry of the bath, the temperature and the time of immersion.

The Economics Chemical polishing has tended to replace electro-polishing due to lower costs. The process is cheap to set up and operate: no power supply or expensive racking are needed, and the capital investment and labor costs are low.

Typical Products Jewelery, razor parts, automotive trim, fountain pens, searchlight reflectors, architectural trim, household appliances, thermal reflectors for components of space vehicles.

The Environment The chemicals involved here are aggressive but manageable. Disposal of spent chemical polishing fluids requires a recycling loop.

Competing Processes Mechanical polishing; electro-polishing; vapor metallizing; electro-plating; electro-less plating.

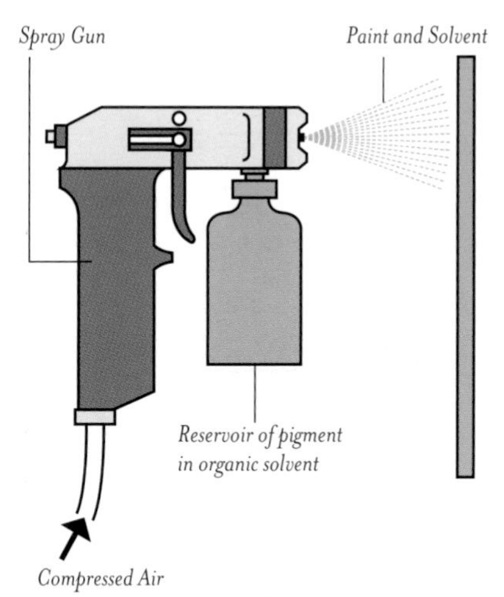

Attributes of Solvent-based Painting

Surface hardness, Vickers	10–16
Coating thickness, μm	10–1000
Curved surface coverage	Good
Processing temp., C	10–100

Solvent-based Painting

What is it? Watching paint dry is a synonym for boredom. But modern paints are far from boring. In solvent-based painting, coloring materials (pigments) are suspended, together with binding agents (resins), in a volatile organic solvent (VOC). When spread thin over a surface, the solvent evaporates; the resins hold the pigments in place to form a decorative and protective coating. Some few paints are not much more than this today. New developments now give formulations that dry in seconds, have fade-resistant colors, soft textures, visual effects, powerful protective qualities, and much more. But – and here it comes – there is a problem. Solvent-based paints are environmentally bad, so bad that their very future is under threat.

Functions Color, texture, feel; protection against corrosion and bacteria; wear protection.

Design Notes Solvent-based paints give the smoothest, most uniform coating and the greatest control of color – the automobile industry and most product designers insists on them. Metallic paints mix flake aluminum powder in the coating; the trick is to have the coating thin enough that the metal flakes lie in a plane so that the color does not "flip" when viewed from different angles. Color is determined by the differential absorption and reflection of the various wavelengths of light; the color seen is that at the least absorbed wavelength from the angle of view.

Technical Notes Paints are applied by brushing, dipping or spraying, and can be applied to virtually any surface provided it is sufficiently clean.

The Economics Painting is cost effective. The equipment costs are low for non-automated painting, but can be high if the equipment is automated. Paints are a $75 billion per year industry.

Typical Products About half of all paints are used for decorating and protecting buildings, the other half for manufactured products, most particularly cars and domestic appliances; marine applications create important market for high-performance corrosion and anti-fouling formulations; "printers inks" are paints that play a central role in publishing and packaging.

The Environment Emissions from the evaporating solvents from solvent-based paints (VOCs) are toxic, react in sunlight to form smog and are generally hostile to the environment. Auto manufacturers and others are under increasing pressure to meet demanding environmental standards. The solvents must now be recaptured, burnt or recycled. There is growing incentive to replace them by water-based paints (but they dry slowly) or dry polymer coatings (but they cannot yet offer the same surface quality).

Competing Processes Water-based painting; powder coating; electro-painting.

Water-based Painting

What is it? Water-based (or latex) paints are synthetic resins and pigments, plus coalescing agents, that are kept dispersed in water by surfactants. They dry by evaporation of the water; the coalescing agents cause the particles of resin to fuse together (coalesce) as the water evaporates to form a continuous coating.

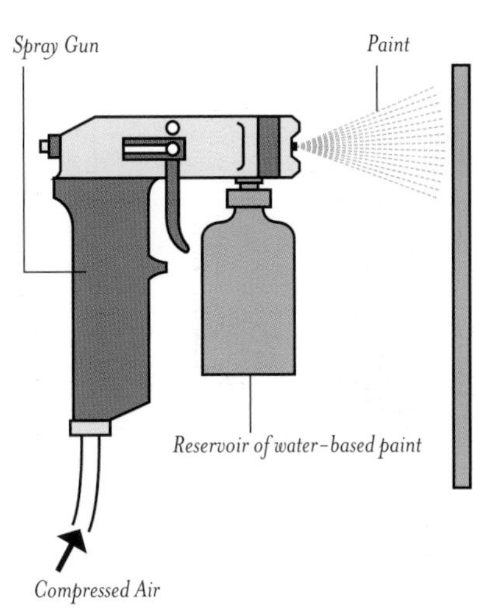

Functions Color, texture, feel; protection against corrosion, fungal and bacterial attack; wear protection.

Design Notes Water-based paints are based on acrylic, urethane, polyvinyl acetate (PVA) or epoxy dispersions. Acrylic emulsions are used in exterior applications where their non-yellowing characteristics, as well as excellent weatherability, are outstanding. Water-based urethanes are suited to uses where good flexibility and toughness are important, such as leather and polymer coatings, but their major drawback is their high cost. PVA and epoxies give good weathering resistance. Water-based paints take longer to dry than many organic solvent-based paints, and give a surface finish that is less good.

Attributes of Water-based Painting

Surface hardness, Vickers	10–16
Coating thickness, μm	10–1000
Curved surface coverage	Good
Processing temp., C	10–100

Technical Notes Water-based paints must be protected from freezing and applied at a minimum temperature of 10 C. Humidity and temperature control are critical for the drying time. A heat cure is sometimes necessary. Pigments must be compatible with water. Metallic particles are usually coated before being mixed into the paint to prevent chemical reaction with water. Many conventional binders (alkyls, acrylics, and epoxies) can be made water soluble by chemically attaching polar groups such as carboxyl, hydroxyl, and amide. Dispersions are very small particles of binders, less than 0.1 microns diameter, dispersed in water. Emulsions, or latex, differ from dispersions by having much larger particle size on the order of 0.1 micron or more. They are made by precipitation in water and therefore do not need to be dispersed mechanically.

The Economics Solvent-based paint systems can usually be converted to water-based paint systems with a limited capital investment. The cost of water-based paints depends on the type; they can be less expensive than the solvent-based equivalent.

Typical Products Flooring, wood lacquers, decorative paints, automotive coatings, polymer coatings, inks, adhesives.

The Environment Water-based paints reduce VOC emissions and worker exposure to toxic pollutants, and present no fire hazard. They are displacing solvent-based paints, although these dry fast and give a better surface finish with better adhesion.

Competing Processes Solvent-based painting; powder coating; electro-painting.

Electro-painting

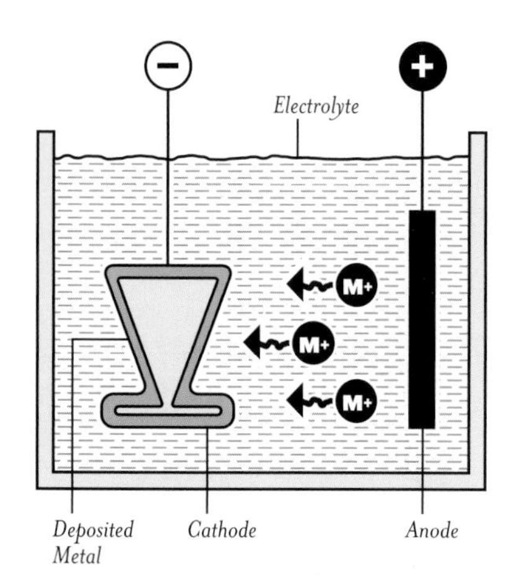

Attributes of Electro-painting

Surface hardness, Vickers	5–15
Coating thickness, μm	10–30
Curved surface coverage	Good
Processing temp., C	25–230

What is it? In electro-painting the component forms the cathode in an electrolytic cell. The anode is inert – usually graphite or stainless steel – and the electrolyte is an aqueous solution in which the pigment is emulsified. An alkaline environment builds up at the cathode, causing the lacquer to coagulate. The charged particles of pigment are dragged by the electric field onto the surface of the component where they form a strongly bonded film. The film is relatively thin (10–30 microns), but gives good corrosion resistance, particularly if baked at 150–230 C after deposition. Other layers of paint are applied on top. The principle can be compared with electro-plating, and is widely used in the automobile industry.

Functions Color, reflectivity; corrosion resistance.

Design Notes The process is largely used to give uniform undercoats on large, complex components. It lends itself to large volume production of products coated with the same color; examples are the electro-painting of a white paint on metal window and door frames and undercoating of auto body shells. The paint layer is distributed evenly over the surface, even in those areas that would otherwise be inaccessible. The main disadvantage is the requirement of a single, uniform color.

Technical Notes Electro-paint films are deposited from a dip tank. The bath consists of resin concentrate and pigment concentrate mixed with de-ionized water and small amounts of solubilizers and de-foamers. The concentration of non-volatile solids in the bath varies from about 10 to 20%, depending on type and composition. This process sometimes uses autophoretic paints; these are water-reducible paints deposited on metal surfaces by the catalytic action of the metal on the paint materials in the bath. Currently, only ferrous alloys activate the electro-paints available commercially. Tubular automotive frames are coated by this method, because the entire length of the tubing can be coated inside and out with equal ease.

The Economics The tooling costs are low; the capital cost medium.

Typical Products Undercoats for body lacquering of car bodies, single-color coatings for appliances, architectural panels and frames.

The Environment The process poses no particular environmental hazards, apart from the disposal of spent bath fluids.

Competing Processes Solvent and water-based painting; powder coating.

Powder Coating

What is it? Dust particles stick to the face of a television set because of a difference in electrical charge between the dust and the screen. This effect is used for practical purposes in electrostatic powder coating. It is an efficient, widely used process for applying decorative and protective finishes to metallic or conducting components. The powder is a mixture of finely ground pigment and resin that is sprayed through a negatively charged nozzle onto a surface to be coated. The charged powder particles adhere to the surface of the electrically grounded component. The charge difference attracts the powder to the component at places where the powder layer (which is insulating) is thinnest, building up a uniform layer and minimizing powder loss. The component is subsequently heated to fuse the layer into a smooth coating in a curing oven. The result is a uniform, durable coating of high quality and attractive finish.

In polymer flame coating, a thermoplastic in powder form (80–200 μm) is fed from a hopper into a gas-air flame that melts the powder and propels it onto the surface to be coated. The process is versatile, can be mechanized or operated manually, and can build up coatings as thick as 1 mm. A wide range of thermoplastic powders can be used and the process is cheap. The disadvantages: line-of-sight deposition, and surface finish that is inferior to other processes.

A hot donut, dunked into fine sugar, emerges with a crisp sweet skin. Fluidized-bed coating works in much the same way. The component, heated to 200–400 C, is immersed for 1 to 10 seconds in a tank containing coating powder, fluidized by a stream of air. The hot component melts the particles, which adhere to it, forming a thick coating with excellent adhesion. In electrostatic bed coating the bed is similar but the air stream is electrically charged as it enters the bed. The ionized air charges the particles as they move upward, forming a cloud of charged particles. The grounded component is covered by the charged particles as it enters the chamber. No preheating of the component is required but a subsequent hot curing is necessary. The process is particularly suitable for coating small objects with simple geometries.

Functions Color, texture, feel; protection against aqueous and organic corrosion; wear protection.

Design Notes Powder coating applications are limited to those that will sustain the processing temperatures required for polymer melting, curing, and film formation – but that includes all metals. The process can be easily automated, and can create thin films (50 μm), with good edge cover. Thick films can be applied in a single application, and the coating toughness is generally better than its liquid-based counterparts. Powder coatings are available in a wide variety of glosses and textures.

Technical Notes Electrostatic powder coating is routinely applied to components of steel, aluminum, magnesium, brass, copper, cast iron and most metallic alloys. It can be applied to non-metals if their surface is first made conducting. The development of powder that can be cured at lower temperatures has allowed powder coating to be applied to non-metal surfaces such as ceramics, wood and polymers. Coating materials

Electrostatic Powder Coating

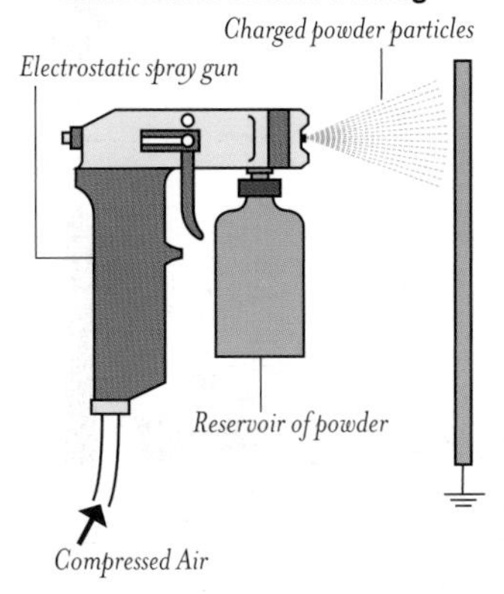

Polymer Flam Coating

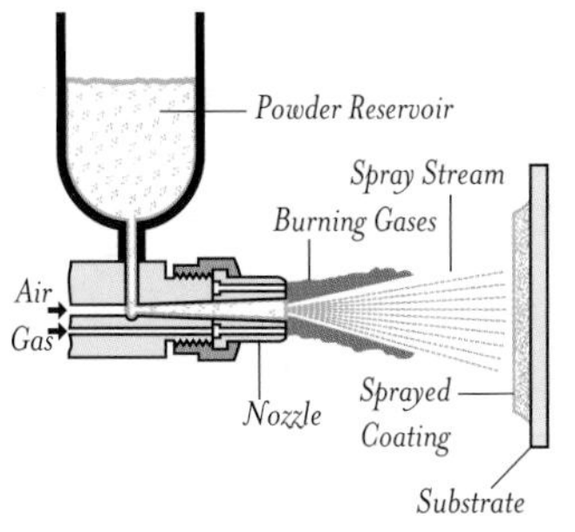

Fluidized Bed Coating

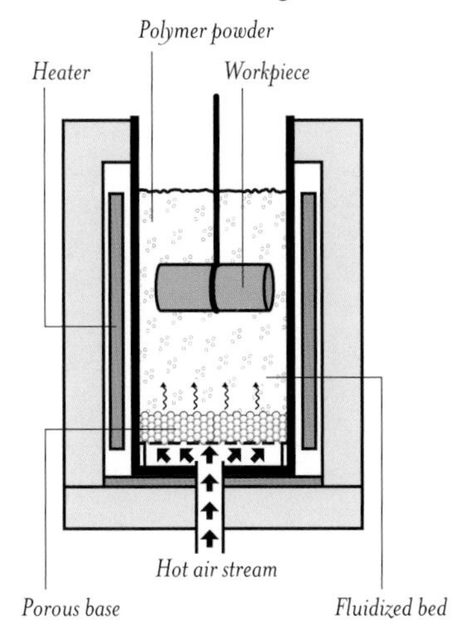

Attributes of Powder Coating

Surface hardness, Vickers	10–16
Coating thickness, μm	50–2000
Curved surface coverage	Good
Processing temp., C	125–400

are typically nylons, polyesters, polyethylene, polypropylene, polyvinyl-chloride (PVC), polyvinylidene fluoride (PVDF) and ethylene acrylic acid (EAA) co-polymers.

Almost all metals, ceramic and even wood can be coated by polymer flame spraying – unlike electrostatic powder coating, it is not necessary for the surface to be conducting. Typical coatings are low and high density polyethylene, polypropylene, thermoplastic polyesters, and polyamides (Nylon 11). Pre-heating of the surface to 150–250 C may be necessary.

Fluidized-bed coating works well with nylons, PVC, acrylics, polyethylene, polypropylene, silicones, EVA and polystyrene. The process gives good coverage, particularly of irregular shapes, allowing simultaneous coating of external and internal surfaces. Thick coatings are possible in a short time, with little waste of powder.

The Economics Powder coating is the fastest-growing finishing technique, now representing over 10% of all industrial finishing. For electrostatic polymer coating, the capital cost is medium, the tooling cost low, and the equipment is not portable – a workshop process. The coating rate is between 5 and 20 µm/s. For polymer flame spraying, the capital cost is low, the tooling cost is low and the equipment is portable. There is no limit on the size of component. The coating rate is between 10 and 30 microns/second. For fluidized-bed coating, the capital cost is higher than that for polymer flame spraying; the tooling costs are low. The coating rate is between 20 and 50 microns/second.

Typical Products Automotive industry: wheels, bumpers, hubcaps, door handles, decorative trim and accent parts, truck beds, radiators, filters, numerous engine parts. Appliances: front and side panels of ranges and refrigerators, washer tops and lids, dryer drums, air-conditioner cabinets, water heaters, dishwater racks, cavities of microwave ovens; powder coating has replaced porcelain enamel on many washer and dryer parts. Architecture: frames for windows and doors and modular furniture. Consumer products: lighting fixtures, antennas, electrical components, golf clubs, ski poles and bindings, bicycles, computer cabinets, mechanical pencils and pens.

The Environment The process uses no solvents and releases almost no volatile organic compounds (VOCs), so operators are not exposed to high levels of toxicity. Air pollution, fire hazards, and volatile organic compounds are low. Excess or over-sprayed powders are recycled, blended, and re-sprayed with virgin material, which increases utilization up to 99% by weight.

Competing Processes Anodizing, organic and water-based painting.

Enameling

What is it? Enameling is painting with glass. It creates coatings of exceptional durability: the enameled death-masks of the Pharaohs of Egypt look as bright, vibrant and perfect today as they did when they were created over 3000 years ago (Tutankhamen, BC 1358–1340). A thin layer of glass powder with binder and coloring agent is applied to the object to be enameled by painting, spraying or screen printing; the layer is then fused to the object, generally made of cast iron, pressed steel, copper, silver or even – in the case of Tutankhamen – of gold, creating a continuous, strongly-bonded coating of colored glass. It is a hot process – the object being enameled must be heated to the melting temperature of the glass powder, limiting it to use on metals and ceramics.

Spray Coat

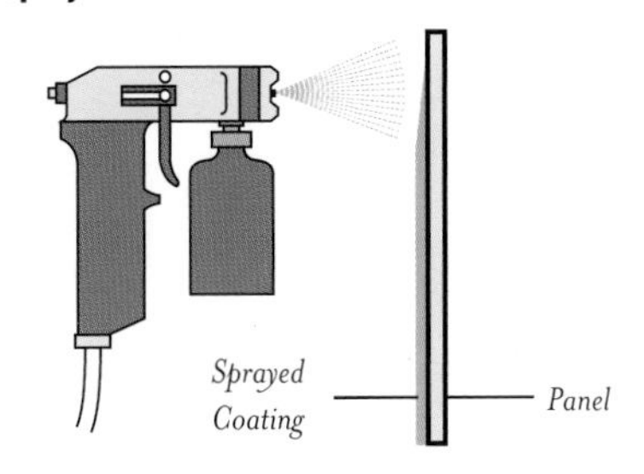

Fuse and Bond

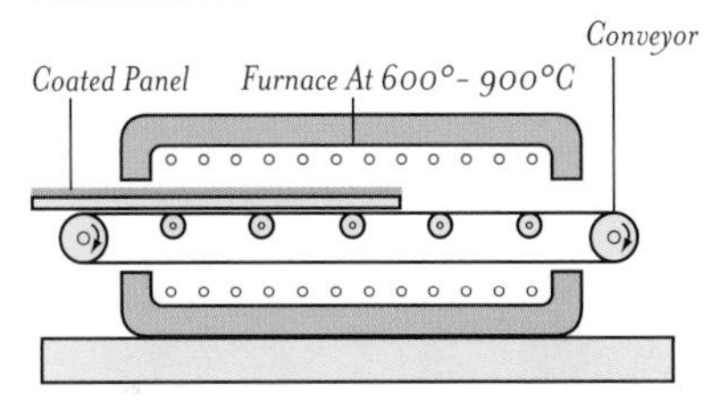

Attributes of Enameling

Surface hardness, Vickers	7K–10K
Coating thickness, µm	500–1000
Curved surface coverage	Good
Processing temp., C	600–900

Functions Color, texture, reflectivity; corrosion resistance; scratch resistance.

Design Notes Enamel coatings have exceptional hardness, wear and corrosion resistance, and can be made in an unlimited range of colors.

Technical Notes The glass used for enameling must flow easily, wet and bond to the metal surface, and have an expansion coefficient that differs only slightly from that of the underlying metal. This is achieved by using a glass rich in boric oxide (up to 35%), which also gives strong bonds and lustrous colors. In practice enameling usually marries two different kinds of glass: the base coat includes cobalt and nickel oxides, which help to form a strong bond with the metal; the second coat carries color and decoration. The result is extremely durable coating that is able to withstand thermal shock.

The Economics Enameling is not cheap – it is both slow and energy intensive. Its choice relies on the balance between quality and cost – its durability and aesthetic qualities are unequalled.

Typical Products In the home: baths, washing machines, dishwashers, heaters, stovetops, fireplaces, gas and electric cookers, clock faces and cook ware. In places you can't see: hot water services, storage tanks, car exhaust systems, printed circuits and heat exchangers. In places you can see: street signs, railway signs and murals. In the building industry: interior and exterior architectural panels, fascias, spandrels and partitions.

The Environment The process requires high temperatures, but is otherwise environmentally benign. Enameled surfaces are easily cleaned (even graffiti can be removed without damage), scratch resistant and exceptionally hygienic.

Competing Processes Solvent and water-based painting; powder coatings – though none give coatings of comparable hardness and durability.

Acid Etching

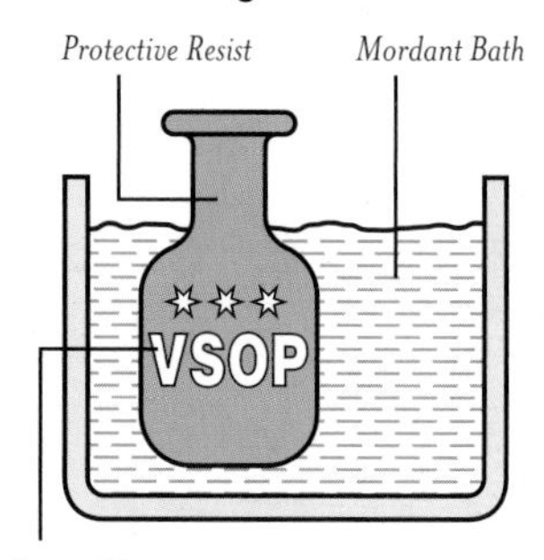

Sand Etching

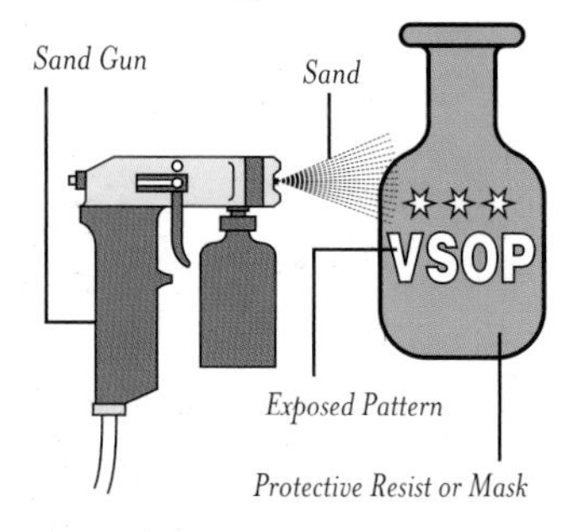

Attributes of Etching

Surface hardness, Vickers	7K–10K
Coating thickness, μm	500–1000
Curved surface coverage	Depends on process
Processing temp., C	18–60

Etching

What is it? There are three basic processes for etching. Chemical etching is widely used to decorate glass, metal and polymer objects. The areas or patterns to be etched are exposed to an acid or "mordant;" the areas that are not to be etched are protected with a wax resist. Glass used to be etched by hydro-fluoric acid, a particularly nasty chemical that gives off poisonous fumes; today this has largely been replaced by proprietary etch-gels, by sand etching or by engraving with a diamond tool. Copper, zinc, nickel and steel are etched in various combinations of sulfuric, nitric and hydrochloric acids.

Sand etching makes use of a jet of angular sand particles in a pressurized gas stream to erode the surface where it is not protected by a mask. The method works well with metals, ceramics and polymers.

Electro-etching is a non-toxic way of etching of metals. Here the acid bath is replaced by one of copper sulfate (for etching copper) or zinc sulfate (for etching zinc) that never has to be changed. The object, protected by wax resists where necessary, is made the anode (+) of an electro-polishing bath, cutting into the areas of surface not protected by the resist.

Functions Color, texture, patterning, decorating.

Design Notes Etching is widely used to pattern glass and metal. The freedom of design is limited only by the technique for creating the wax pattern of resistance.

Technical Notes Techniques exist to etch glass, tile, wood, stone, metals, polymers. The acids used to chemical etching, are by their nature, particularly aggressive, requiring special containers, handling and disposal. Sand etching is less demanding, but less controllable.

The Economics Chemical etching is slow and expensive. Sand etching and electro-etching are faster and cheaper.

Typical Products Decoration of glass, metal, polymer or wood objects. Art: jewellery, ornaments, clocks, bookmarks, photographs etched into metal, medallions, street art, event posters, light switches. Advertising: plaques, business cards, graphic plates for brochures, awards, cover elements for annual reports, logos or tags to be placed on products, elevator button plates, signs to go on doors, name-plates, insignias, face plates, office decorations.

The Environment Chemical etching presents major environmental problems associated with toxic fumes, aggressive chemicals, and the difficulty of disposing of spent baths. Sand etching and electro-etching have none of these problems.

Competing Processes Texturing.

Texturing

What is it? Texture can be created in many ways: by casting or molding in a patterned tool; or by rolling between patterned rolls ("Diamond plate," used for stair treads, is made in this way). By crimping, pressing of sheet between patterned dies: pre-plated sheet, up to 0.8 mm thick, can be crimped to give sparkling reflective patterns; it is used, too, to increase bending stiffness and strength, allowing lighter gauge material for boxes and cans. By sand blasting, or abrasive polishing, the microscopic roughening of the surface in a random or patterned way (as in "brushed aluminum"). By laser texturing, usually applied to rolls or dies; the pulsed laser beam creates minute craters on the roll surface in a precisely controlled pattern that is then transferred to the component during the rolling operation. By electro-texturing, a process like electro-discharge machining (EDM), but using shaped graphite electrodes to smooth or give texture to the surface of the component. And by chemical engraving ("etching" and "chemical polishing") either of molds, rolls and dies, or of the component itself.

Functions Color, texture, feel; friction (grip).

Design Notes Texturing is important in design for both technical and aesthetic reasons. It can improve product appearance, giving a visually interesting surface as well as hiding minor flaws, scratches and sink-marks. It can improve grip and it contributes to the tactile qualities of a product.

Technical Notes Virtually any material can be textured by one or another of the methods described above. Texturing by patterned rolls or crimping is limited to simple shapes (like sheet); electro-texturing, chemical engraving and patterned tool methods allow greater freedom of shape.

The Economics The cost depends on the process. Some, like sand blasting and pattern-rolling are fast and cheap. Others, like laser texturing and chemical engraving are slower, and consequently more expensive.

Typical Products There are two very broad classes of application: to impart visual features or to create tactile characteristics on a product.

The Environment Texturing processes do not present any particular environmental hazards.

Competing Processes Etching.

Attributes of Texturing

Surface hardness, Vickers	Depends on substrate
Coating thickness, μm	50–1000
Curved surface coverage	Depends on process
Processing temp., C	18–30

Appendix: Exercises for the Eye and Mind

Here are some things to try. Most can be done individually; 4 and 10 require a group. Some need a supply of materials.

1. Materials in the in the real world
Examine the use, in products and structures, of one material class from the following list.

Woods	Polymers
Ceramics	Metals
Glass	Textiles

Seek examples of their use by visiting shops, observing products and structures in the street, stations, and museums. Prepare a commentary, with sketches rather than photographs, bringing out the way the material has been used, the forms that it has allowed, and the associations and perceptions its use has created.

2. Product anatomy
Dissect a product of your choice, examining how materials have been used, doing your best to identify the material and the way it was processed. Examine how the design choices used for the interior of the product have influenced its external form. How far does the external form express the function of the product?

3. Product redesign
Dissect a product, identifying the function both of the systems it contains and of the individual component. Explore alternatives – how could one or more functions be achieved in other ways? Could components be redesigned using other materials? What changes would these alterations allow in the number of parts in the product and in the design of its external form?

4. Associations of materials
What do you think of when you see something made of gold? You might – because of its expense – associate it with wealth and luxury; or perhaps – because it is a visible symbol of riches – with extravagance and power; or perhaps – because of its total resistance to chemical attack – with stability and permanence. What associations would you attach to the following materials? Ask: "What do you think of when you see something made of...?"

Machined steel	Nylon
Stainless steel	Diamond
Rusted steel	Polished cherry wood
Brushed aluminum	Plastic with a wood finish
Lead	Marble
Polyethylene	Glass

5. Creating associations with materials

You are asked to design a rack for holding CDs in a den, study or home, using the choice of material to prompt associations in the mind of the purchaser or user. What materials (colors, finishes, styles) would you choose to suggest that the rack is...

State of the art, technically advanced
Environmentally friendly
Easy to use, non-threatening
Desirable to children
Durable, robust

6. Material metaphors

An "iron" is used for smoothing crumpled clothing. It derives its name from the material of which it is made. A "glass," similarly, is a product that shares its name with its material. What – if you could name it, would be...

an aluminum?
a polycarbonate?
a bamboo?
a polyethylene?
a nickel?
a copper?
a zinc?
a nylon?

7. Making a mood board

Identify a product and the environment in which it will be used. Assemble images, samples, sketches and other concept-suggesting material, and assemble them into a mood board.

8. Product Evolution

Explore the evolution of a product, documenting its design history (examples were given in Chapter 3). Report on the ways in which the choices of material, processing and surface treatment have been used to meet technical and aesthetic requirements and to create associations and perceptions. Start with your own observations, then seek information from design-history books and contemporary magazines and catalogs. Products that are well documented in this way include

Radios
Vacuum cleaners
Skis
Bicycles
Mobile phones
Personal computers
Pens
Cars

9. Variations on a theme

Sketch designs for one product as it would appear if made from each of several given materials. For example, design a desk made of metal, of a polymer, of glass and of wood.

10. Inspiration from materials

Brainstorm on the uses to which a common material, or one that is new or unusual (if available), could be put. Both functional and aesthetic solutions are possible. Take one of the ideas emerging from the brainstorm and develop it further, trying out the material in the chosen role. Describe the features of the idea and how the material contributes to these features. (The project requires materials samples and is best done with a group of individuals attempting the same result.)

11. Chromatic series

Develop a chromatic series of materials, as Itten suggested to his students in the Bauhaus course. The most recognizable chromatic series is that of color: red, orange, yellow, green, blue, violet. This series provides a limited structure and a visual representation of a wide spectrum of colors. The same is needed to guide designers across the wide range of available materials.

12. Significant associations

Analyze the responses to Question 4 (assuming that responses are available from – say – 10 participants).

- Examine the responses, judging the degree to which participants agree on associations
- Use statistical methods to identify association-choices with a significant degree of common meaning
- Use cluster analysis to group materials with similar associations.

13. Major projects

Develop a design for a project. Research the background (clarifying the task), devise concepts, develop these though visualization by sketching, and materialization by model building. Estimate the cost.

Appendix: Selected Material Maps

Chart 1 – Elastic Modulus/Density 316
Chart 2 – Strength/Density 318
Chart 3 – Fracture Toughness/Elastic Modulus 320
Chart 4 – Elastic Modulus/Strength 322
Chart 5 – Loss Coefficient/Elastic Modulus 324
Chart 6 – Thermal Expansion/Thermal Conductivity 326

The six charts shown here are maps of technical attributes. They show the islands and continents, so to speak, that are occupied and the oceans in between that are empty. Parts of these oceans can be filled by making composites, sandwich structures and other configured materials (foams, for instance) that combine the properties of two or more single materials – there is still uncharted water here. Other parts cannot be reached at all, for fundamental reasons to do with the physics of the way in which atoms are bonded in solids.

The charts form part of a larger set that can be found in the companion text (Ashby, 1999); they were constructed using the CES 4 (2002) software. Some of the labels have been shortened for the sake of space: thus tsPolyester is thermosetting polyester, tpPolyester is thermoplastic and elPolyester is elastomeric; fPU is a flexible polyurethane foam, rPU is rigid foam, sPU a structural foam, ocPU is open cell, and ccPU is closed cell.

Chart 1 – Elastic modulus, E, and Density, ρ

If a solid is deformed elastically, it springs back to its original shape when released. The atoms of solids are held together by atomic bonds – think of them as little springs linking the atoms. If the springs are hard to stretch, the solid is stiff; if they are easy to stretch, the solid is compliant. Young's modulus is a measure of the stiffness or compliance of the bonds. Metals and ceramics are stiff; polymers are much more compliant. Foams and elastomers are even more compliant, and for a new reason: their structure allows additional deflection (cell wall bending, molecular rearrangement) which, nonetheless, is recovered when the loads on the material are released.

Inter-atomic spacings in materials do not vary very much: they are all within a factor of 3 of 0.3 nm (3×10^{-10} m). The densities of materials – the mass per unit of their volume – varies far more widely, simply because some atoms are heavy (copper, iron – and thus steel) and others are light (hydrogen, carbon – and thus polymers like polyethylene, $(CH2)_n$).

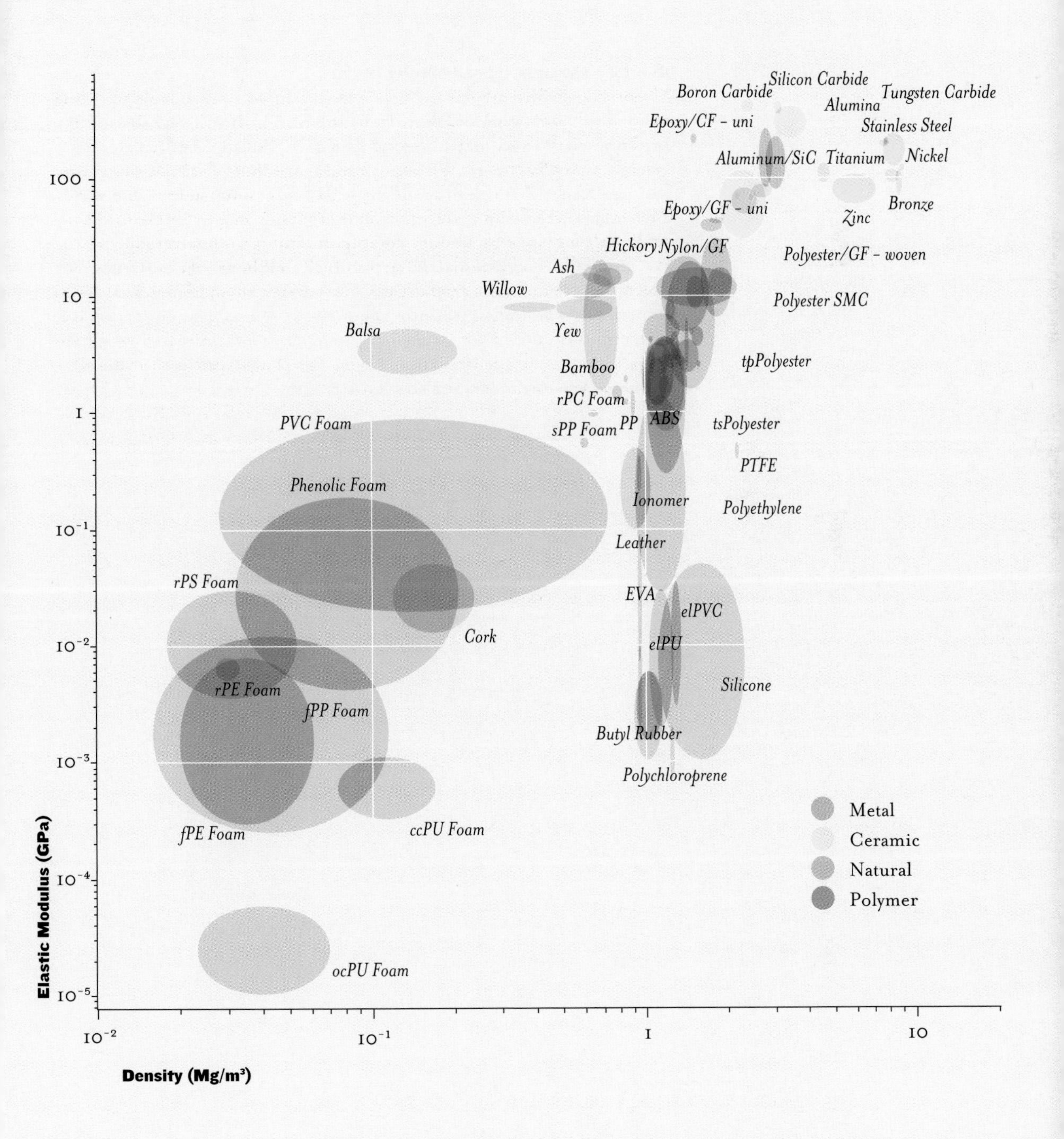
Silicon Carbide
Boron Carbide
Tungsten Carbide
Alumina
Epoxy/CF - uni
Stainless Steel
Aluminum/SiC
Titanium
Nickel
Epoxy/GF - uni
Bronze
Zinc
Hickory
Nylon/GF
Polyester/GF - woven
Ash
Willow
Polyester SMC
Balsa
Yew
Bamboo
tpPolyester
rPC Foam
sPP Foam
PP
ABS
tsPolyester
PVC Foam
PTFE
Phenolic Foam
Ionomer
Polyethylene
Leather
rPS Foam
EVA
elPVC
Cork
elPU
Silicone
rPE Foam
fPP Foam
Butyl Rubber
Polychloroprene
fPE Foam
ccPU Foam
ocPU Foam
Metal
Ceramic
Natural
Polymer
Elastic Modulus (GPa)
Density (Mg/m³)
100
10
1
10⁻¹
10⁻²
10⁻³
10⁻⁴
10⁻⁵
10⁻²
10⁻¹
1
10

Chart 2 – Strength, σ_f, and Density, ρ

Strength is different from stiffness – think of it as a measure of the force needed to break atomic bonds. In brittle materials (glasses, ceramics) breaking means just that: the solid fractures. But in ductile materials (metals, many polymers) the bonds break, the atoms or molecules move, and new bonds form, leaving the material just as it was before, but with a new shape. This ability to reform bonds gives ductility, and thus the ability to be shaped by rolling, drawing, molding, and stretching.

"Strength," here, means the 0.2% offset yield strength for metals. For polymers it is the stress at which the stress-strain curve becomes markedly non-linear – typically a strain of about 1%. For ceramics and glasses, it is the compressive crushing strength; remember that this is roughly 15 times larger than the tensile fracture strength. For composites it is the tensile strength. For elastomers it is the tear strength.

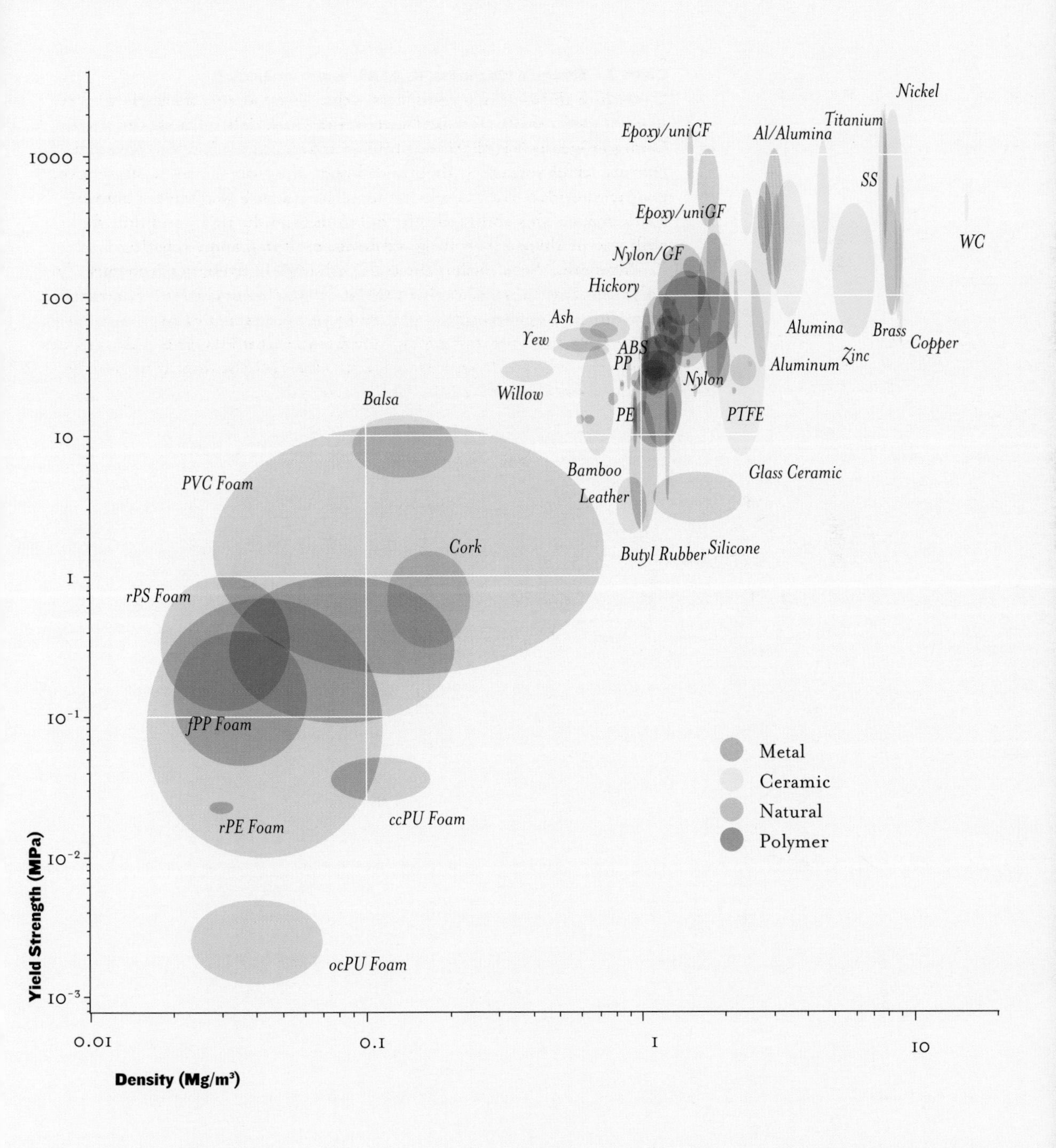
Nickel
Titanium
Epoxy/uniCF
Al/Alumina
SS
Epoxy/uniGF
WC
Nylon/GF
Hickory
Ash
Yew
Alumina
Brass
Copper
ABS
PP
Zinc
Aluminum
Nylon
Willow
Balsa
PE
PTFE
Bamboo
Glass Ceramic
PVC Foam
Leather
Cork
Butyl Rubber
Silicone
rPS Foam
fPP Foam
Metal
Ceramic
Natural
Polymer
rPE Foam
ccPU Foam
ocPU Foam
Yield Strength (MPa)
1000
100
10
1
10^{-1}
10^{-2}
10^{-3}
0.01
0.1
1
10
Density (Mg/m³)

Chart 3 – Fracture toughness, K_{IC}, and elastic modulus, E

If you nick the edge of a cellophane wrapper or scratch the surface of a plate of glass, it will crack or tear from the nick or scratch. If you do the same to copper or steel, it won't. That is because the first two have low fracture toughness, K_{IC}, the property that measures the resistance to the propagation of a crack. Materials with low fracture toughness are strong when they are perfect (that's why it's so difficult to get the cellophane wrapper off a new CD or packet of biscuits), but as soon as the surface is damaged they break easily. Materials with high fracture toughness are crack tolerant; they still carry loads safely even when cracked. The chart shows fracture toughness, K_C, plotted against modulus E. The "toughness," G, is related to K_{IC} by $G = K_{IC}^2/E$ – shown as a set of diagonal contours.

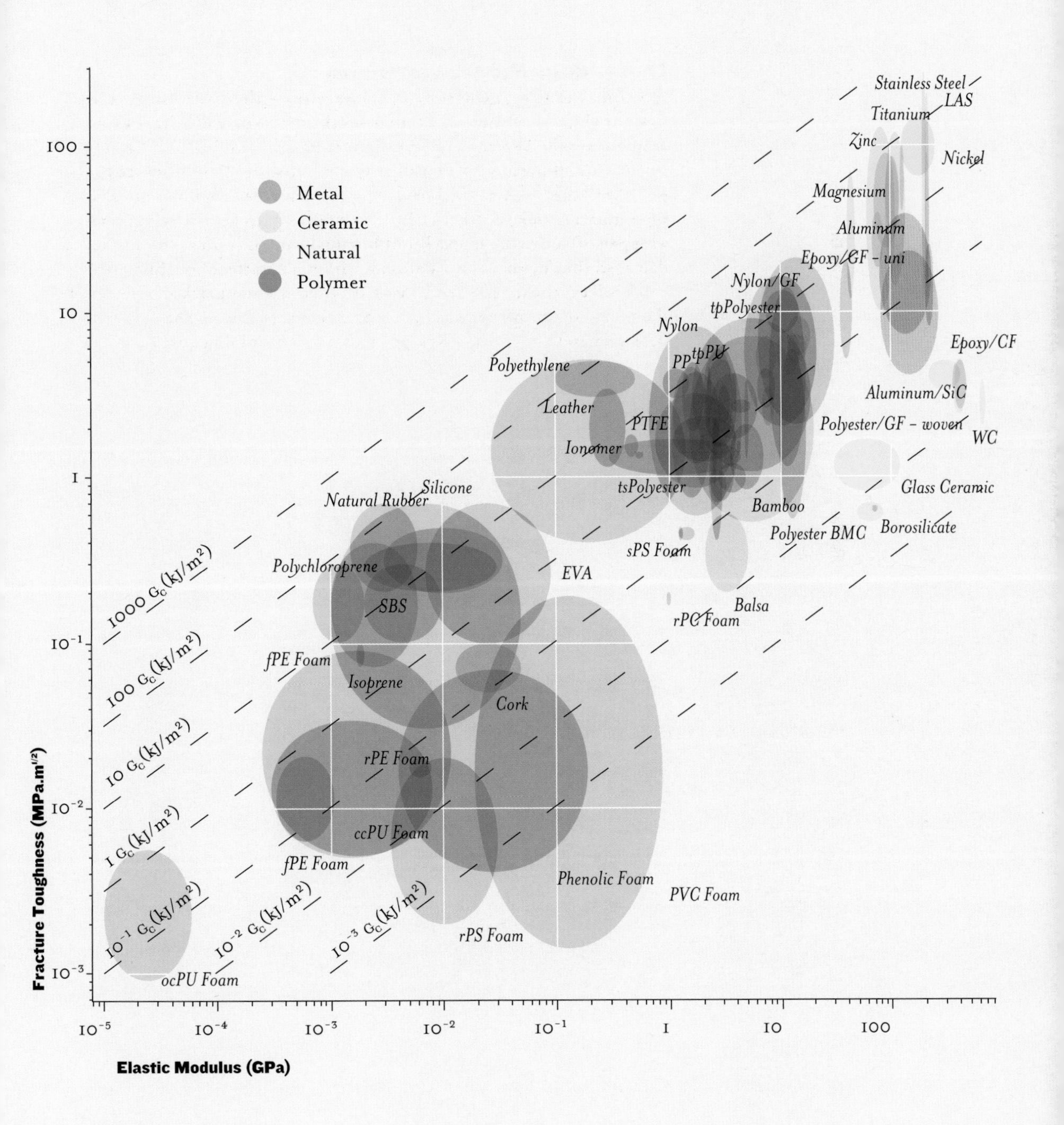

Metal
Ceramic
Natural
Polymer
Stainless Steel
LAS
Titanium
Zinc
Nickel
Magnesium
Aluminum
Epoxy/GF - uni
Nylon/GF
tpPolyester
Nylon
tpPU
PP
Epoxy/CF
Polyethylene
Leather
PTFE
Aluminum/SiC
Polyester/GF - woven
WC
Ionomer
tsPolyester
Glass Ceramic
Silicone
Natural Rubber
Bamboo
Polyester BMC
Borosilicate
sPS Foam
Polychloroprene
EVA
SBS
Balsa
rPC Foam
1000 G_C(kJ/m²)
100 G_C(kJ/m²)
fPE Foam
Isoprene
Cork
10 G_C(kJ/m²)
rPE Foam
1 G_C(kJ/m²)
ccPU Foam
fPE Foam
Phenolic Foam
PVC Foam
10^{-1} G_C(kJ/m²)
10^{-2} G_C(kJ/m²)
10^{-3} G_C(kJ/m²)
rPS Foam
ocPU Foam
Fracture Toughness (MPa.m$^{1/2}$)
100
10
1
10^{-1}
10^{-2}
10^{-3}
10^{-5}
10^{-4}
10^{-3}
10^{-2}
10^{-1}
1
10
100
Elastic Modulus (GPa)

Chart 4 – Elastic Modulus, E, and Strength, σ_f

There are no new properties here. The value of the chart is that many designs depend on one or another combination of E and σ_f. Thus resilience – the ability to bend without damage (like leather) is measured by the combination σ_f/E, shown as one set of contours on the chart. The ability to store and return elastic energy (something rubber is good at) is measured by σ_f^2/E (not shown). Materials with large values of these combinations perform well in these functions.

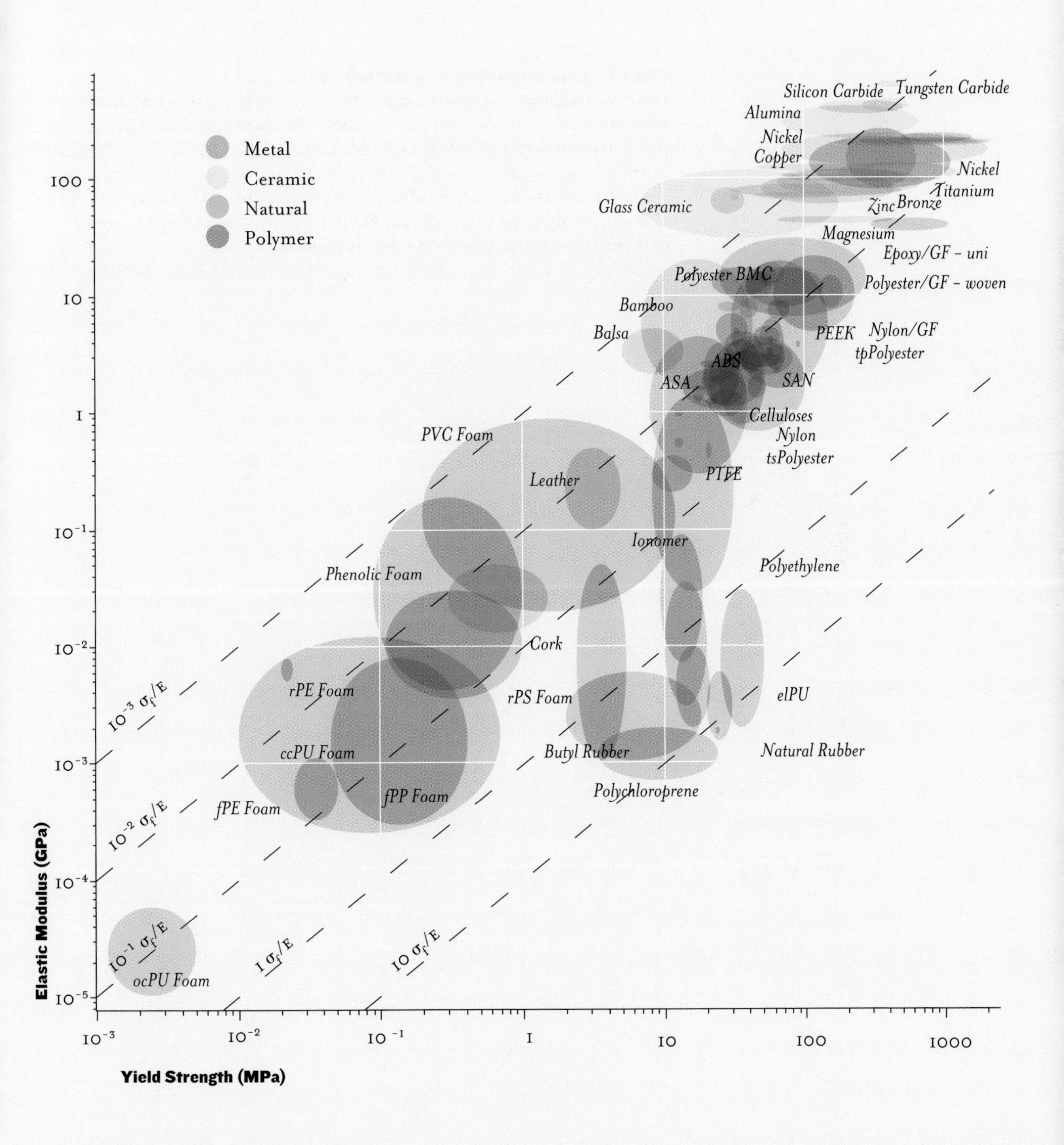

Metal
Ceramic
Natural
Polymer
Silicon Carbide
Tungsten Carbide
Alumina
Nickel
Copper
Nickel
Titanium
Glass Ceramic
Zinc
Bronze
Magnesium
Epoxy/GF - uni
Polyester BMC
Polyester/GF - woven
Bamboo
Balsa
PEEK
Nylon/GF
tpPolyester
ABS
ASA
SAN
Celluloses
Nylon
tsPolyester
PVC Foam
Leather
PTFE
Ionomer
Polyethylene
Phenolic Foam
Cork
rPE Foam
rPS Foam
elPU
ccPU Foam
Butyl Rubber
Natural Rubber
fPE Foam
fPP Foam
Polychloroprene
ocPU Foam
$10^{-3}\ \sigma_f/E$
$10^{-2}\ \sigma_f/E$
$10^{-1}\ \sigma_f/E$
$1\ \sigma_f/E$
$10\ \sigma_f/E$
Elastic Modulus (GPa)
Yield Strength (MPa)

Chart 5 – Loss Coefficient, η, and Elastic Modulus, E

If a hard steel ball is dropped onto the surface of a material, it bounces back. The height to which it bounces is a measure of the loss of energy in the material: the higher it bounces, the less is the loss. Low loss-coefficient materials return almost all the energy of the ball so that it will bounce many times (as many as 1000) before it comes to rest. High loss materials bring the ball to rest almost immediately. Choose low loss materials for springs, vibrating reeds, strings for musical instruments and precision suspension systems; choose high loss materials for vibration damping.

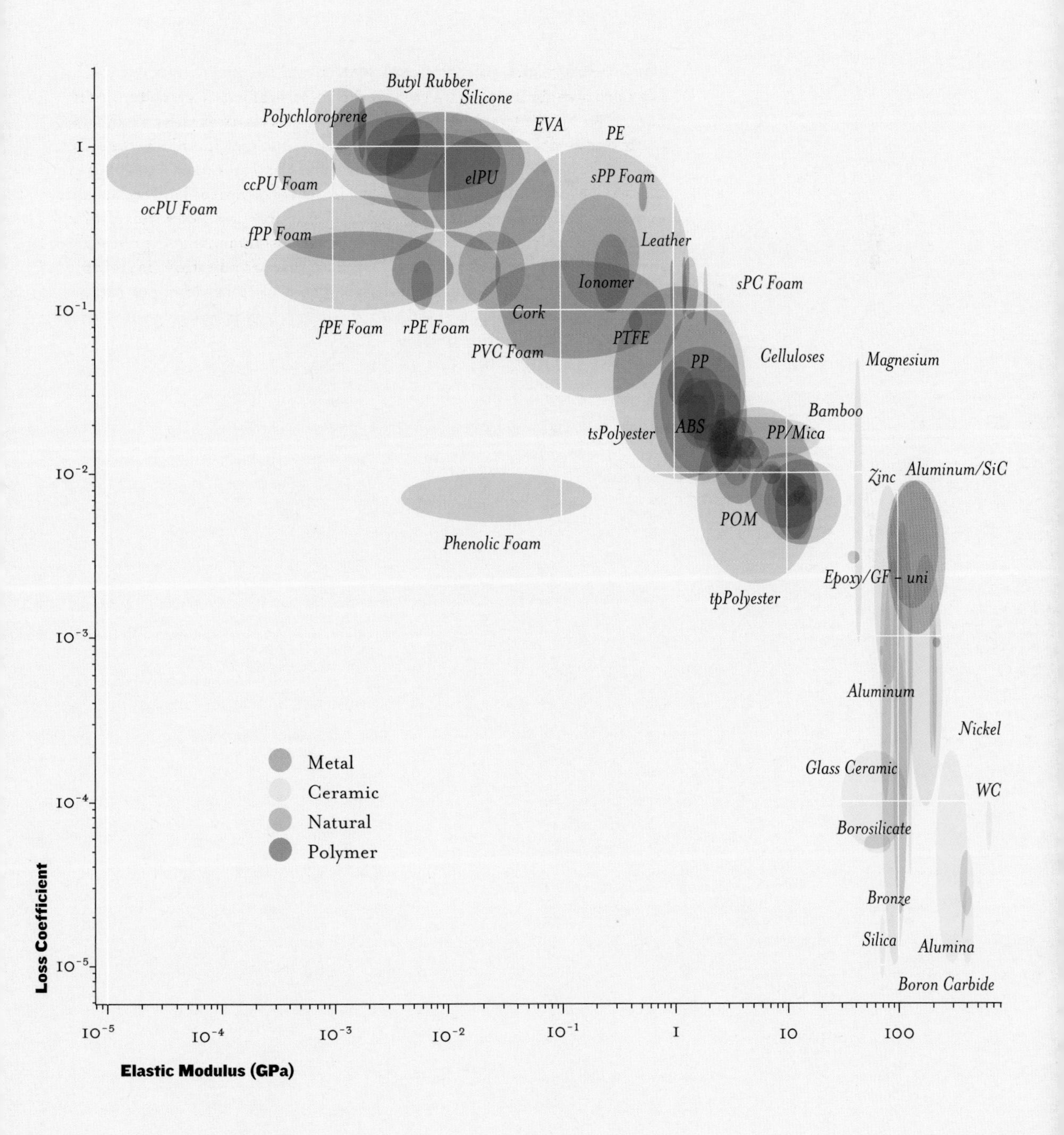
Butyl Rubber
Silicone
Polychloroprene
EVA
PE
ccPU Foam
elPU
sPP Foam
ocPU Foam
fPP Foam
Leather
Ionomer
sPC Foam
Cork
fPE Foam
rPE Foam
PTFE
PVC Foam
PP
Celluloses
Magnesium
Bamboo
tsPolyester
ABS
PP/Mica
Zinc
Aluminum/SiC
POM
Phenolic Foam
Epoxy/GF - uni
tpPolyester
Aluminum
Nickel
Glass Ceramic
WC
Borosilicate
Bronze
Silica
Alumina
Boron Carbide
Metal
Ceramic
Natural
Polymer
Loss Coefficient
Elastic Modulus (GPa)
1
10^{-1}
10^{-2}
10^{-3}
10^{-4}
10^{-5}
10^{-4}
10^{-3}
10^{-2}
10^{-1}
1
10
100

Chart 6 – Thermal Expansion Coefficient, α, and Thermal Conductivity, λ

Expansion and conductivity are the two most important thermal properties of materials for product design. They vary greatly: the chart shows that metals and certain ceramics have high conductivities together with low expansion. Polymers and elastomers have conductivities that are 1000 times less, and they expand at least 10 times more. It is important to match expansion coefficients when joining different materials that will be used at high temperature; if this is not done, damaging stresses appear when the temperature is changed. High thermal conductivity is valuable in heat exchangers and heat sinks, and when a uniform temperature is required (as in a cooking vessel). Low conductivity is valuable when the aim is to insulate against heat.

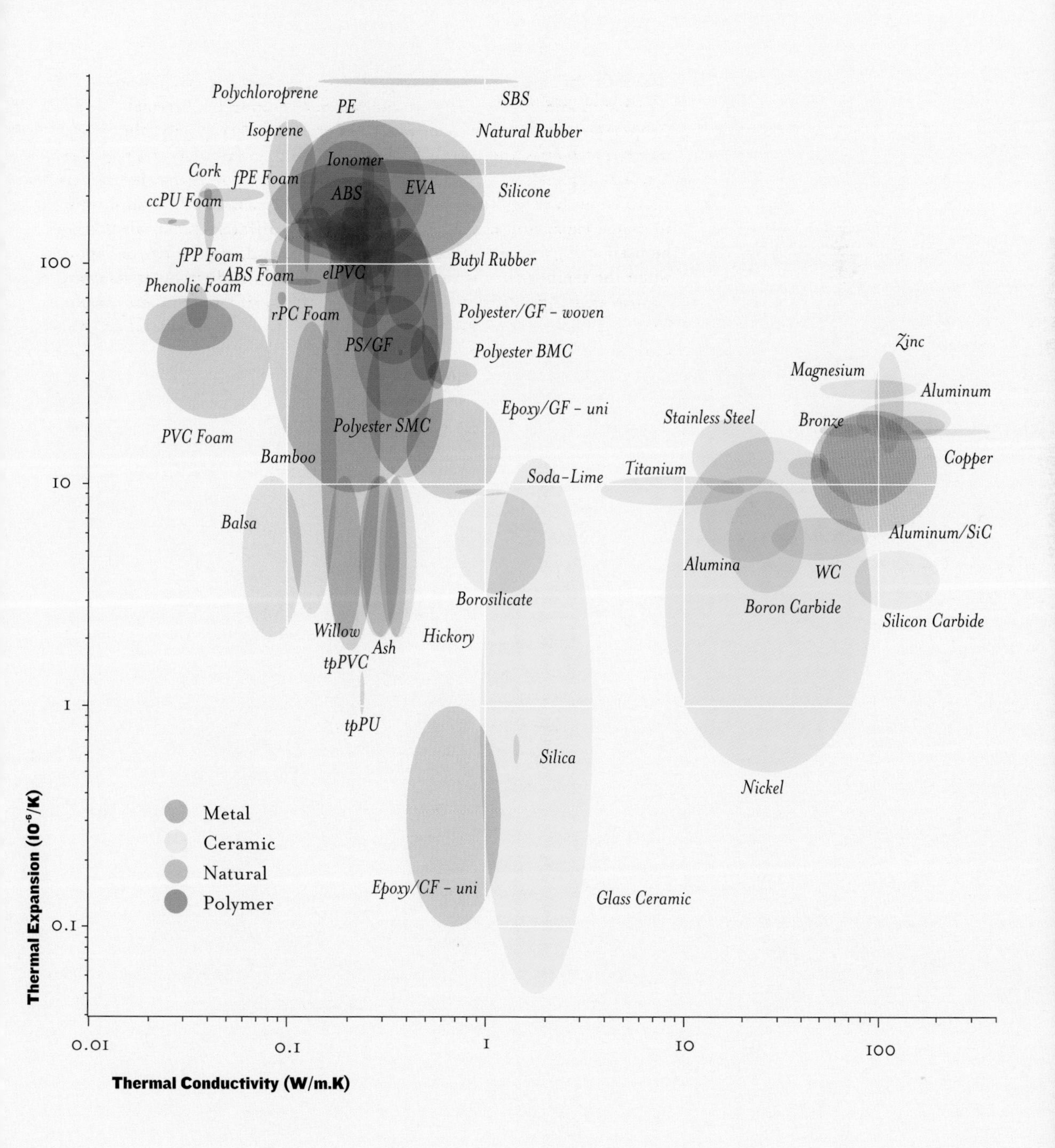
Polychloroprene
PE
SBS
Isoprene
Natural Rubber
Ionomer
Cork
fPE Foam
EVA
Silicone
ABS
ccPU Foam
fPP Foam
Butyl Rubber
ABS Foam
elPVC
Phenolic Foam
rPC Foam
Polyester/GF - woven
PS/GF
Polyester BMC
Zinc
Magnesium
Aluminum
Epoxy/GF - uni
Stainless Steel
Bronze
Polyester SMC
PVC Foam
Copper
Bamboo
Titanium
Soda-Lime
Balsa
Aluminum/SiC
Alumina
WC
Borosilicate
Boron Carbide
Silicon Carbide
Willow
Hickory
Ash
tpPVC
tpPU
Silica
Nickel
Metal
Ceramic
Natural
Polymer
Epoxy/CF - uni
Glass Ceramic
Thermal Expansion (10⁻⁶/K)
100
10
1
0.1
0.01
0.1
1
10
100
Thermal Conductivity (W/m.K)

Index

A

Abrasion resistance, 78, 79, 86, 151–2, 292
Acetal, 195
Acoustic attributes, 68, 72–3, 86
Acoustic design, 61
Acrylate elastomers, 208
Acrylics, 37, 193, 249, 303
  adhesives, 263
Acrylonitrile-butadiene-styrene (ABS), 191, 256–7
Actuator materials, 59
Adaptation, 32, 66
Adhesives, 262–5
Advertising, 16, 63, 161
Aerospace materials, 113, 159–61
Aesthetics
  appeal, 1, 7, 14–17, 18–19, 34, 63, 92, 157, 294–7
  attributes, 29, 49, 68–73, 74, 121, 169
  modeling, 85–7
Alloys, 176, 183–4, 219–25, 235, 276–82, 304–6
  see also Steels
Alternatives, 127, 147, 149–51, 311
Alumina, 226
Aluminum
  anodizing, 298
  electro-plating, 295
  foam, 163, 164, 233
  vapor metalizing, 293
Aluminum alloys, 219, 246, 250
Amorphous metals, 234
Anaerobic adhesives, 265
Analysis in material selection, 124–6, 129–30, 131, 132, 134–9, 141–2, 144, 171
  case studies, 141–2, 144, 147–8
Anodizing, 298
Anthropomorphics, 57–8
Antiques, 63–4
Aramid fibers, 230
Arc welding, 278–80
Archaeology, 110–13
Architecture, 1, 16, 17, 40, 99–101
Art Deco, 112
Assembly, 10, 143, 158, 269
Associative attributes, 1, 22, 27, 29, 30, 38, 41, 43, 44, 63, 73, 81
  material selection, 311–12, 313
  new materials, 162
Atomic structure, 50–1, 315, 316, 318
Axisymmetric shapes, 238, 244, 251

B

Backpacks, 144–5
Bacteria resistance, 302–3
Bakelite, 21, 37, 112, 207, 264
Ballistic particle manufacture, 257
Balsa, 231
Bamboo, 231
Bang & Olufsen, 19–20
Batch sizes, 236, 248
Bending, 99–101, 144
Bicycles, 108–9, 125, 130, 135–7, 144–5
Bio-active (passive) surfaces, 10
Bio-mechanics, 57–8
Biodegradation, 180–1
Blends, 53
Blind stapling, 267
Blow molding, 242
Bookshops, 118–19, 121
Boron carbide, 226
Borosilicate glass, 228
Bottle openers, 110
Boundary conditions, 7, 12
Brain, 29–31, 44, 51, 56
Brainstorming, 42, 165, 313
Brand identity, 15–16
Brass, 223–4, 255, 276, 295
Brazing, 276
Brick, 99, 100
Brightness of sound, 72, 73, 86
Brittleness, 151–2
Bronze, 223–4, 255, 276, 295
Browsing, 118–19, 128, 129, 142, 148, 171
Bubble analogy of design, 34, 35, 129–30
Bulk forming, 250–1
Bulk shapes, 238, 240, 244, 246, 250, 255–7
Business case, 13
Business strategy, 8
Butt joints, 260, 262, 266, 270, 272–3, 276, 279, 281–2
Butyl rubbers (NR), 208
By-products, 11–12, 65

C

CAD files, 43, 256–7
Cadmium, 295
Calendering, 183
Cameras, 20–1
Carbon fiber-reinforced polymers (CFRP), 109, 144–5, 182–3, 214–15
Carbon fibers, 229
Carbon steels, 216, 251
Cars, 9, 64, 144
Case studies, 141–55
Cast iron, 99, 100
Casting, 246–9
Caulking gum adhesion, 262
CD cases, 146–8
CD players, portable, 18, 19
Cell biology, 10
Cellular phones, 18–19, 20
Celluloid, 37
Celluloses, 198
Centrifugal casting, 249
Ceramics, 22, 51–2, 54–5, 75, 176
  adhesion, 263–4
  cements, 265
  coating, 306
  diffusion bonding, 283
  extrusion, 251
  lass ceramics, 228
  material maps, 316–27
  polishing, 299
  powder sintering, 255
  profiles, 184–5, 226–32
Chairs, 101–8, 145
Change driving forces, 9, 64
Characterization, 27, 49, 56, 157
Chemical etching, 309
Chemical polishing, 301
Children, 58
Child's car seat, 123–4
Chlorinated elastomers, 210
Chromatic series, 313
Chromium plating, 295
Circular shapes, 238
Classic products, 16
Classification
  information, 117–20

materials, 50–2, 54–5
polymers, 181
processes, 98–91
Clustering, 51–3, 54, 72, 73, 111, 112
Coating, 302–7
Cobalt alloys, 295
Cold drawing, 251
Cold rolling, 250
Coldness, 68, 69, 85, 151–2
Collages, 42, 106–7, 113
Collections, 18, 41, 128
Color, 19–20, 32, 59, 70–1
chromatic series, 313
surface processes, 288–96, 298–308
Commercialization, 13, 14, 157, 161
Communication, 58–60, 161–2, 164–5
Competition, 13, 23
Complex shapes, 238, 240, 247–9, 252, 257
Composites, 9, 50, 53, 55, 157, 176
adhesion, 263–4
injection molding, 240
metal-matrix, 184, 255, 282
polymer–matrix, 145, 159, 181, 182–3, 214–15, 254, 264
Composition, 51
Compression, 99–101, 255, 318
Compression molding, 244
Computer-aided design (CAD), 43, 256–7
Computer mouse, 19
Conceptual design, 13, 32, 33, 35, 36, 49, 102–7, 141, 170
Concrete, reinforced, 100, 101
Conducting materials, 62–3, 277, 293–4, 297
Conservation, 64, 66
Constraints, 103, 124, 125, 127, 129, 134–7
case studies, 141–2, 147, 149, 151–2, 154
Consumers, 1–2, 7, 15, 161
expectations, 34, 141
influences, 63, 106, 114
product design, 15, 17, 18, 22, 32, 38, 170
Control variables, 134–5
Controls, 15, 19, 58–9
control-loops, 59
Copper alloys, 223–4, 250
Copper plating, 295
Cork, 62, 231
Corporate identity, 15–16, 19–20
Corrosion resistance, 293–4, 297–8, 302–7
Costs, 9, 13, 33, 64, 123, 313
material selection, 137–9, 142–4, 147, 157–8
material-cost sensitivity, 159–60
see also Shaping, profiles; Surface processes, profiles
Cracking, 146–7
Craft-based design, 16, 20, 79, 81, 157
Creativity, 2, 12, 28, 30, 40, 41, 128, 131, 169
aids, 42–3
Critical properties, 34
Crystallization, 177
Cubic printing, 290
Culture, 1, 27, 66
Curing, 179
Curiosity, 129
Curved shapes, 238, 244, 252–4
Customers, *see* Consumers
Cut–seal process, 275
Cyanoacrylate adhesives, 263
Cylinders, joining, 260

D

Damping, 62, 72, 73, 86, 149–50, 324
Data capture, 49, 56
Databases, 120–1, 125–31, 142, 144, 147, 150, 151, 155
Datasheets, 51, 56, 161, 162
Decibels (dB), 61
Decoration, 1, 292, 308
Deductive reasoning, 28, 30, 124, 131
Deformation mechanism maps, 30, 31, 51–3, 315–27
Density, 51–3, 85, 136–9, 143–4, 150, 153–5, 316–19
Deposition prototyping, 257
Design
definition, 27
detail, 32–3, 34, 35, 37–9
features list, 102–8, 112
integrity, 79–80
intentions, 105, 121, 122, 123, 129, 170
types, 32
Design brief, 8, 32, 44, 103, 129, 141, 142
Design life, 12, 16, 63–4
Design process, 7–17, 23, 27, 32–45
bubble analogy, 34, 35
Desire (want), 8, 35
Desks, 142–5
Development, 14, 32, 33–4, 35, 36, 162, 174
see also Research and development
Die casting, 91, 109, 246
Die forging, 250
Differentiation, 15, 17–18, 20–1, 153
Diffusion bonding, 283
Digital cameras, 21
Digital watches, 9, 17
Direct extrusion, 251
Disabled people, 58
Discarding products, 11–12, 13, 16, 63–5, 66, 91
Display technology, 158
Disposal, 8, 11–12, 20, 49, 67
Dissimilar material joining, 266–9, 273, 276–7, 282–3, 326
Distance-maps, 53–5
Drape thermoforming, 253
Ductile-to-brittle transition, 152, 318

E

Ebonite, 37
Eco-design, *see* Environmental impact
Ecology, 7
Economic conditions, 7, 8, 64, 236
see also Shaping, profiles; Surface processes, profiles
Economical design, 17, 169
Efficiency, 9, 17, 65

Elastic limit, 136
Elastic modulus, 51–2, 143–4, 153–5, 179, 316–17, 320–5
Elastomers, 51–2, 54–5, 176
  injection molding, 240
  joining, 263–4, 269
  material maps, 326–7
  profiles, 179, 208–11
Elderly people, 58, 122
Electric motors, 21–2
Electrical attributes, 50, 51
Electro-etching, 309
Electro-less plating, 297
Electro-painting, 304
Electro-plating, 294
Electro-polishing, 300
Electron beam welding, 274–5
Electronic controls, 58–9, 67–8
Electronic messaging, 59
Electronic products, 65
Electronic structure, 50–1, 183
Electrostatic powder coating, 305–6
Elegance, 15, 16, 17
Emissions, 11–12
Emoticons, 59
Emotions
  appeal, 2, 16, 19, 41, 64
  attributes, 29, 44
Enameling, 307
Energy
  consumption, 11–12, 64–8
  conversion, 12
  efficiency, 65
  loss, 68, 149–50, 324–5
  reduction, 12
Engineering design, *see* Technical design
Engineering education, 2, 132
Engineering materials, 9, 21, 50–6, 169
Engraving, 308
Environmental impact, 7, 8, 9, 11–13, 16, 63–8
  legislation, 12, 49
  pernambuco, 149–50
  polymers, 180–1
  see also Profiles
Epoxies, 206, 263, 303
Ergonomics, 57, 58, 92, 106
Etching, 309
Ethylene-propylene elastomers, 209
Ethylene-vinyl-acetate elastomers (EVA), 209, 211
Eutectic composition, 277
Evaporative sand casting, 247
Evolution of material use, 20–2, 174, 176, 312
Exhibitions, 40, 76, 162
Expanded foam molding, 243
Expectations, 12, 16, 34, 141
Expressive features, 92–4
Extrusion, 251, 255
Extrusion blow molding, 242

F

Fabrics, 100, 101, 263–4
Failure analysis, 34
Fans, 21–2
Fashion, 17, 19, 21, 40
Fasteners, 266–9
Fatigue resistance, 299
Features list, 102–8, 112
Feedstock, 11–12, 66
Feel, 302–3, 305–6
  see also Touch
Female characteristics, 57
Fiber-reinforced composites, 52, 53, 159
  adhesion, 263–4
  resin transfer molding, 245
  profiles, 182–3, 214–15
  carbon fiber-reinforcement, 109, 144–5, 182–3, 214–15
  glass fiber-reinforcement, 53, 144, 159, 182, 214, 215, 244–5
Fiberglass, 214
Fibers, 62, 180, 182, 229, 266
Flanges, 260
Flat shapes, 238, 244, 252
Flexible manufacturing methods, 9
Flexible materials, 59, 153
Flexural modulus, 149–50
Fluidized-bed coating, 305–6
Fluoro-carbon elastomers, 210
Foams
  expanded foam molding, 243
  metals, 163, 164–5, 184, 233
  polymers, 51–2, 55, 62, 164, 179, 212–13
Forces for change, 7–17
Form, 99–114, 236, 311
Formal attributes, 111
Forming, 250–3
Fracture toughness, 87, 153–5, 320–1
Free-market economies, 8
Frequencies, 61, 72–3, 86, 150
Friction reduction, 299
Friction welding, 282
Fuel consumption, 64, 67
Functional materials, 8, 10, 14, 181
Functionality, 1–5, 8, 9, 65, 106
  and form, 113–14
  material selection, 58–9, 129, 134–6
  and personality, 1–3
  product design, 15, 20, 35, 38
Fungal resistance, 303
Furniture, 66–7, 101–8, 142–5
Fused deposition modeling, 257

G

Gas metal arc (MIG) welding, 279
Glass ceramics, 228
Glass fiber-reinforced polymers, 53, 144, 159, 182, 214, 215, 244–5
Glasses, 55, 75, 176, 185–6
  enameling, 307
  etching, 308
  fibers, 229
  profiles, 228
Glaze bonding, 283
Global growth rate, 11, 23, 64
Gold, 37, 295
Governments, 14
Graphs, 30, 31
Green sand casting, 247

H

Hairdryers, 21–2
Hand lay-up methods, 254
Handling materials, 41
Hardness, 68–9, 70, 78, 79, 85, 86, 293, 297, 307
Harley Davidson motorcycle, 28
Health concerns, 18, 61
Hearing, 68, 72–3
Hemp, 230

High temperature materials, 10
Historical overview, 1
Hollow shapes, 238, 241–2, 246, 251–2, 256–7
Hot bar welding, 271
Hot die forging, 250
Hot gas welding, 270
Hot isostatic pressing, 255
Hot-melt adhesives, 263, 265
Hot plate welding, 272
Hot rolling, 250
Hot stamping, 291
Houses, 66
Human factors, 11, 12, 56–8

I

Ice axes, 151–2
Icons, 59
Identification of concept, 103–6, 141, 170
Images, 29–31, 41–2, 44
Imagination, 16, 30
Imide-based adhesives, 263–4
Impact extrusion, 251
Impact resistance, 87, 153–5
Impact strength, 262
Improvisation, 43
Impulse welding, 271
In-mold decoration, 292
Indexing, 44–5, 113, 117–20, 121, 128, 131
Indicators, 59
Indirect extrusion, 251
Inductive reasoning, 28, 124, 131
Industrial design, 7, 8, 14–17, 23, 27, 162
  balance, 16–17
  environmental impact, 65–6
  resources, 2–3
Inertial friction welding, 282
Information
  access, 2–3, 117, 162
  classification, 117–20
  display, 31, 59, 60
  indexing, 44–5, 113, 117–20, 121, 128, 131
  management, 58–60
  new materials, 161–3
  retrieval, 45, 117
  services, 41, 162
  structuring, 117–39, 141–2, 170–1
  use, 141
Injection blow molding, 242
Injection molding, 90, 91, 109, 240, 255, 292
Inline skates, 153–5
Innovation, 2, 9, 10, 20, 40, 43, 127
  new materials, 157–66
Inspiration, 2, 9, 10
  material selection, 124, 128–9, 132, 141, 147–8, 313
  in nature, 43–4
  sources, 39–45, 103, 107, 169
Insulation
  electrical, 21, 22, 623
  sound, 61–2
  thermal, 62, 68
Insulin pump, 126
Intellectual property, 13
Intentions, 105, 121, 122, 123, 129, 170
Interfaces, 15, 18–19, 56–7
Internet searching, 45, 128
Investment, 7, 8, 13–14, 158
Investment casting, 248
Ion plating, 293
Ionomers, 197
Iridium, 37
Irregular surfaces, 289
ISO Standards, 49
Isocyanate-based adhesives, 263–4
Isolation, 61
Isoprene, 208

J

Joining, 9, 89, 91, 92–3, 108–9, 183
  profiles, 258–83
Joint geometry, 258, 260
Juxtaposition, 42

K

Kevlar, 214, 215, 230

L

Laminated object manufacture, 256
Lamination, 183
Lap joints, 260, 262, 266–8, 270–1, 273–7, 279, 281
Laser-based processes, 158
  prototyping, 256
  welding, 274–5
Laws of mechanics, 10
Lay-up methods, 254
Leather, 231
Legislation, 12, 49, 57
Lifestyle, 11, 64, 106
Lifetime, 9, 11–13, 16, 63–4, 65, 66–7, 68
Light management, 63
Lightweight materials, 9, 10, 12, 58, 67–8
  bicycles, 108–9, 136
  inline skates, 153–5
  office furniture, 142–5
Limited editions, 19
Loads, 99–101, 136
Logic, 28, 30
Logos, 23, 151
Loss coefficient, 149–50, 324–5
Lost wax process, 248
Low alloy steels, 218, 251

M

Magazines, 40, 76–7, 128, 147–8
Magnesium alloys, 220, 246, 250
Magnetic permeability, 297
Male characteristics, 57
Manual metal arc welding, 278
Manufacturing methods, 9
Mapping, 30, 31, 51–5, 70–2, 79, 80, 143, 146, 154, 164, 315–27
Market forces, 7, 8–9, 159
Market saturation, 8, 15, 17, 18, 20
Market share, 2, 22
Marketing, 63
Mass production, 16, 122, 123, 142, 145, 240
Material-cost sensitivity, 159–60
Material indices, 135–8, 142
Material/process matrix, 107, 109, 110
Material property charts, 30, 31
Materialization, 107–8, 170, 313
Materials
  classification, 50–2, 54–5
  consumption, 64–7, 158–9
  cost sensitivity, 159–60

development, 14, 174
evolution in use, 20–2, 174, 176, 312
and form, 114, 311
information services, 41, 162
information structure, 117–39, 141–2, 170–1
lifecycle, 11–12, 66
multi-dimensional, 9, 49–87
new, 157–66
profiles, 161, 163–4, 174–235
properties, 34, 50, 125, 142
reduced use, 12
sample collections, 41, 128
selection, 33–4, 49, 56, 101–2, 108–9, 117, 141–2, 169–71
analysis, 124–6, 129–30, 131, 132, 134–9, 141–2, 144, 147–8, 171
charts, 30–31, 51–55, 70–72, 79–80, 143, 146, 154, 315–327
inspiration, 124, 128–9, 132, 141
similarity, 124, 127–8, 129–30, 131, 141–2, 149–51, 171
synthesis, 124, 126–7, 129–30, 131, 132, 141–2, 144–5, 151–2
Materials science, 30–1, 50
Mathematical procedures, 28, 29–30
MDS (multi-dimensional scaling), 52–5, 72, 74, 79, 80, 137, 138, 143, 154, 164
material maps, 51–3, 316–27
Mechanical attributes, 50, 51, 54–5, 169
Mechanical fasteners, 267–9
Mechanical loads, 9, 99–101
Mechanical polishing, 299
Mechanical properties, 250
Medical products, 158, 159–60
MEMS devices, 10
Metal arc welding, 278–9
Metal-matrix composites, 184, 255, 282
Metal oxides, 185–6
Metalizing, 293–8
Metals, 51–2, 54–5, 74–5, 176
adhesion, 263–4
amorphous, 234
anodizing, 298
brazing, 276
casting, 246–8
coating, 302–7
diffusion bonding, 283
etching, 308
extrusion, 251
foams, 163, 164–5, 184, 233
forming, 250
material maps, 316–27
plating, 294–7
polishing, 299–301
powder shaping, 255
profiles, 183–4, 216–25
refractory, 255
welding, 273–5, 278–82
Micro electro-mechanical systems, 10
Micrographs, 30, 31
Micron-scale components, 10
MIG welding, 279
Military vehicles, 112–13
Mimic materials, 10, 81
Mind, *see* Brain
Mind-mapping, 42
Miniaturization, 10, 12, 20
Mobile phones, 18–19, 20
Modeling, 33, 43, 85–87, 313
Molding, 58, 90, 91, 109, 180, 183, 240–5
Mood boards, 41–2, 106–7, 113, 312
Muffling, *see* Damping
Multi-dimensional materials, 9, 49–87
Multi-dimensional scaling, *see* MDS
Multi-objective optimization, 134, 139
Museums of Design and Applied Art, 39, 40

N

Naming, 18, 312
Nanoscale assembly techniques, 10
Natural materials, 208, 231–2
material maps, 316–27
Natural resources, 64
Nature as inspiration, 43–4
Need (want), 7, 8, 9, 32, 36, 102
New materials, 8, 157–66, 233–5
Nickel alloys, 222
Nickel plating, 295, 296
Nitrile elastomers, 208
Noise management, 61–2
Non-axisymmetric shapes, 238
Non-circular shapes, 238
Nuclear industry, 159
Nylon, 53, 153–4, 192, 256–7, 263

O

Objectives, 124, 129, 134–6, 141
Observation, 28–9
Obsolescence, 63, 65
Office furniture, 105–6, 142–5
Opaque materials, 68, 69–70, 71
Open molds, 254
Optimization, 7, 12, 32–3, 54
Optimum design, 134–9
Ores, 11–12, 64, 66
Osmium, 37
Oxides, 185–6

P

Pad printing, 289
Painting, 302–4
Parallel features, 238
Particulate-reinforced polymers, 53
Passive products, 66–7
Pattern recognition, 51
Patterning, 292, 308
Pens, 35–9
Perception, 27, 28–9, 44, 64, 66, 92, 162, 169–71
material selection, 121, 126, 129, 150, 311
vocabulary, 76–8
Performance, 13, 15, 17, 20, 33, 125, 134–5, 158–9
Pernambuco, 149–50
Personal computers, 18
Personality, 2, 73–6, 105–7
material selection, 122, 170
new materials, 49
of materials, 73–76
product design, 14, 15, 19, 27, 38

Personalization, 66
Phase diagrams, 30, 31
Phenolic adhesives, 263–4
Phenolic resins, 207, 263
Physical attributes, 50, 51
Pictograms, 59
Pitch, 72–3, 86
Plastic deformation, 250
Plastics, *see* Polymers
Plates, joining, 260, 270, 272–3
Plating, 293–8
Plug-assisted thermoforming, 253
Polishing, 299–301
Polyamide (PA), 192
Polybutadiene elastomers, 209
Polycarbonate (PC), 194
Polychloroprene, 211
Polyesters, 181, 182, 204–5
Polyetheretherketone (PEEK), 199
Polyethylene (PE), 153–4, 181–2, 188, 212
Polymer flame coating, 305–6
Polymer-matrix composites, 145, 159, 181, 182–3, 214–15, 254, 264
Polymers, 21–2, 176
  adhesion, 263–4
  aesthetic attributes, 75, 80–1
  casting, 249
  classification, 181
  coatings, 306
  etching, 308
  extrusion, 251
  fiber-reinforced, 53, 109, 144–5, 159, 182–3, 214–15, 264
  foams, 51–2, 55, 62, 164, 179, 212–13
  material maps, 316–27
  plating, 294–7
  profiles, 177–83, 188–215
  technical attributes, 51–3, 54–5
  welding, 271, 273–5
Polymethylmethacrylate (PMMA), 193, 249
Polyoxymethylene (POM), 195
Polypropylene (PP), 51, 189
Polystyrene (PS), 146, 147, 181, 182, 190, 212–13, 243
Polysulphide elastomers, 209
Polytetrafluoroethylene (PTFE), 196
Polyurethane (PU), 181, 182, 201–2, 213, 241, 263–4
Polyvinylchloride (PVC), 146, 147, 148, 181, 182, 200
Population growth, 12, 65
Portable CD players, 18, 19
Powder coating, 305–6
Powders, 185, 255
Power-beam welding, 274–5
Power-density, 10, 22
Precision engineering, 93, 145
Preconceptions, 127–8, 141, 153
Press forming, 252
Pressing and sintering, 255
Pressure bag molding, 254
Pressure die casting, 246
Pressure thermoforming, 253
Printing, 288–92
Prismatic shapes, 238, 250–2, 255
Processes, 9, 89–94, 121, 122, 142, 158
Product archaeology, 110–13
Product design, 27, 49, 170–1, 311
  influences, 7–25
  information structure, 120–32, 161
Product family, 15–16
Product life, 9, 11–13, 16, 63–4, 65, 66–7, 68
Production, 8, 9, 13
  see also Mass production
Products
  attributes, 120–1, 311
  collections, 18
  development, 32, 33–4, 35, 36
  differentiation, 15, 17–18, 20–1, 153
  personality, 2, 14, 15, 19, 27, 38, 49, 73–6, 105–7, 122, 170
  requirements, 32, 162, 170
Profiles
  joining, 258–83
  materials, 161, 163–4, 174–235
  shaping, 236–57
Projection spot welding, 281
Prototyping, 33, 34, 43, 158, 236, 256–7
Pultrusion, 145, 183
PVD metalizing, 293

Q

Quality, 9, 19, 49
Quality of life, 16

R

Rapid prototyping, 158, 256–7
Re-flow soldering, 277
Reasoning, 38–9
Recognition, 30
Recycling, 11–12, 65, 66, 68, 122, 180
Reflectivity, 63, 71, 288–91, 293–4, 298–301, 304, 307
Refractory metals, 255
Regulations, 12
Reinforced composites, *see* Fiber-reinforced composites; Particulate-reinforced polymers
Reinforced concrete, 100, 101
Remote controls, 19
Renewable materials, 12
Research and development, 9, 14, 157, 161, 313
Resilience, 78, 80, 87, 146–8
Resin transfer molding, 245
Resistance welding, 281
Reuse of materials, *see* Recycling
Rhodium plating, 296
Risk, 14, 157–8, 159
Rivets, 267
Roll forming, 252
Rotational molding, 241
Rubbers, 208

S

Safety, 15, 51
Sample collections, 41
Sand casting, 247
Sand etching, 309
Satisfaction, 49
Saturated markets, 8, 15, 17, 18, 20
Scale, 10
Scarf joints, 260, 276–7
Schematics, 31
Scientific research, 7, 8, 161
Scratch resistance, 307
Screen printing, 288
Screws, 268
Sculptural solutions, 2

Sealants, 262, 264
Seam resistance welding, 281
Selection, 117–31, 141–55
  by analysis, 124–126, 134–139
  by inspiration, 128
  by similarity, 127–28
  by synthesis, 126–127
Selective laser sintering, 256
Sensors, 59
Serendipity, 119
Services, 12, 41, 162
Sewing, 266
Shape-memory alloys, 184, 235
Shape rolling, 250
Shapes, 58, 180, 236, 238
Shaping, 9, 89–91, 92, 158
  profiles, 236–57
Shear strength, 262, 273, 277
Sheet forming, 252–3
Shell casting, 247
Shell-like structures, 100, 101
Shielded metal arc welding, 278
Shrinkage, 177–8, 185, 249
Sight, 68, 69–71
Silica glass, 228
Silicate sand casting, 247
Silicon carbide, 2, 227
Silicones (SIL), 203, 263–4
Silk screen printing, 288
Silver plating, 296
Similarity in material selection, 124, 127–8, 129–30, 131, 171
  case studies, 141–2, 149–51
Sintering, 255
Size, 58
Sketching, 43, 104, 107, 109, 312, 313
Sleeve joints, 260, 277
Smell, 68
Snap fits, 269
Social groups, 15, 22, 49, 77, 153
Soda-lime glass, 228
Sodium silicate, 247
Softness, 68–9, 70, 85
Software tools, 43, 45, 56, 136–7, 155, 171, 315
Soldering, 277
Solid shapes, 238, 243, 250–1, 256–7
Solvent-based painting, 302
Solvent resistance, 293
Sound
  absorption, 61–2
  brightness, 72, 73, 86
  levels, 61–2
  pitch, 72–3, 86
  reduction, 61–2
  warning devices, 59
Space, 15
Specific heat, 85
Specifications, 8, 32, 33, 37, 103, 141, 171
Spinning, 252
Sports equipment, 58, 158, 159–60
Spot welding, 281
Spray adhesion, 262
Spray-up methods, 254
Sputtering, 293
Stainless steels, 217, 251, 255
Stamping, 291
Stand-by modes, 67–8
Standard Design Person, 57–8
Standards, 49
Staples, 267
Statistical analysis, 49
Steels, 100, 101, 183, 216–19, 250
Stereolithography, 256
Stick diagrams, 104
Stiffness, 10, 51–3, 78, 80, 85, 86, 108–9, 136, 143, 149, 153
Stitching, 266
Stone, 99, 100
Strength, 10, 62, 108–9, 136, 137, 149–50, 318–19, 322–3
  strength/modulus ratio, 22, 269
Stress analysis, 101
Structural adhesives, 262
Structural materials, 8–10, 14
Studs, 260, 267
Styles, 29, 38, 77–8, 112–13
Stylistic attributes, 111
Styrene–butadiene elastomers, 208, 211
Substitutes, 127, 147, 149–51
Sumarian plates, 111
Super glue, 263
Superplastic alloys, 184
Suppliers, 41, 161, 162
Support structures, 10
Surface models, 33, 43
Surface printing, 60
Surface processes, 9, 89, 91, 92, 94
  profiles, 284–309
Surfaces
  profiling, 60
  reflectivity, 63
  texture, 20, 32, 63, 165, 288–96, 299–303, 305–9
Sustainability, 7, 8, 11–13, 149–50
Swatch, 17–18
Swiss watch industry, 17–18
Symbolic objects, 1
Synthesis in material selection, 124, 126–7, 129–30, 131, 132
  case studies, 141–2, 144–5, 151–2
Synthetic adhesives, 262

T

Tactile appeal, 2, 9, 19, 22, 27, 58, 162, 284
  see also Touch
Taste, 68
Teaching resources, 2, 132
Technical attributes, 27, 34, 50–6, 89–92, 113, 315–27
Technical data, 49, 51, 56, 89, 90, 91
  see also Profiles
Technical design, 15, 27, 114
  balance, 16–17
  engineering dimension, 89–92, 134–9, 169
  resources, 2–3
Technology, 7, 8, 9–10, 20–1
  attributes, 111
  coupling, 127
Tensile steel cables, 100, 101
Tension, 99–101, 318
Tests, 49
Textile adhesion, 263–4
Texture, 20, 32, 63, 165, 288–96, 299–303, 305–9
Thermal attributes, 50, 51, 54–5
Thermal conductivity, 62, 69, 85, 151–2, 241, 326–7
Thermal diffusivity, 85
Thermal expansion coefficient, 326–7

Thermal management, 10, 12, 22, 62–3
Thermoforming, 253
Thermoplastics, 37, 54–5
- adhesives, 265
- casting, 249
- coatings, 306
- elastomers, 210
- molding, 240–4
- powder coating, 305
- profiles, 177–8, 197
- prototyping, 256–7
- thermoforming, 253
- welding, 270–2

Thermosets
- adhesives, 263–4
- profiles, 178, 206
- shaping, 240–1, 244–5, 249, 254

Thin-walled shapes, 238, 241–2, 244, 246, 252–4
Thinking processes, 1, 28–31, 38–9
Threaded fasteners, 268
Three-dimensional surface models, 33, 43, 256–7
TIG welding, 280
Tightening, 268
Time-dependent properties, 27, 157
Tin plating, 296
Tinning, 277
Titanium alloys, 221
Tongue and groove joints, 260
Torch welding, 278
Touch, 59–60, 61, 68–9, 70, 85, 302
- *see also* Tactile appeal

Toughness, 87
Toxicity, 49, 295
Trade-off analysis, 138–9, 143–4
Tradeshows, 40
Transfers, 290
Translucent materials, 68, 69–70, 71, 148
Transparent materials, 6, 69–70, 71, 146–8, 249, 274
Transverse features, 238
Trends, 19, 31, 63
Tungsten carbide, 226
Tungsten inert gas (TIG) welding, 280
Two-dimensional devices, 10
Typefaces, 59
Typology, 111

U

Ultrasonic welding, 273
Utility, 8, 35, 38, 56–63, 67

V

Vacuum bag molding, 254
Vacuum hot pressing, 255
Vacuum thermoforming, 253
Value capture, 13, 14, 19
Vapor metalizing, 293
Venn diagrams, 30–1
Verbal reasoning, 29–30
Viability, 13
Vibrations, 61–2, 86, 324
Violin bows, 149–51
Visual appeal, 2, 9, 19, 20, 22, 27, 38, 284
Visual attributes, 68, 69–71, 162, 165
Visual displays, 59, 60
Visual thinking, 29–31, 42
- *see also* Mapping

Visualization, 105, 106–7, 170, 313
Vulcanization, 179

W

Walkman, 9, 18, 19
Want, *see* Desire; Need
Warmth, 68, 69, 85
Waste products, 11–12
Water-based painting, 303
Waterproof cameras, 20
Wavelengths, 61
Wear resistance, 293–4, 297–9, 302–3, 305–7
Weight, 51–2, 79, 136, 143–4, 149
Welding, 270–83
White finger, 61
Wood, 51–2, 55, 73–4, 163, 164, 231–2
- adhesion, 263–4
- coating, 302–3, 306
- in construction, 99, 100
- environmental impact, 149–50

Workshops, 164–5
Wristwatches, 9, 17–18, 20
Writing, 35–9
Wrought iron, 100, 101, 250

Y

Yearbooks, 40
Yield strength, 87, 135, 136–9, 269, 318–19, 322–3
Young's modulus, 136

Z

Zinc alloys, 225, 246
Zinc plating, 296